Handbook of Neuroemergency
Clinical Trials

HANDBOOK OF NEUROEMERGENCY CLINICAL TRIALS

Edited by

WAYNE M. ALVES, Ph.D.
Valeant Pharmaceuticals International
Costa Mesa, California

BRETT E. SKOLNICK, Ph.D.
Novo Nordisk Pharmaceuticals, Inc.
Princeton, New Jersey

AMSTERDAM • BOSTON • HEIDELBERG • LONDON
NEW YORK • OXFORD • PARIS • SAN DIEGO
SAN FRANCISCO • SINGAPORE • SYDNEY • TOKYO

Academic Press is an imprint of Elsevier

Elsevier Academic Press
30 Corporate Drive, Suite 400, Burlington, MA 01803, USA
525 B Street, Suite 1900, San Diego, California 92101-4495, USA
84 Theobald's Road, London WC1x 8RR, UK

This book is printed on acid-free paper. ♾

Library of Congress Cataloging-in-Publication Data
Handbook of neuroemergency clinical trials / [edited by] Wayne M.
Alves, Brett E. Skolnick.
        p. ; cm.
   Includes bibliographical references.
   ISBN 0-12-648082-6 (alk. paper)
     1. Nervous system–Diseases–Handbooks, manuals, etc.
2. Neuropharmacology–Handbooks, manuals, etc. 3. Clinical trials–
Handbooks, manuals, etc.
     [DNLM: 1. Central Nervous System Diseases–drug therapy.
2. Neuroprotective Agents–therapeutic use. 3. Clinical Trials–methods.
4. Emergency Medicine–methods. 5. Neuropharmacology–methods.   QV
76.5 H2355 2006]   I. Alves, Wayne M.   II. Skolnick, Brett Evan.
   RC346.H2355 2006
   616.8′0461–dc22

British Library Cataloguing in Publication Data
A catalogue record for this book is available from the British Library

ISBN 13: 978-0-12-088523-7
ISBN 10: 0-12-648082-6

For information on all Elsevier Academic Press publications
visit our website at www.books.elsevier.com

Printed in the United States of America
05   06   07   08   09   10   9   8   7   6   5   4   3   2   1

Contents

11. Data Safety and Monitoring Board: Role in Acute Neurological Trials
BRETT E. SKOLNICK

12. Role of a Project Medical Officer in Acute Neuroemergency Clinical Trials
JOSEPH A. KWENTUS

13. Ethical Considerations in Neuroemergency Clinical Trials
WAYNE M. ALVES

14. Industry Perspective on Drug Development
THOMAS C. WESSEL AND CHRISTOPHER GALLEN

Foreword

"Where tireless striving stretches its arms towards perfection..."

–Rabindranath Tagore

Clinical trials in emergency neurological disorders are not suited for the faint of heart. Most conditions under study have devastating consequences, the nervous system is not very forgiving of insults, the potential impact of any single therapy is limited, the therapeutic window of opportunity is generally brief and difficult to define, obtaining consent for participation can be a challenge, especially if the patient is cognitively impaired and unable to consent personally, patient accrual is usually limited at any single medical center, and uniformity of management is often elusive. Why then the interest?

Acute neurological disorders affect a large number of people in all age groups. The public health impact of these conditions therefore certainly cannot be ignored. It has been argued that the outcome from some of these diseases is determined *ab initio* and that any intervention is likely to be futile. However, the fact that the outcomes from many of the diseases have improved substantially over the past few decades argues against this nihilistic posture. It is fairly clear that improvements can, have and will be made. And the humanistic, intellectual and commercial pay-off of such advances is impossible to resist.

In this book Drs. Skolnick and Alves have brought together some of the most experienced investigators to create an invaluable road-map through this poorly charted and perilous territory. The result is a concise and easily digestible how-to manual that is a must-read for anyone who is in the field, or contemplating entry. Many lessons that have been learned the hard way from previous trials never make it into the published literature for a variety of reasons. Thus errors can be made over and over again. Therefore the greater value of this book may be in teaching us what not to do.

The editors have extensive real-world experience in the trenches of clinical trial design and implementation. The authors are leading experts in their respective disciplines. Together they have created a volume that investigators in the field of neurological emergency trials would ignore at their own peril. As we enjoy one of the most exciting eras in brain research, it is our hope that the information contained in this volume will serve as the foundation for many exciting breakthroughs in the not-so-distant future.

Raj K. Narayan, MD, FACS
Mayfield Professor and Chairman
The University of Cincinnati

and

The Mayfield Clinic
Cincinnati, Ohio

Acknowledgments

The editors express their sincere appreciation to all of the patients and their families who participated in the numerous neuro-emergency clinical trials that served as the basis for this volume. Special acknowledgment goes to all of the clinician-scientists who have devoted much of their careers in the search for safe and effective therapies.

Special thanks to:

Ruth, Marianne, Amy, Timothy, Jesse, and Shannon

and

John A. Jane, Sr. M.D., Ph.D. and Neal F. Kassell, M.D., Clinical Trialists
sine qua non

WMA

This handbook was inspired by the dedication of the many clinician-scientists with whom I have had the opportunity to interact with and learn from over the past 20 years. Special acknowledgment to Howard I. Hurtig, M.D., and Mathew B. Stern, M.D. and my many other colleagues at the University of Pennsylvania who provided the foundations for my first explorations into the area of acute stroke in the early 80's. Finally, to my coworkers at Novo Nordisk who have enabled me to continue to apply these experiences in the CNS arena.

And a special thanks to my family:

Mary Ann Crawford and Maxwell Skolnick for their continued support.

BES

Introduction

Wayne M. Alves and Brett E. Skolnick

During the 1990s, scientific advances in understanding the mechanisms and pathophysiology of acute central nervous system injury, especially the neurochemical cascade associated with secondary brain injuries that occur most prominently with stroke and trauma, were offset by a history of disappointing results from phase III clinical trials of an unprecedented number of novel neuroprotective drugs. Novel compounds were "tested" and seemingly just fell by the wayside. The list of apparently ineffective compounds includes free radical scavengers, calcium channel blockers, and glutamate N-methyl-D-aspartate receptor antagonists along with many other classes of molecular targets. Were these disappointments reflective of failure of our therapeutic hypotheses or our inability to provide a level playing field to test the safety and efficacy of novel drugs?

The focus of this volume is the "state of the practice" of clinical trials in acute neuroscience populations, or "neuroemergencies" (1). Acute aspects of chronic neurological disorders, in so far as they pose special difficulties for evaluating novel therapies focused on the acute features of those diseases, are also relevant topics (e.g., drugs for acute exacerbations of multiple sclerosis or neuromuscular disorders). The book is intended to focus on novel therapies and the unique challenges their intended targets pose for the design and analysis of clinical trials.

We entered the 1990s as the clinical epidemiology of acute neuroemergencies was becoming well understood. High incidence, potentially devastating consequences, and recognition of the complexity of damage and outcome made these patients the sickest of the sick, with little or no effective treatments beyond supportive management and improved neurosurgical and neurointensive care management. This was combined with an unparalleled optimism regarding the potential of novel neuroprotective compounds. The Decade of the Brain provided disappointment as a legacy of failed clinical trials emerged. Table 1 lists some of the molecular and cellular targets for compounds that either failed or for which uncertain results

TABLE 1. Molecular and cellular targets of compounds in development for neuroemergencies since the 1990s

Neuroprotectants
    Antioxidants/free radical scavengers
    N-methyl-D-aspartate antagonists
    Glycine antagonists
    AMPA/kainite antagonists
    Polyamine antagonists
    Free fatty acid inhibitors
    Adenosine antagonists
    Bradykinin antagonists
    Cholecystokinin B antagonists
    Neurokinin receptor antagonists
    γ-Aminobutyric acid agonists
    Calcium channel blockers
    Calcium-dependent protease (calpain) inhibitors
    Sodium channel blockers
    Lactate buffers/inhibitors
    Nitric oxide synthase antagonists
    Nonpsychotropic cannabinoids
    Opiate receptor antagonists
    Endothelin receptor antagonists
    Apoptosis inhibitors
    Gene expression regulators
    Intracellular adhesion molecule inhibitors

Thrombolytics and antifibrinolytics
    Recombinant tissue plasminogen activator
    Streptokinase
    Prourokinase
    Fibrinogen-clearing enzyme
    Tranexamic acid
    Antifibrinolytic (e.g., Ancrod)

Anticoagulants and antiplatelets
    Low-molecular-weight heparin
    Heparinoids
    Antiplatelet agents (e.g., Ticlopidine)

were obtained during the past 15 years. Although disappointments have been many, attempts to organize a consortium to handle the complexity of neuroemergency clinical trials offer hope (2).

Although neuroemergencies have a fairly high incidence, they are relatively rare compared with non–central nervous system diseases. They carry with them significant risk for devastating complications and long slow recovery. These are complex diseases and disorders with no singular recovery patterns. In some cases,

similar injuries appear to have different outcomes, whereas in other cases the same outcomes result from quite different injuries. Morbidity is often underestimated, and factors of lifestyle and life cycle are important in both etiology and recovery. As such, not all the sequelae are directly attributable to injury *per se*, as indirect effects on important life domains are important and sociological factors contributing to outcomes lurk in the background.

The most significant emergent hypothesis of the 1990s regarding the potential of novel neuroprotective agents for neuroemergencies explicitly recognized that overlapping pathological processes in the early days postinsult led to irreversible cell damage or cell death, that early treatments were needed to interrupt a "secondary cascade," and if successful we might observe improved cerebral metabolism with better clinical outcomes. The challenge was to find the ideal therapeutic milieu in which recovery could occur (3). It was left for us to test this hypothesis with new chemical entities with the potential to interrupt the secondary injury cascade.

By the mid-1990s over 100 new chemical entities were under development for a number of neurological disease indications, including about a dozen for traumatic brain injury. Yet we still have no approved drugs for traumatic brain injury, and only a single compound (recombinant tissue plasminogen activator) has been approved for use in ischemic stroke. This disappointing experience made it clear that safe and effective drugs would be hard to come by and success at best would be incremental. The problem, we are coming to understand, is how to find a level playing field to fairly demonstrate the safety, efficacy, and effectiveness of novel drugs targeted for neuroemergencies. Given that we have a need to recognize the multiplicity of damage and outcomes in clinical trials, the need to understand factors that influence outcomes, and our past clinical trials failures, what do we have to offer?

The purpose of this volume is to explore the issues we face and the strategies that might lead to future success in developing drugs for neuroemergencies, which remains a critical area of unmet medical need. In retrospect, in our evaluation of past neuroemergency development programs, we are tempted to attribute our failures to skipping steps in the drug development process. This does not mean we should not strive to be creative in defining the "optimal" drug development paradigm for specific neuroemergency indications. The conventional drug development process is a staged sequential process that commercial scientists have long sought to reengineer and streamline. But the answer goes beyond simply the logistics of drug development. Our ability to define relevant treatment populations and measure the effects of treatment interventions is equally important. Improved disease classifications based on pathology and the use of continually improving imaging methods, improved endpoint measurement and analysis, identification of "leveraged" in vivo models to provide for better proof-of-concept studies, development of surrogate endpoints, and innovative clinical trials methodologies all can contribute to future success.

The minimal target criteria for a successful neuroprotectant are not difficult to describe. It must be safe, reach an intended action site (i.e., cross the blood–brain barrier), have an expected neurochemical effect, produce an expected neurophysiological effect leading to functional changes, and thereby improve clinical outcomes. The issue is how to demonstrate this in the conpara of adequate and well-controlled clinical trials.

Criticisms of previous neuroemergency development programs include bias in treatment group assignment due to imbalance in important covariates, inability to use classical statistical tests procedures, not addressing treatment delays, and difficulty in obtaining informed consent in many indications (see Chapter 13).

CURRENT STATUS OF TREATMENT OF NEUROEMERGENCIES

The brain is a small somewhat round object weighing approximately 3 pounds. As an organ, it has unique vulnerabilities. Its energy requirements demand a constant blood supply providing glucose and oxygen substrates. The brain is the organ most prone to spontaneous hemorrhage and second most prone to symptomatic ischemic infarction. Cerebral arteries are thinner and less elastic than in other systems of the body. Injury produces not only neurophysical impairments, but also changes in intellectual, emotional, and personality function (3).

Although the mechanisms of damage (e.g., infarction, hemorrhage, contusion, or edema) in neuroemergencies are limited, they seldom occur in isolation. It is often the case in the individual patient that several pathophysiological mechanisms are combined (1,2). This multiplicity of pathways for damage and outcome may be a major contributing reason for the failure of phase III clinical trials. The characteristic mechanisms of acute brain injury, listed below, are limited in that they tend to occur in combination with each other to create in each instance complexity of damage and outcome:

- Brain edema
- Hemorrhage
- Ischemia and brain swelling
- Hydrocephalus
- Neurotransmitter failure
- Toxic substances that cross blood–brain barrier
- Infection or inflammation
- Brain atrophy

Numerous reasons have been offered for the failure of clinical trials in acute neuroscience disorders in the 1990s, including whether the underlying therapeutic hypothesis is flawed, the nature of acute neuroscience populations, whether the drug is able to cross the blood–brain barrier, study design considerations, the clinical populations actually enrolled in the trials, and failure to control relevant disease cofactors. Especially relevant is the adequacy of brain penetration of the investigative agents tested in terms of optimizing dosage and the dosing regimens used. To address these issues and shortcomings, academic–industry collaboration has tried to define optimal preclinical and clinical strategies for drug development in ischemic stroke (6,7).

ACUTE NEUROCLINICAL TRIALS

There is a relatively limited history of drug development in acute neuroscience populations. As mentioned earlier, the diseases are fairly "rare" and require more research sites for sufficient enrollment. Individual practice variations in hospital-based settings (e.g., emergency departments or neurological intensive care units) contribute to a plethora of examinations, drugs, and supportive interventions. Treatment decisions are often idiosyncratic and there are few gold standards. Consequently, subjective definitions and perceptions are very important in guiding treatment decisions. Guideline statements are becoming more robust regarding treatment options but are still limited by the number of level I studies (8,9). This poses considerable challenges for the design, conduct, and analysis of randomized clinical trials. Because gold standards are few, often there is a lack of consensus on measurement of damage and outcome that contributes to large

case report books. Because trials are large, they take time to conduct and analyze, and there is a danger that the rate of change in standards of clinical care could out-run our ability to prove efficacy. An example is the evolution of HHH therapy for the management of clinical vasospasm as a complication of subarachnoid hemorrhage as various pharmacological interventions were being tested. The fact that rescue therapies could have been efficacious (albeit risky and expensive) meant it was difficult to compare endpoints.

Many steps might be contemplated in improving neuroemergency trials, including:

- Identification of leveraged *in vivo* models that may reduce the inherent complexity of damage and outcome of neuroemergencies
- Improved efforts to understand underlying mechanisms of action
- Improved measurements of disease burden and/or activity
- Improved outcomes measurement
- Identifying procedures for handling the inherent overlap of various outcomes domains
- Identifying procedures for handling spillover and swamping effects of major prognostic factors
- Achieving agreement on how to order competing sets of explanatory variables in outcomes models
- Clinical phenotyping of treatment populations to avoid including patients with excessively good or excessively poor prognosis
- Focus on clinical benefit and crisper endpoint assessment
- Improved assessment of intermediate effects (i.e., biomarkers or mechanistic endpoints) as supportive evidence
- Consider "novel" approaches to neuroemergency trials design and randomization strategies.

PURPOSE OF THIS VOLUME

Modern clinical drug development involves complex interactions among scientific, medical, commercial, regulatory, and manufacturing issues (10). This volume is intended to provide developers of novel therapies with a more complete understanding of the scientific and medical issues of relevance in designing and initiating clinical development plans intended for acute neuroscience populations. We hope that we can provide an understanding of the pitfalls associated with drug development in neuroemergencies as well as a single source for the best information available regarding how to approach and solve the issues that have plagued drug development since the early 1990s.

We asked authors to include disorders generally requiring emergency care or intensive care in highly specialized clinical settings (e.g., neurological intensive care units). The authors could include discussion of drug development for disorders where the brain is a component (e.g., HIV-1 infection or sickle cell crises) and clinical development is primarily focused on brain protection in the setting of chronic disorders. Authors also could include neuroprotection in the compara of systemic disease (e.g., brain protection in coronary artery bypass graft surgery or out-of-hospital cardiac arrest). Device trials (e.g., endovascular obliteration of cerebral aneurysms) and brain access technologies where relevant could also be discussed. Out of sheer practicality, we excluded systemic complications in the compara of neuroemergencies (e.g., neurogenic cardiovascular disorders or respiratory syndromes), except as they are relevant to understanding the nature of the acute central nervous system disease and have implications for clinical drug development program. We also excluded evaluation of neurosurgical interventions *per se*, and drug development for disorders where the brain is a disease component but the therapeutic focus is on the systemic disease itself (e.g., HAART in HIV-1 infection as opposed to a drug focused on HIV-1–associated cognitive impairments).

The mandate to authors was to focus on relevant aspects of their respective disease areas that bore importance to the design and analysis of clinical trials. This could include the following:

- Brief overview of disease epidemiology and natural history
- Current management guidelines relevant for drug development
- Recent successes and disappointments of novel drugs
- Consensus regarding "failed trials" and how we might solve trials design and analysis problems
- Advances in preclinical evaluation of novel therapies
- Current "state of the practice" in the design and analysis of randomized clinical trials
- "Gold" and "silver" measures for diagnosis, definition of subpopulations, and outcomes assessment
- Biological markers and surrogate endpoints
- Emergent clinical technologies and methodologies relevant for future clinical trials (pros and cons). Examples include censoring excessively good or poor prognoses, shift analyses over a range of outcomes categories, and strategies for improving interrater reliability in outcome assessment.

No single volume can do justice to the complexity of drug development in acute neuroscience populations. Our hope is simply to stimulate discussion focused on providing solutions to the problems that have plagued the search for safe and efficacious drugs/biologics in the acute neurological area in the hope that investigators will be able to provide the level playing field that has eluded us for so many years.

References

1. Cruz J, ed. Neurologic and Neurosurgical Emergencies, 1st ed. Philadelphia: W.B. Saunders, 1998.
2. Barsan WG, Pancioli AM, Conwit RA. Executive summary of the National Institute of Neurological Disorders and Stroke conference on Emergency Neurologic Clinical Trials Network. Ann Emerg Med 2004;44:407–412.
3. Becker DP, Gudeman SK. Textbook of Head Injury. Philadelphia: W.B. Saunders, 1989.
4. DiMasi JA, Hansen RW, Grabowski HG. The price of innovation: new estimates of drug development costs. J Health Econ 2003;22:151–185.
5. DiMasi JA, Hansen RW, Grabowski HG, Lasagna L. Research and development costs for new drugs by therapeutic category. A study of the US pharmaceutical industry. Pharmacoeconomics 1995;7:152–169.
6. Stroke Therapy Academic Industry Roundtable (STAIR). Recommendations for standards regarding preclinical neuroprotective and restorative drug development. Stroke 1999;30:2752–2758.
7. Stroke Therapy Academic Industry Roundtable II (STAIR II). Recommendations for clinical trial evaluation of acute stroke therapies. Stroke 2002;33:639–640.
8. Broderick et al., 1999, ICH.
9. Acute Ischemic Stroke: Adams, 2005.
10. Steiner J. Clinical Development: Strategic, Pre-Clinical, Clinical, and Regulatory Issues. Buffalo Grove, IL: Interpharm Press, 1997.

Contributors

Numbers in parentheses indicate the chapter to which the author has contributed.

Wayne M. Alves (4, 10, 13), Valeant Pharmaceuticals International, Costa Mesa, CA 92626

Laura J. Balcer (6), University of Pennsylvania School of Medicine, Philadelphia, PA 19104

Don Berry (9), University of Texas M.D. Anderson Cancer Center, Houston, TX 77030

Christopher Bladin (1), Monash University, Melbourne, Australia

Stephen Davis (1), University of Melbourne, Melbourne, Australia

Aaron S. Dumont (2), University of Virginia, Charlottesville, VA 22908

Christopher Gallen (14), Neuromed Pharmaceuticals Inc., Conshohocken, PA 19428

Romergryko G. Geocadin (8), Johns Hopkins Hospital, Baltimore, MD 21287

Daniel F. Hanley (8), Johns Hopkins School of Medicine, Baltimore, MD 21287

Susan T. Herman (5), Hospital of the University of Pennsylvania, Philadelphia, PA 19104

Michael D. Hill (3), University of Calgary, Calgary, Alberta T2N 2T9 Canada

Neal F. Kassel (2), University of Virginia, Charlottesville, VA 22908

Christopher C. King (7), University of Pittsburgh Medical Center, Pittsburgh, PA 15213

Michael Krams (9), Pfizer CNS Clinical, Groton, CT 06340

Joseph Kwentus (12), University of Mississippi Medical Center, Jackson, MS 39216

Lawrence F. Marshall (4), University of California San Diego, San Diego, CA 92093

Petter Müller (9), University of Texas M.D. Anderson Cancer Center, Houston, TX 77030

Edwin Nemoto (7), Presbyterian University Hospital, Pittsburgh, PA 15213

Tom Parke (9), Tessella Support Services, Abingdon OX14 3PX, United Kingdom

Nader Pouratian (2), University of Virginia, Charlottesville, VA 22908

Brett E. Skolnick (11), Novo Nordisk, Inc., Princeton, NJ 08540

Madhura A. Tamhankar (6), Scheie Eye Institute, Philadelphia, PA 19104

Lisa L. Travis (15), Schering-Plough Research Institute, Kenilwort, NJ 07033

Thomas C. Wessel (14), Sepracor, Inc., Marlborough, MA 01752

1

Acute Ischemic Stroke

Christopher Bladin and Stephen Davis

Stroke is one of the most devastating diseases of Western society. In most countries, stroke is the third most common cause of death and the leading cause of adult neurological disability (1). The social and psychological costs are enormous, and the health economic costs run into billions of dollars. Developing a successful and reliable acute treatment for stroke remains an elusive "Holy Grail." Fortunately, significant advances over the past decade indicate a breakthrough is not too far away.

STROKE THROMBOLYSIS

Intravenous tissue plasminogen activator (tPA) was approved for use in acute stroke in the United States in 1996 after publication of the landmark National Institute of Neurological Disorders and Stroke (NINDS) study (2). Approval for the use of tPA in acute stroke has occurred in many regions, including Canada (1999), Europe (2002), and Australia (2003). Acute stroke treatment guidelines, including use of tPA, have been published by a number of organizations, including the American Heart Association (3) and the Canadian Stroke Consortium (4). However, the benefits and risks of tPA in acute stroke are still the subject of much debate (5,6).

Differences in trial methodology and outcome measures and conflicting results from various thrombolytic trials have made interpretation of the literature difficult and controversial for many. In addition, there have been claims of financial conflicts of interest in those devising these guidelines as well as concerns about inappropriate conclusions being drawn from the original NINDS publication. The *British Medical Journal* website has posted the many contributions to this often heated debate (5).

As a consequence, many neurologists and emergency medicine physicians have unfortunately expressed reluctance to use tPA in acute stroke. The knowledge base is therefore small, and only a few centers have depth of experience with stroke thrombolysis, further hindering the more widespread use of tPA. To fully understand the issues involved in the use of tPA in stroke, it is worth undertaking a brief overview of the seminal trials undertaken so far and following this with discussion on the phase IV (postmarketing) studies of tPA in acute stroke, otherwise known as "tPA use in the real world" (7).

Stroke tPA Thrombolysis Trials

As mentioned previously, the NINDS study was first published in 1995 (2). Acute ischemic stroke patients were treated with tPA within 3 hours of symptom onset, and results indicated that those receiving this treatment achieved greater neurological recovery and experienced less disability than patients who received placebo. The tPA dose used was 0.9 mg/kg (maximum dose, 90 mg), and half of the patients were treated within 90 minutes of stroke onset. Patients in this study had moderately severe strokes with a median baseline score (National Institutes of Health Stroke Scale [NIHSS]) of 14 for tPA-treated patients and 15 for the placebo group. There was a strict protocol for managing hypertension, and all patients were admitted to the intensive care unit for the first 24 hours. Outcome measures were based on a "global" outcome score. This was a composite endpoint based on four disability scales (Barthel, Glasgow Outcome Scale, Rankin, and NIHSS) to detect a consistent and persuasive difference in the proportion of patients achieving a favorable outcome. At 3 months, each of the four primary outcome scales and the combined global tested statistics showed a statistically significant benefit for the use of tPA. In summary, 42% of the tPA-treated patients and only 26% of the placebo-treated patients had regained functional independence at 3 months. Overall, six patients (95% confidence interval, 5–11) had to be treated for one additional patient to recover self-care independence, and nine patients (95% confidence interval, 5–25) had to be treated for one additional patient to achieve full neurological recovery (7). The beneficial effects occurred in patients with all subtypes of stroke, including lacunar infarction. Further analysis of the NINDS data set revealed that the benefits were sustained at 1 year with no additional increase in mortality (6).

The occurrence of intracranial hemorrhage is the complication of most concern with tPA. These may be either asymptomatic (usually of small size) or larger symptomatic intracranial hemorrhages with

clinical deterioration and possible impact on eventual outcome. In the NINDS study, symptomatic intracranial hemorrhages occurred in 6.4% of the tPA-treated patients and in 0.6% of placebo-treated patients ($p < 0.01$) (8). Most tPA-related hemorrhages occurred within the first 24 hours, and nearly half were fatal. The risk factors for developing intracerebral hemorrhage included increased stroke severity (NIHSS score) and hyperglycemia. Although the European tPA trials (9,10) suggested that baseline computed tomography (CT) findings of early cerebral edema with mass effect predicted hemorrhagic transformation with tPA, reanalysis of the NINDS trial did not suggest any major association (11). Despite the 10-fold difference in rate of symptomatic intracranial hemorrhage, the all-cause mortality rate was 17% for tPA-treated patients and 21% for placebo-treated patients (not statistically significant), with no increase in mortality attributable to tPA within the first week or even within the first 3 months.

Another argument that has been put forward is that some patients are "rescued" from death due to stroke only to be left with severe disability. However, the improved outcome in tPA-treated patients was not associated with an increase in the number of patients surviving with severe disability (2).

The NINDS tPA Controversy

The NINDS trial has undergone considerable scrutiny and interpretation since its publication (2). An imbalance in baseline stroke severity between the tPA and placebo treatment groups has been the primary focus of discussion (5,12,13) When the baseline NIHSS scores were divided into quintiles (0–5, 6–10, 11–15, 16–20, >20), it was found that imbalances existed in the mildest and most severe stroke groups. Of the 58 patients in the 0–5 NIHSS group, 42 (72%) were from the tPA treatment group, versus

16 (28%) from the placebo treatment group. Among the 140 patients in the >20 NIHSS group, 63 (45%) were from the tPA treatment group, versus 77 (55%) from the placebo treatment group. The imbalance in baseline stroke severity generated concerns that the treatment benefit reported in favor of tPA may have been explained by the excesses of both mild strokes allocated to tPA and more severe strokes allocated to placebo.

To determine whether the baseline stroke severity imbalance affected the outcome of the trial, the NINDS appointed an independent committee made up of three biostatisticians and three stroke clinicians to reanalyze the NINDS trial data. In addition to the issue of baseline stroke severity imbalance, the committee was asked to determine whether eligible stroke patients may not benefit from tPA given according to the protocol used in the trials. After performing extensive analyses, the committee reported that the baseline stroke severity imbalance did not affect the outcome of the study (14). Indeed, they confirmed on multivariate analysis evidence of a statistically significant tPA treatment effect. Exploratory analyses did not identify any group of acute ischemic stroke patients who would be harmed by receiving tPA. Specifically, there was no evidence that either baseline NIHSS or time from stroke onset to treatment modified the t-PA treatment effect.

Studies on tPA in acute stroke were also undertaken in Europe. The two studies performed were the European Cooperative Acute Stroke Studies, ECASS (9) and ECASS II (10). In the first ECASS study, the dose of tPA was higher than that used in the NINDS trial, at 1.1 mg/kg with a maximum dose of 100 mg. The other difference was that the window for administration of tPA was broader at 6 hours and the median time to treatment was 4 hours. There was a 21% incidence of intracranial hemorrhage in the tPA-treated patients.

There were a number of possible causes for this, including the longer treatment window, the greater dose of tPA, and, perhaps most importantly, the inclusion of large numbers of patients (almost one in five) with protocol violations. These deviations mainly consisted of the failure to recognize changes on the pretreatment CT that should have excluded the patient from the study. As a consequence, there was no statistically significant difference in primary outcome where tPA-treated and placebo groups were based on the intention to treat analysis (9). In a reanalysis of the ECASS data (9), excluding patients who were inappropriately included in the study, the proportion of patients with minimal or no disability (modified Rankin scale of 0 or 1) at 3 months was significantly greater in the treatment group than in the control group (41% vs. 29%, $p < 0.05$).

With the many lessons learned during the first ECASS trial, ECASS II was undertaken in the late 1990s (10). The tPA dose was reduced to 0.9 mg/kg, as in the NINDS trial. Investigators were extensively trained to recognize the CT abnormalities of early ischemic stroke, in particular focusing on the exclusion of patients with more than one-third of the middle cerebral artery (MCA) territory involved in the ischemic process on the initial CT. Strict blood pressure controls were also implemented. The primary outcome measure was defined as the proportion of patients with a favorable outcome based on the modified Rankin scale score of 0 or 1 at 3 months, again in keeping with the NINDS trial. Based on this outcome measure, there was no significant difference between tPA treatment and placebo, although the distribution of modified Rankin Score (mRS) scores revealed a benefit in favor of tPA treatment. A post-hoc analysis was then undertaken, in which patient outcomes were dichotomized as either a good outcome, as indicated by independence in self-care (mRS score, 0 to 2), or a bad outcome, as indicated by death or dependence (mRS score, 3 to 6). A significantly greater proportion of the tPA-treated patients achieved independence at 3 months (54% vs. 46%, $p = 0.024$) (10). From this post-hoc analysis it was determined that 12 patients had to be treated for one additional independent survivor. Intracranial hemorrhage was more common in the tPA-treated patients (9%) than in the placebo-treated patients (3%), but again there was no difference in mortality between the two groups.

One of the important points with ECASS II was that the time window remained at 6 hours, and the results indicated that tPA reduced disability without increasing the mortality rate. It should be emphasized that the primary outcome of the study was negative, and that a positive result was only achieved after a post-hoc analysis with reconfiguring of the methodological definition of "favorable outcome."

Another trial testing the effects of tPA on stroke was the Alteplase Thrombolysis for Acute Interventional Therapy in Ischemic Stroke study (15). This differed from other tPA studies in that it experienced a number of protocol changes due to publication of the NINDS trial and had two time windows: part A (<3 hours) and part B (3–5 hours). For part B there was no clinical benefit and there was an increase in mortality. Although the numbers in part A were small, there was a benefit in this cohort (16). The completion of four randomized controlled trials using tPA in acute ischemic stroke enabled several meta-analyses to be undertaken (17–19). These meta-analyses revealed an overall benefit for tPA treatment, regardless of the modified Rankin grading used to define outcomes of minimal disability/fully independent living, with no increase in mortality.

Guidelines for early management of acute ischemic stroke have now been extensively published and are freely available on the Internet (3,20). The recommendations are based on evidence-based practice and recommend the use of tPA in acute ischemic stroke in carefully selected patients, who can be treated within 3 hours of onset of ischemic stroke.

Phase IV Data: Postmarketing Studies in tPA Stroke Thrombolysis

There is now a large body of international experience in the use of tPA in acute ischemic stroke. Many centers have published data allowing perspective on the use of tPA in routine clinical practice (7).

The largest published experience is the Standard Treatment with Alteplase to Reverse Stroke study, which is a prospective review of the management of stroke with tPA in 24 academic and 33 community centers (21). The results from this phase IV study compared very favorably with those from the NINDS study. The characteristics of the patient population treated with tPA were similar to those in the NINDS study. The 1-month outcome indicated that 43% were independent and 35% had minimal or no disability. The rate of symptomatic intracranial hemorrhage was low at 3.3% compared with 6.4% in the NINDS study (21).

In Canada, after the approval for the use of tPA in acute stroke, a national database was required to be established as part of regulatory conditions for approval of tPA. The Canadian Activase for Stroke Effectiveness Study was established to collect data prospectively from academic and community hospitals across Canada (22). The results from this database indicated that patients receiving tPA were older than those in the NINDS cohort with a similar stroke severity. Again, the symptomatic intracranial hemorrhage rate was low (4.4%), and the 3-month outcomes were favorable and comparable with the NINDS data (7,22). Similar findings have been published from centers in Germany, the United States, and Australia (23–25).

An important message that came from the many postmarketing studies was the considerable danger in violating the tPA stroke protocol. This refers to treatment of patients who do not meet the eligibility criteria or are given additional treatments that deviate from published guidelines. The incidence of symptomatic intracranial hemorrhage in these patients is often much higher, with a greater risk of death or dependency. In one study from Indianapolis, tPA-related intracranial hemorrhage occurred in 38% of patients with protocol violations but only in 2% of those patients without protocol violations (26). A widely publicized report from Cleveland, largely based at community hospitals, highlighted the problems that can occur (27). In this cohort of 70 patients from 29 hospitals, the rate of symptomatic intracranial hemorrhage and in-hospital mortality was around 16%. This cohort had the highest reported rate of protocol violations, with 50% of patients deviating from national treatment guidelines. A quality improvement program with frequent educational sessions was then initiated to address these problems. A subsequent follow-up report in which stroke tPA guidelines were adhered to resulted in a significant improvement in outcome at the same group of hospitals (28).

Phase IV data should always be interpreted with caution, because there are often many differences between stroke centers (e.g., university hospitals vs. community hospitals), in patient demographics, in baseline stroke severity, and in the rate of protocol deviations. Adequate and complete follow-up, overall accuracy, and degree of completion of database sets are also very important.

Ongoing tPA Trials and Future Directions

There are still many challenges ahead for the proper administration of tPA in acute stroke. The poor public recognition of stroke symptoms, delays in transportation, and limitations of a 3-hour time window present considerable obstacles for patient recruitment (29). It is estimated that tPA treatment currently reaches only 2–3% of the North American stroke population (30). For example, in the NINDS trial, over 17,000 patients were screened, but only 624 eligible subjects were recruited. Most of those excluded were ineligible because of the time that had elapsed since stroke onset. Time delays from stroke onset to presentation continue to be frustrating. Public awareness of the symptoms of stroke is poor, and emphasizing the need to act quickly requires considerable education. Studies have shown that one-third of the general public cannot name a single warning sign of stroke (29). Prehospital stroke screening tools have been developed (e.g., the Los Angeles Pre-hospital Stroke Scale and the Cincinatti Pre-hospital Stroke Scale) to facilitate paramedic diagnosis of tPA-eligible stroke patients and to prenotify emergency departments of impending arrival (31,32). It is agreed that tPA should only be administered by physicians with expertise in acute stroke with strict adherence to published treatment guidelines. These may be neurologists or emergency medicine physicians with an interest in stroke. Similarly, nursing protocols, particularly for management of the post-tPA stroke patient, need to be closely followed. Comprehensive registries of tPA use, including hemorrhage rates, are used in Canada and Europe. Quality assurance via web-based database programs can simplify data collection and allow for more accurate postmarketing surveillance. Web-based programs have also been used to help improve skills in the radiological diagnosis of stroke on CT (e.g., www.neuroimage.co.uk) (33).

The target of acute stroke therapies, such as tPA, is the ischemic penumbra, a zone of incomplete ischemia where neurons are hypoxic and functionally inactive but still viable. The ischemic

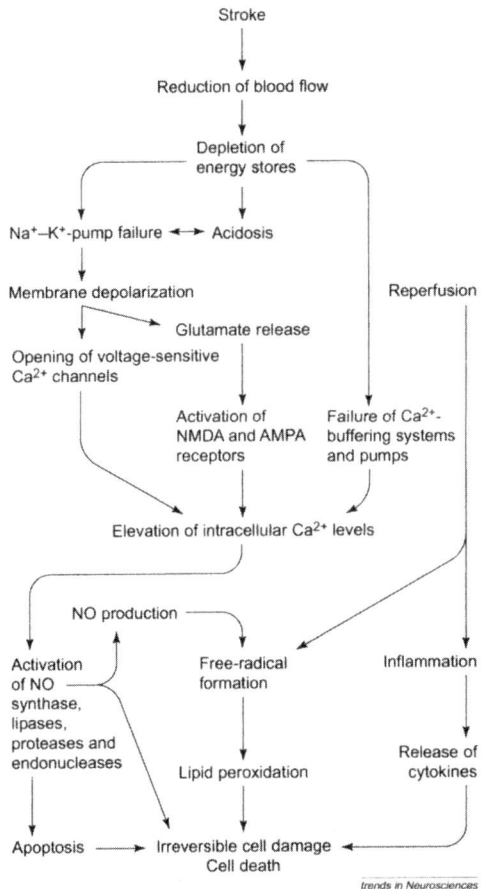

FIGURE 1.1   Neurotoxic cascade in the ischemic penumbra. A complex neurotoxic cascade is triggered by a focal deficit in brain perfusion. Key events are uncontrolled neuronal depolarizations, an overexcitation in glutamate receptors, a buildup of intracellular Ca^{2+} levels, the generation of free radicals, the stimulation of several catabolic systems, and the induction of inflammation. AMPA, ; NO, nitric oxide. (Adapted from ref. 44.)

penumbra is a dynamic time-based condition in which brain parenchyma undergoes necrosis over hours to days due to a cascade of biochemical events termed the ischemic cascade (Fig. 1.1). It has been suggested that the critical time for intervention, based on stroke studies using magnetic resonance imaging (MRI), may be around 4.5 hours, with earlier use of reperfusion strategies leading to greater tissue salvage (34). As the ischemic process unfolds, there is a progressive decrease in cerebral blood flow. When this falls from the normal levels of 50 to 55 ml/100 mg/min to below 8 to 10 ml/100 mg/min there is rapid neuronal cell death. However, between this ischemic core and the normally perfused brain at the periphery there exists zones of moderately reduced blood flow, the extent of which depends on collateral supply from surrounding arteries.

There are a number of ongoing trials to determine whether tPA can be used beyond the currently accepted 3-hour time window, targeting the penumbral region, which has been shown to last many hours in some patients (35). These include clinical trials such as ECASS 3 (tPA vs. placebo 3–4 hours after stroke onset) and the International Stroke Trial 3 (tPA vs. placebo 0–6 hours after onset). There has only been one intraarterial phase III trial of thrombolysis. The Prolyse in Acute Cerebral Thromboembolism trial used a 6-hour time window and focused on patients with MCA territory infarction, randomized to prourokinase versus placebo (36). The trial was positive, but another definitive trial is needed before this approach can be licensed.

There are also MRI-based trials, using combined perfusion-weighted imaging (PWI) and diffusion-weighted imaging (DWI), aimed at determining whether PWI–DWI mismatch can be used to select treatment responders beyond 3 hours. This MRI signature, where the PWI boundary represents tissue at risk and the DWI lesion represents tissue usually destined to infarct, is postulated to represent the ischemic penumbra, that is, critically hypoperfused but potentially recoverable brain tissue. These research trials include Echoplanar imaging thrombolysis evaluation trial (EPITHET) in Australasia and the Diffusion weighted imaging evoluation for understanding stroke evolution (DEFUSE) in the United States (34,37). The pilot EPITHET data suggested that tPA delivered to acute stroke patients with MRI

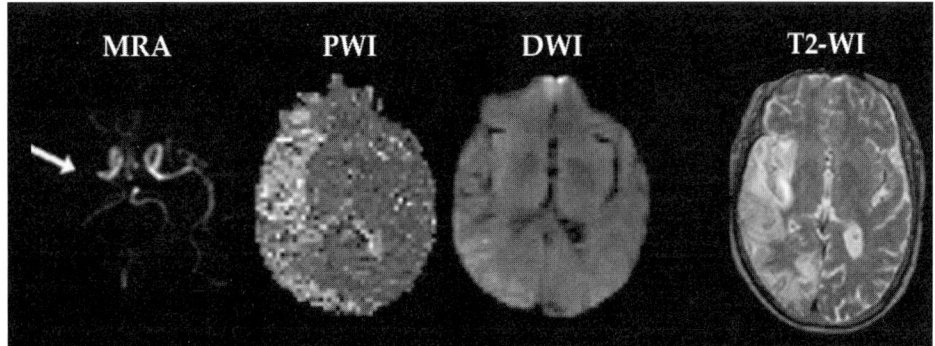

FIGURE 1.2 Acute right MCA infarct, with occlusion of proximal MCA on magnetic resonance angiography (MRA). Acute PWI–DWI mismatch is shown. In the absence of thrombolysis, an extensive infarct is present on outcome T2-weighted MRI (T2-WI). Rapid reperfusion has been shown to salvage tissue in the mismatch region.

mismatch enhanced brain tissue reperfusion and penumbral salvage (Fig. 1.2). Using this MRI-based approach, another thrombolytic agent, desmoteplase, has been tested 3–9 hours after stroke onset. Phase II studies have shown enhanced reperfusion to the ischemic region, very few intracerebral hemorrhages, and promising outcome data. Phase III studies will follow (38).

ANTITHROMBOTIC DRUGS

Acute aspirin is now used almost routinely in acute ischemic stroke, based on the two megatrials, the International Stroke Trial and Chinese acute stroke trial (CAST), which showed that aspirin given within 48 hours modestly reduced poor outcomes at 6 months (39–41). The exception is for patients who have received tPA, where antiplatelet therapy is contraindicated for 24 hours after tPA administration. These trials suggested that for every 1000 patients treated, poor outcomes could be reduced in about 10 patients. There is controversy whether this is truly an acute benefit or in fact a secondary prevention effect. Another acute antiplatelet approach, using the intravenous glycoprotein 2b/3a antagonist abciximab, has

yielded promising results in a phase II trial and is being studied further (42).

In contrast, formal anticoagulation with heparin or low-molecular-weight heparin or heparinoids has not been beneficial in acute stroke, In most trials, a small reduction in acute recurrent stroke has been offset by a greater increase in rate of intracerebral hemorrhage (40,41,43). These agents are still sometimes recommended in patients with minor ischemic stroke and a very high risk of recurrent embolism (e.g., prosthetic valve disease). They are also used in cerebral vein thrombosis, extracranial arterial dissection, and in low dose for deep vein thrombosis prophylaxis.

NEUROPROTECTIVE DRUGS IN ACUTE ISCHEMIC STROKE

The use of tPA in acute ischemic stroke realistically represents only one part of the treatment equation. Apart from restoring blood flow to ischemic brain tissue, the cellular and pathobiochemical consequences of prolonged cerebral ischemia also need to be dealt with. In the ischemic penumbra, oxygen delivery becomes insufficient to allow normal levels of oxidative metabolism. This produces lactic acidosis and impedes the production of ATP, the energy source of cellular ionic pumps.

Failure of the sodium-potassium pump results in rapid loss of potassium from the neurons with extensive neuronal depolarization. Voltage calcium channels are opened, leading to an extracellular build-up of excitatory amino acids that over-stimulates receptors. One of the principal excitatory amino acids is glutamate, with stimulation of the *N*-methyl-D-aspartate (NMDA) and AMPA receptors, a key component of the excitotoxic process. There is failure of the mitochondrial energy systems, resulting in elevated calcium levels, the generation of free radicals, and formation of excessive amounts of nitric oxide. Lipid peroxidation occurs, resulting in membrane damage, and there is sustained activation a large range of calcium-dependent enzymes (e.g., lipases, proteases, endonucleases, etc.) with further impairment of cellular function and membrane structure. As a background to this there is also damage to cellular DNA via endonucleases or free radicals. This triggers a complex self-destructive process involving gene expression, known as apoptosis, or programmed cell death. Mitochondrial dysfunction is thought to be a key element in the initiation of apoptosis with release of caspases, cytochrome c, and mitochondrial apoptosis-inducing factor (44).

The restoration of blood flow, for example via thrombolysis, may potentially exacerbate this process. Oxygen can enhance the biochemical reactions that generate free radicals. Inflammatory processes also play a roll with up-regulation of endothelial adhesion receptors and other chemoattractants resulting in invasion of leucocytes and macrophages and the release of metalloproteinases and cytotoxic cytokines (such as tumor necrosis factor and interleukins).

Potential Targets for Neuroprotection

The many stages of the neurotoxic pathway outlined in Figure 1.1 would seem to offer a wealth of opportunities to impede or alter the course of this dynamic process. In animal studies, reduction of infarct size of 50% or even greater has been demonstrated with strategies that attenuate the excitotoxic cascade, reduce free radical toxicity, diminish the various inflammatory responses, and finally curb progressive neuronal cell death by apoptosis (44–46).

Neuroprotective agents can target one or more of these processes. Following are some examples of areas that have been pharmacologically targeted and the results of trials that have ensued.

1. Sodium channel blockers: The anti-epileptic medication phenytoin blocks voltage-dependent sodium channels and reduces infarct size in permanent and reperfusion models of focal brain ischemia (47). Fosphenytoin is a prodrug of phenytoin that was evaluated in a phase III trial; enrollment was halted because of lack of efficacy in an interim analysis (48).

2. Calcium channel blockers: The prominent roll of calcium in the excitotoxic process has led to many attempts to develop therapies to inhibit voltage-sensitive calcium channels. Nimodipine and flunarizine are calcium channel blockers that have been shown to reduce infarct size in animal models of permanent and transient focal cerebral ischaemia (49). Nimodipine has been demonstrated to be of significant benefit in subarachnoid hemorrhage but has been less than impressive in trials for acute ischemic stroke, with worsened outcome when administrated intravenously. One particular problem was the marked hypotensive effect of intravenous administration. A meta-analysis of studies using oral nimodipine with a broad 12-hour time window suggested a possible benefit for the drug. However, the prospective, randomized, controlled trial, VENUS (Very Early Nimodipine

Use in Stroke), was stopped due to a failure of benefit (50).

3. Glutamate inhibition: Blockade of the NMDA receptor was an early target in the development of neuroprotective agents with promising data from animal studies. Early studies of various NMDA receptors were stopped in phase I and II development because of unacceptable neuropsychiatric side effects (51). Some agents (e.g., selfotel, aptiganel, eliprodil) were studied in phase III trials but were terminated prematurely because of poor benefit (44,52–54). Selfotel was thought to have neurotoxic effects, with increased early mortality (52).

4. γ-Aminobutyric acid (GABA) agonists: The use of drugs that cause activation of GABA receptors have been proposed as a strategy to counteract the actions of glutamate. Clomethiazole is an antiepileptic drug that causes neuronal hyperpolarization by enhancing the activity of GABA and $GABA_A$ receptors (44). Again, different ischemic animal stroke models indicated a neuroprotective effect for clomethiazole, but phase III clinical trials failed to show any benefit (55). A post-hoc analysis did indicate that patients with severe stroke may receive some benefit, leading to the establishment of the Clomethiazole Acute Stroke Study Ischemia. This study concluded with a negative result (56).

5. Nitrogen oxide inhibitors and free radical scavengers: The rise in intracellular calcium levels in ischemia leads to activation of nitric oxide synthase and nitric oxide production, a cytotoxic free radical. Down-regulation of the nitric oxide synthase pathway has been investigated using drugs such as lubeluzole, a purported neuroprotective agent that acts by reducing nitric oxide-related neurotoxicity. Initial animal stroke studies and small low-dose phase II human studies indicated reduced mortality in ischemic stroke (44,57). However, a number of phase III trials, some using a double-dose regimen producing plasma concentration equivalent to the levels associated with neuroprotection in rats, failed to produce any benefit in over 3000 patients (57,58).

Free radical scavengers have also been investigated. Tirilazad is a nonglucocorticoid lipid peroxidation inhibitor that acts as a free radical scavenger. As with many other neuroprotective agents, this drug demonstrated reduced cerebral infarct volume in animal models but in phase III studies in humans did not result in any improvement in functional outcome (59). Concerns were then raised about the dose being inappropriately low, but subsequent trials with higher doses of tirilazad were stopped prematurely because of safety problems (60).

6. Anti-inflammatory agents: The inflammatory process is an integral part of neurotoxicity in acute ischemic stroke. Adhesion molecules such as endothelial adhesion molecule-1 are rapidly expressed in the zone of focal cerebral ischemia attracting leucocytes into the region of ischemia and with cytokine release further enhance developing necrosis. Enlimomab, a mouse monoclonal antibody against endothelial adhesion molecule-1, was demonstrated to be effective in animal models in reducing infarct volume in reperfusion models but not with permanent MCA occlusion models (61,62). Phase III studies indicated that there is indeed a neurological deterioration in patients receiving this treatment, quite possibly related to the mouse antigens provoking an inflammatory response (44).

Other agents that have been studied include those acting at the cell membrane level such as G_{M1}-ganglioside, citicoline, and piracetam (63–65). As with other agents, results in phase III trials have largely been disappointing.

G_{M1}-ganglioside licenses have been suspended due to concerns over possible occurrence of Guillain-Barré neuropathy.

STUMBLING FROM THE BENCH TO BEDSIDE

Despite extensive research with many different compounds, which have demonstrated promising results in animal stroke models, all phase III clinical trials conducted so far indicate that these drugs have failed to live up to their initial promise (46,51,52,66,67). Many compounds that interfere with the excitotoxic pathway have been demonstrated to be neuroprotective in preclinical models of stroke. Safety in subsequent phase I and II clinical trials led to phase III, randomized, double-blind, placebo-controlled, efficacy trials. The resources required to complete such a trial are prodigious, often estimated to be over 50 million U.S. dollars. Despite these efforts, all phase III trials have so far failed to demonstrate the efficacy of neuroprotective agents. The reader looking for further detailed analysis on the many trials conducted in this area is directed to a number of excellent published reviews and commentaries (44–46).

Failure of Clinical Trials in Neuroprotective Therapies

The common theme in all trials of neuroprotective therapies is the failure of promising results in animal models of ischemic stroke to be replicated in phase III clinical trials. There are many possible reasons for this (44,54,68), including difficulty in translating tightly structured animal stroke models to complex human clinical scenarios and inadequate penetration of neuroprotective agents into the poorly perfused hemisphere and ischemic penumbra. It has also been suggested that the secondary and reperfusion biochemical cascade may be responsible for relatively little additional tissue damage compared with that due to the initial hypoxic ischemic insult (69). Other possibilities include the relatively long time windows used in many neuroprotective trials, adverse effects of some drugs such as hypotension and sedation, inadequate sample size, other trial design issues, and the relative lack of neuroreceptors in white matter.

Stroke Heterogeneity

Stroke in the human brain is much less predictable than in the animal model: The etiology, location, and severity of ischemic stroke in human subjects is very heterogenous. All ischemic stroke animal models are based around permanent or reversible models of ischemia. In reversible cerebral ischemia the vessel is occluded, usually with a silicon or cotton thread for variable time periods ranging from several minutes to several hours. The occlusive device is removed and cerebral perfusion is allowed to reoccur. Laser Doppler flowmetry is now considered an essential part of all animal studies to ensure that reperfusion has indeed occurred. In permanent ischemia models, the occlusion is left *in situ* with no reperfusion allowed to occur. A number of neuroprotective agents have demonstrated efficacy in one of these ischemic models but not in the other.

However, the animal model is set in a highly contrived and stabilized environment using animals of similar age and standardized amounts of focal cerebral ischemia induced by a reproducible intervention. In humans, both types of focal brain ischemia can, and often do, occur, producing a potentially salvageable ischemic penumbra; reperfusion after transient occlusion further adds to the evolving neurotoxicity. Collateral circulation influences infarct and penumbral size (70), with considerable variability in the human model of collateral flow around the circle of Willis that is often only partially complete. Variability in the location of the embolic occlusion (e.g., proximal or distal MCA) also leads to variability in the ability of the collateral branches, such as those through the lenticulostriate arteries, to sustain cerebral parenchymal blood flow.

The human stroke model has patients of different ages with variable comorbidities, all of which can heavily affect the outcome of death or disability. The nature of ischemic stroke is also variable, ranging from large cortical infarctions through to small lacunar infarction. Variations in collateral circulation may further alter the size of the ischemic penumbra. Many patients show spontaneous reperfusion in the early stages of stroke and consequently have a better clinical outcome. At the time of stroke, systemic factors such as blood pressure, body temperature, oxygenation, and glucose levels can also affect eventual outcome and can potentially override any beneficial effect of a neuroprotective agent (see later).

Optimal Therapeutic Doses of Neuroprotective Agents

Determining the correct balance between the adverse and beneficial effects of neuroprotective drugs is often difficult. In many studies, doses of neuroprotective drugs that limit infarct size in animals are associated with adverse effects that can limit tolerable doses. Psychiatric side effects were the main reason for the premature termination of a number of trials with NMDA receptor antagonists, and adverse hemodynamic consequences have limited the efficacy of drugs such as nimodipine (71). Conversely, suboptimal doses of neuroprotective agents may be used in phase III trials because of undue concern regarding some safety aspects. In phase III trials of lubeluzole, concerns about the QTc interval were possibly misinterpreted from

the phase II trial data, leading to an incorrect dosage regimen (58,72). In other examples, the length of administration of neuroprotective agent may not be sufficient; clomethiazole was administered for only 24 hours, although it had been demonstrated that excitatory amino acid levels in the ischemic penumbra could remain significantly elevated for at least 6 days after the onset of stroke. Concerns of excessive sedation shortened the protocol for administration of this medication to a probable inappropriate time period.

Combination Neuroprotective Therapies

Developing therapies that target the hypoxic brain cell is only the first step; appropriate delivery to this site is clearly required and, in many ways, can only be done after vascular reperfusion has occurred. Combination with a thrombolytic drug is the next natural progression to this process, at which point the combination of a "clot buster and a cell saver" is really more likely to succeed.

Most, if not all, neuroprotective drugs developed so far target one specific aspect of the ischemic cascade pathway. To expect a definitive result from this limited approach is in many ways unrealistic, and a multimodal therapy, targeting multiple areas of the ischemic cascade, would be more practical and likely to succeed. An example of this is the novel agent AM-36, an arylalkylpiperazine with combined antioxidant and Na^+ channel blocking actions (73). Individually, these properties have been shown to confer neuroprotection in a variety of *in vitro* and *in vivo* animal models of stroke. Preliminary studies have demonstrated that AM-36 is neuroprotective *in vivo* and protects against both neuronal damage and functional deficits even when administered up to 180 minutes after induction of stroke. In fact, the greatest protection was found when administration was delayed by 180 minutes after stroke (73). This multimodal approach to neuroprotective therapy clearly holds great promise.

Physiological Modification of the Ischemic Environment

There is increasing interest in physiological approaches that impact on the acute ischemic process. These strategies include careful maintenance of euglycemia in hyperglycemic stroke patients, aggressive treatment of fever, varied ways of inducing hypothermia, and potential manipulation of blood pressure in the acute setting.

About one-third of acute stroke patients present with hyperglycemia. These include known and newly diagnosed diabetic subjects, but also those with stress hyperglycemia. These groups can be distinguished by measurements of acute blood glucose and HbA1C to determine whether the hyperglycemia has predated the stroke. Regardless of etiology, hyperglycemia independently predicts higher mortality and worse functional outcome (74). Although the precise mechanism is unclear, higher lactate levels have been identified in animal and human studies in the ischemic region, consequent upon activation of the glycolytic pathway in hyperglycemia in anaerobic regions (75). A large phase III trial is investigating whether dextrose-insulin infusions improve outcome (76). In the meantime, stroke guidelines currently advocate avoidance of glucose-containing solutions in acute stroke and correction of hyperglycemia.

In both animal and human stroke, fever is independently associated with a worse outcome (77). Fever is known to accentuate the neurotoxic cascade after acute ischemic stroke, as demonstrated in animal models. There is a consensus that fever should therefore be aggressively treated. Furthermore, small phase II studies indicated that mild to moderate hypothermia, aiming at a core temperature of about 33°C, might be beneficial (78). Techniques to induce

hypothermia include external cooling blankets, cooled intravenous fluids, and intravenous heat-exchange catheters that can rapidly and precisely induce and maintain hypothermia and subsequent rewarming. Larger trials are planned.

In ischemic brain, perfusion pressure is important, but the normal autoregulation (where cerebral blood flow is maintained constant despite wide fluctuations in blood pressure) is lost in acute stroke. There is enormous uncertainty about the optimal approach to blood pressure management in acute stroke, and this remains a great challenge in stroke trials. Strategies have ranged from pressor therapies to elevate perfusion pressure to a variety of blood pressure–lowering therapies (79). It is recognized that marked acute hypertension increases the risk of hemorrhagic transformation in ischemic stroke treated with tPA, so that there is an recommended upper limit set of 180/100 mm Hg. Conversely, rapid blood pressure lowering is generally considered hazardous. Elevated blood pressure levels in acute stroke tend to spontaneously lower over the first 24–48 hours. Current trials include the use of glyceryl trinitrate (GTN) paste to lower blood pressure, being tested in efficacy of nitric oxide in stroke (ENOS) (80).

FUTURE STROKE TRIALS: THE STROKE THERAPY ACADEMIC INDUSTRY ROUNDTABLE

The difficulty with current animal models and the poor rate of progression from phase II to phase III studies, coupled with the failure of many phase III studies, led to the development of the Stroke Therapy Academic Industry Roundtable (STAIR) (68). This collaborative group is actively examining preclinical issues and trial design to ensure the optimal development of new acute stroke therapies. There has been particular scrutiny of phase II and phase III studies with regard to trial design and appropriately valid outcome measures/endpoints that are easy to measure, reproducible, valid, clinically meaningful, and resistant to bias (68).

Stroke is a heterogenous entity, and a "one size fits all" approach to treatment is not appropriate. The challenge is to reliably identify the different stroke subgroups and tailor the therapy accordingly. Experimental data in animal stroke models suggest that the treatment window may be as long as 8 to 12 hours. Imaging and biochemical studies in acute stroke patients suggest that a similar time window may be present in selected stroke patients. The STAIR group is examining the incorporation of imaging methods into trial methodology. Imaging techniques such as multimodal MRI and CT using diffusion and perfusion imaging offer an opportunity to greatly assist patient selection (68,81) and may allow the expansion of the therapeutic time window beyond 3 hours, with some trials investigating treatment up to 9 hours (82).

CONCLUSIONS

Treatment of stroke in the new millennium offers much promise, with a renewed interest and vigor for the management of acute ischemic stroke with both clinical therapies and pharmacotherapies. Yet despite all this good work, one of the key difficulties still remains targeting the acute stroke patient as quickly as possible to administer these treatments. Our excellent stroke research will have only minimal public health impact if only a small percentage of the stroke population is able to receive and benefit from these new therapies. Education is an integral part of this process, both to our colleagues and to the general public. In many ways, this represents equally as great a challenge to stroke physicians and stroke researchers.

References

1. Warlow CP, Sudlow C, Dennis M, Wardlaw J, Sandercock P. Stroke. Lancet 2003;362:1211–1224.
2. Group NIoNDaSrSS. Tissue plasminogen activator for acute ischemic stroke. N Engl J Med 1995;333:1581–1587.
3. Adams HP, Adams JR, Brott TG, del Zoppo GL, Furlan AJ, Goldstein LB, et al. Guidelines for the early management of patients with ischemic stroke, a scientific statement from the Stroke Council of the American Stroke Association. Stroke 2003;34:1056.
4. Norris JW, Buchan A, Cote R, Hachinski V, Phillips SJ, Shuaib A, et al. Canadian guidelines for intravenous thrombolytic treatment in acute stroke: a consensus statement of the Canadian Stroke Consortium. Can J Neurol Sci 1998;25:257–259.
5. Lenzer J. Alteplase for stroke: money and optimistic claims buttress the "brain attack" campaign. Br Med J 2002;324:723–729.
6. Marler JR, Tilley BC, Lu M, Brott TG, Lyden PC, Grotta JC, et al. Early stroke treatment associated with better outcome: the NINDS rt-PA stroke study. Neurology 2000;55:1649–1655.
7. Gladstone DJ, Black SE. Update on intravenous tissue plasminogen activator for acute stroke: from clinical trials to clinical practice. Can Med Assoc J 2001;165:311–317.
8. Group TNt-PSS. Intracerebral hemorrhage after intravenous t-PA therapy for ischemic stroke. Stroke 1997;28:2109–2118.
9. Hacke W, Kaste M, Fieschi C, Toni D, Lesaffre E, von Kummer R, et al. Intravenous thrombolysis with recombinant tissue plasminogen activator for acute hemispheric stroke: the European Cooperative Acute Stroke Study (ECASS). JAMA 1995;274:1017–1025.
10. Hacke W, Kaste M, Fieschi C, von Kummer R, Davalos A, Meier D, et al. Randomized double-blind placebo-controlled trial of thrombolytic therapy with intravenous alteplase in acute ischemic stroke (ECASS II). Lancet 1998;352:1245–1251.
11. Patel S, Levine S, Tilley B, Grotta J, Lu M, Frankel M, et al. Lack of clinical significance of early ischemic changes on computed tomography in acute stroke. JAMA 2001;286:2830–2838.
12. Mann H. Alteplase for stroke. Uncertainty remains about efficacy. Br Med J 2002;324:1581.
13. Mann J. The efficacy analysis of the NINDS trial is flawed. Br Med J 2002;324:723–729.
14. Ingall TJ, O'Fallon WM, Louis TA, Hertzberg V, Goldfrank LR, Asplund K. Initial findings of the rt-PA acute stroke treatment review panel. Cerebrovasc Dis 2003;16(suppl 4):125.
15. Clark W, Wissman S, Albers G, Jhamandas J, Madden K, Hamilton S. Recombinant tissue-type

plasminogen activator (alteplase) for ischemic stroke 3 to 5 hours after symptom onset: the ATLANTIS Study: a randomized controlled trial. Alteplase Thrombolysis for Acute Noninterventional Therapy in Ischemic Stroke. JAMA 1999;282:2019–2026.

16. Albers G, Clark W, Madden K, Hamilton S. ATLANTIS trial: results for patients treated within 3 hours of stroke onset. Alteplase Thrombolysis for Acute Noninterventional Therapy in Ischemic Stroke. Stroke 2002;33: 493–495.

17. Wardlaw JM, del Zoppo G, Yamaguchi T. Thrombolysis for acute ischaemic stroke [Cochrane review]. In: The Cochrane Library. Issue Oxford: Update Software, 2000.

18. Wardlaw JM, Sandercock PAG, Warlow CP, Lindley RI. Trials of thrombolysis in acute ischemic stroke: does the choice of primary outcome measure really matter? Stroke 2000; 31:1133–1135.

19. Steiner T, Bluhmki E, Kaste M, Toni D, Trouillas P, von Kummer R, et al. The ECASS 3-hour cohort. Secondary analysis of ECASS data by time stratification. European Cooperative Acute Stroke Study. Cerebrovasc Dis 1998;8:198–203.

20. Committee EIECaW. The European Stroke Initiative recommendations for stroke management: update 2003. Cerebrovasc Dis 2003;16:311–318.

21. Albers GW, Bates VE, Clark WM, Bell R, Verro P, Hamilton SA. Intravenous tissue-type plasminogen activator for treatment of acute stroke: the Standard Treatment with Alteplase to Reverse Stroke (STARS) study. JAMA 2000;283:1145–1150.

22. Hill MD, Buchan AM, Investigators C. The Canadian Activase for Stroke Effectiveness Study (CASES): interim results [abstract]. Stroke 2001;32:323.

23. Schmulling S, Grond M, Rudolf J, Heiss WD. One-year follow-up in acute stroke patients treated with rtPA in clinical routine. Stroke 2000;31:1552–1554.

24. Chiu D, Kreiger D, Villar-Cordova C, Kasner SE, Morgenstern LB, Bratina PL, et al. Intravenous tissue plasminogen activator for acute ischemic stroke; feasibility, safety and efficacy in the first year of clinical practice. Stroke 1998;29:18–22.

25. Szoeke C, Parsons M, Butcher K, Baird T, Mitchell P, Fox S, et al. Acute stroke thrombolysis with intravenous tissue plasminogen activator in an Australian tertiary hospital. Med J Australia 2003:324–328.

26. Lopez-Yunez A, Bruno A, Williams L, Yilmaz E, Zurru C, Biller J. Protocol violations in community-based rtPA stroke treatment are associated with symptomatic intracerebral hemorrhage. Stroke 2001;32:12–16.

27. Katzan IL, Furlan AJ, Lloyd LE, Frank JI, Harper DL, Hinchey JA, et al. Use of tissue-type plasminogen activator for acute ischemic stroke: the Cleveland area experience. JAMA 2000;283:1151–1158.

28. Katzan IL, hamer MD, Furlan AJ, Hixson ED, Nadzam DM. Quality improvement and tissue-type plasminogen activator for acute ischaemic stroke—a Cleveland Update. Stroke 2003;34:799–800.

29. Greenlund KJ, Neff LJ, Zheng ZJ, Keenan NL, Giles WH, Ayala CA, et al. Low public recognition of major stroke symptoms. Am J Prevent Med 2003;25:315–319.

30. Alberts MJ, Hademenos G, Latchaw RE, Jagoda A, Marler JR, Mayberg MR, et al. Recommendations for the establishment of primary stroke centres. JAMA 2000;283:3102–3109.

31. Kidwell C, Starkman S, Eckstein M, Weems K, Saver J. Identifying stroke in the field. Prospective validation of the Los Angeles Prehospital Stroke Screen (LAPSS). Stroke 2000;31:71–76.

32. Kothari R, Pancioli A, Liu T, Brott T, Broderick J. Cincinnati Prehospital Stroke Scale: reproducibility and validity. Ann Emerg Med 1999; 33:373–378.

33. Wardlaw J, Farrall A. Diagnosis of stroke on neuroimaging. Br Med J 2004;328:655–656.

34. Butcher K, Parsons M, Baird T, Barber A, Donnan G, Desmond P, et al. Perfusion thresholds in acute stroke thrombolysis. Stroke 2003;34:2159–2164.

35. Darby D, Barber P, Gerraty R, Desmond P, Yang Q, Tress B, et al. Pathophysiological topography of acute ischemia by combined diffusion-weighted and perfusion MRI. Stroke 1999; 30:2043–2052.

36. Furlan A, Higashida R, Wechsler L, Gent M, Rowley H, Kase C, et al. Intra-arterial prourokinase for acute ischemic stroke. The PROACT II study: a randomized controlled trial. Prolyse in Acute Cerebral Thromboembolism. JAMA 1999;282:2003–2011.

37. Parsons M, Barber A, Chalk J, Darby D, Rose S, Desmond P, et al. Diffusion and perfusion weighted MRI response to thrombolysis in stroke. Ann Neurol 2002;51:28–37.

38. Fisher M, Davalos A. Emerging therapies for cerebrovascular disorders. Stroke 2004;35:367–369.

39. Chen Z, Sandercock P, Pan H. Indications for early aspirin use in acute ischemic stroke: a combined analysis of 40,000 randomized patients from the chinese acute stroke trial and the international stroke trial. On behalf of the CAST and IST collaborative groups. Stroke 2000; 31:1240–1249.

40. International Stroke Trial Collaborative Group. The International Stroke Trial (IST): a randomized trial of aspirin, subcutaneous heparin, both, or

neither among 19,435 patients with acute ischaemic stroke. Lancet 1997;349: 1569–1581.

41. CAST CASTCG. CAST: randomized placebo-controlled trial of early aspirin use in 20,000 patients with acute ischaemic stroke. Lancet 1997;349:1641–1649.

42. Lapchak P, Araujo D. Therapeutic potential of platelet glycoprotein IIb/IIIa receptor antagonists in the management of ischemic stroke. Am J Cardiovasc Drugs 2003;3:87–94.

43. Berge E, Abdelnoor M, Nakstad P, Sandset P. Low molecular-weight heparin versus aspirin in patients with acute ischaemic stroke and atrial fibrillation: a double-blind randomised study. HAEST Study Group Heparin in Acute Embolic Stroke Trial. Lancet 2000; 355:1205–1210.

44. De Keyser J, Sulter G, Luiten P. Clinical trials with neuroprotective drugs in acute ischaemic stroke: are we doing the right thing? Trends Neurosci 1999;22:535–540.

45. Davis S. Endpoints and statistical concerns for acute stroke therapy trials. In: Fisher MMD, ed. Stroke Therapy, 2nd ed. Butterworth-Heinemann, 2001:123–133.

46. Gorelick PB. Neuroprotection in acute ischaemic stroke: a tale of for whom the bell tolls? Lancet 2000;355:1925–1926.

47. Rataud J, Debarnot F, Mary V, Pratt J, Stutzmann J. Comparative study of voltage-sensitive sodium channel blockers in focal ischaemia and electric convulsions in rodents. Neuroscience 1994;172:19–23.

48. Pulsinelli W, Mann M, Welch M, Sivin J, Biller J, Maisel J, et al. Fosphenytoin in acute ischaemic stroke: efficacy results. Neurology 1999;52(suppl 2):A384.

49. Luiten P, et al. Clinical Pharmacology of Cerebral Ischaemia. Humana Press, 1997:66–99.

50. Horn J. VENUS—Very Early Nimodipine Use in Strokes: final results from a randomised, placebo-controlled trial. Cerebrovasc Dis 1999;9:127.

51. Lees K. Cerestat and other NMDA antagonists in ischemic stroke. Neurology 1997;49(5 suppl 4): S66–S69.

52. Davis SM, Lees K, Albers GW, Diener HC, Markabi S, Karlsson G, et al. Selfotel in acute ischemic stroke: possible neurotoxic effects of an NMDA antagonist. Stroke 2000;31:347–354.

53. Dyker AG, Edwards KR, Fayad PB, Hormes JT, Lees KR. Safety and tolerability study of aptiganel hydrochloride in patients with an acute ischemic stroke. Stroke 1999;30:2038–2042.

54. Lees KR, Diener HC. Neuroprotection in stroke therapy. In: Edvinsson L, Krause D, eds. Cerebral Blood Flow and Metabolism, 2nd ed. Philadelphia: Lippincott Williams & Williams, 2002:452–456.

55. Wahlgren N, Ranasinha K, Rosolacci T, Franke C, van Erven P, Ashwood T, et al. Clomethiazole acute stroke study (CLASS): results of a randomized, controlled trial of clomethiazole versus placebo in 1360 acute stroke patients. Stroke 1999;30:21–28.

56. Lyden P, Shuaib A, Ng K, Levin K, Atkinson R, Rajput A, et al. Clomethiazole Acute Stroke Study in ischemic stroke (CLASS-I): final results. Stroke 2002;33:122–128.

57. Grotta JC. Lubeluzole treatment of acute ischaemic stroke. The US and Canadian Lubeluzole Ischaemic Stroke Study Group. Stroke 1997;28:2338–2346.

58. Diener H, Cortens M, Ford G, Grotta J, Hacke W, Kaste M, et al. Lubeluzole in acute ischemic stroke treatment: a double-blind study with an 8-hour inclusion window comparing a 10-mg daily dose of lubeluzole with placebo. Stroke 2000;31:2543–2551.

59. The RANTTAS Investigators. A randomized trial of tirilazad mesylate in patients with acute stroke (RANTTAS). Stroke 1996;7:1453–1458.

60. Haley ECJ, Lewandowski CA, Tilley BC. High-dose tirilazad for acute stroke (RANTTAS II). RANTTAS II Investigators. Stroke 1998;29:1256–1257.

61. Zhang RM, Chopp MP, Li YM, Zaloga C, Jiang NM, Jones MP, et al. Anti-ICAM-1 antibody reduces ischemic cell damage after transient middle cerebral artery occlusion in the rat. Neurology 1994;44:1747–1751.

62. The Enlimomab Acute Stroke Trial Investigators. The Enlimomab Acute Stroke Trial: final results. Cerebrovasc Dis 1997;7(suppl 4):18.

63. SASS Investigators. Ganglioside GM1 in acute ischemic stroke. The SASS Trial. Stroke 1994;25:1141–1148.

64. Clark WM, Warach S, Pettigrew L, Gammans R, Sabounjian L. A randomized dose-response trial of citicoline in acute ischemic stroke patients. Citicoline Stroke Study Group. Neurology 1997;49:671–678.

65. De Deyn P, Reuck J, Deberdt W, Vlietinck R, Orgogozo J. Treatment of acute ischemic stroke with piracetam. Members of the Piracetam in Acute Stroke Study (PASS) Group. Stroke 1977;28:2347–2352.

66. Lees KR, Asplund K, Carolei A, Davis SM, Diener H, Kaste M, et al. Glycine antagonist (gavestinel) in neuroprotection (GAIN International) in patients with acute stroke: a randomised controlled trial. GAIN International Investigators. Lancet 2000;355:1949–1954.

67. Krams M, Lees KR, Hacke W, Grieve AP, Orgogozo JM, Ford GA, et al. Acute Stroke Therapy by Inhibition of Neutrophils (ASTIN): an adaptive dose-response study of UK-279,

276 in acute ischemic stroke. Stroke 2003;34: 2543–2548.

68. Stroke Therapy Academic Industry Roundtable STAIR II. Recommendations for the clinical trial evaluation of acute stroke therapies. Stroke 2001;32:1598–1606.

69. Heiss WD, Thiel A, Grond M, Graf R. Which targets are relevant for therapy of acute ischemic stroke? Stroke 1999;1999 1486–1489.

70. Liebeskind D. Collateral circulation. Stroke 2003;34:2279–2284.

71. Ahmed N, Nasman P, Wahlgren N. Effect of intravenous nimodipine on blood pressure and outcome after acute stroke. Stroke 2000;31: 1250–1255.

72. Hacke W, Lees K, Timmerhuis T, Haan J, Hantson L, Hennerici M, et al. Cardiovascular safety of lubeluzole (Prosynap(R)) in patients with ischemic stroke. Cerebrovasc Dis 1998;8:247–254.

73. Callaway J, Knight M, Watkins D, Beart P, Jarrott B. Delayed treatment with AM-36, a novel neuroprotective agent, reduces neuronal damage after endothelin-1-induced middle cerebral artery occlusion in conscious rats. Stroke 1999;30:2704–2712.

74. Baird T, Parsons M, Phanh T, Butcher K, Desmond P, Tress B, et al. Persistent post stroke hyperglycemia is independently associated with infarct expansion and worse clinical outcome. Stroke 2003;34:2208–2214.

75. Parsons M, Barber P, Yang Q, Darby D, Desmond P, Gerraty R, et al. Acute hyperglycaemia adversely affects stroke outcome: an MR imaging and spectroscopy study. Ann Neurol 2002;52:20–28.

76. Gray C, Hildreth A, Alberti G, O'Connell J, Collaboration G. Poststroke hyperglyaecemia: natural history and immediate management. Stroke 2004;35:122–126.

77. Reith J, Jorgensen H, Pedersen P, Nakayama H, Raaschou H, Jeppesen L, et al. Body temperature in acute stroke: relation to stroke severity, infarct size, mortality, and outcome. Lancet 1996;347:422–425.

78. Krieger D, De Georgia M, Abou-Chebl A, Andrefsky J, Sila C, Katzan I, et al. Cooling for acute ischemic brain damage (Cool Aid): an open pilot study of induced hypothermia in acute ischemic stroke. Stroke 2001;32: 1847–1854.

79. Bath P, Chalmers J, Powers W, Beilin L, Davis S, Lenfant C, et al. International Society of Hypertension (ISH): statement on the management of blood pressure in acute stroke. J Hypertens 2003;31:665–672.

80. Bath P. High blood pressure as risk factor and prognostic predictor in acute ischemic stroke: when and how to treat it. Cerebrovasc Dis 2004;17(suppl 1):51–57.

81. Fisher M. Recommendations for advancing development of acute stroke therapies: Stroke Therapy Academic Industry Roundtable 3. Stroke 2003;34:1539–1546.

82. Fisher M, Brott TG. Emerging therapies for acute ischemic stroke: new therapies on trial. Stroke 2003;34:359–361.

2

Subarachnoid Hemorrhage

Nader Pouration, Aaron S. Dumont, and Neal F. Kassell

ETIOLOGY, EPIDEMIOLOGY, AND NATURAL HISTORY

Subarachnoid hemorrhage (SAH) can be due to a number of etiologies, including, but not limited to, trauma, intracranial aneurysms, arteriovenous malformations (AVMs), vasculitides, tumor, and coagulopathies. Although trauma is the leading cause of SAH, the most common cause of spontaneous SAH is from intracranial aneurysms. Most investigations of the epidemiology and management of SAH focus on hemorrhages from intracranial aneurysms. This focus is probably due to devastating sequela associated with aneurysmal SAH, including the risk of rerupture and the morbidity and mortality associated with vasospasm. Moreover, traumatic SAH is often trivial (i.e., of minimal volume) or, conversely, so devastating and significant that it cannot be studied in isolation from other significant head injuries (e.g., diffuse axonal injury, edema, increased intracranial pressure). Because most studies of SAH consider aneurysmal SAH, it becomes impossible to discuss the clinical trials for the treatment of SAH without also discussing clinical trials for the prevention and treatment of vasospasm, the leading potentially treatable cause of morbidity and mortality after aneurysm rupture.

Vasospasm is the delayed arterial narrowing that can occur between postbleed days 2 and 14 (with peak incidence at

approximately day 7 or 8), possibly resulting in delayed ischemic neurological deficits (DINDs). Vasospasm accounts for 7% of the mortality and 7% of the morbidity of aneurysmal SAH (1). The exact processes and mechanisms leading to vasospasm remain unclear but appear to be related to multiple parallel yet interacting biochemical cascades (see later).

To gain a better understanding of the implications of clinical trials in SAH and vasospasm, it is important to review the prevalence of aneurysms, the estimated risk of aneurysmal rupture, the morbidity and mortality of aneurysmal rupture, and the incidence of vasospasm as a sequela of aneurysmal rupture. This brief review should provide a framework in which to consider, design, and evaluate the feasibility of proposed interventions.

Despite years of investigation, the precise prevalence of incidental aneurysms is still not definitively known because estimates are based on autopsy studies (2–4) and institutional reviews of cerebral angiograms (5), both of which have intrinsic limitations. Autopsy studies are limited because of selection bias, nonuniformity of aneurysm definition, and lack of interobserver reliability with regards to cerebrovascular dissection. Similarly, prevalence estimates based on review of angiograms are flawed because of limited views on cerebral angiography and selection bias. Nevertheless, when estimates of incidence of aneurysmal SAH and rupture rates (discussed later) are considered, the best estimates suggest that the prevalence of incidental aneurysms is likely approximately 1% (2–6).

Estimating rupture rates has been even more difficult and controversial than estimating prevalence. The preliminary report from the International Study of Unruptured Intracranial Aneurysms (ISUIA) represented the largest study to date, including 2621 patients in 53 participating centers (7). The authors reported that in patients without a history of SAH, aneurysms <10 mm ruptured at a rate of 0.05% annually, aneurysms >10 mm ruptured at a rate of ~1% annually, and aneurysms >25 mm ruptured at a rate of ~6% in the first year. In contrast, in patients with a history of SAH from a different aneurysm, aneurysms <10 mm and >10 mm ruptured at a rate of 0.5% and ~1% annually, respectively. Regardless of size, the authors reported an overall mortality rate of 66% (55% and 83% in those with and without a history of SAH, respectively) when previously unruptured aneurysms hemorrhage. Despite representing the largest sample studied with extensive patient follow-up, the ISUIA study was received with some skepticism because of the unexpectedly low rupture rate reported in patients without a history of SAH with small aneurysms (i.e., 0.05% per year). Most critics blamed surgical selection bias for the surprisingly low rates of rupture and did not believe that the reported rates, at least in that subpopulation, were consistent with known aneurysmal SAH incidence and unruptured aneurysm prevalence (5).

In 2003, the prospective arm of the ISUIA study was published (8). This study confirmed that aneurysm size and location were important predictors of rupture risk. Five-year rupture rates varied between 0% in small aneurysms without a prior history of SAH in the anterior circulation to 50% in large aneurysms (>25 mm) in the posterior circulation. In all, the authors reported 51 aneurysmal ruptures in 6544 patient-years follow-up, corresponding to an overall 0.8% annual risk of rupture, independent of size, location, or history of SAH, and a rate of between 0% and 10% per year, depending on size, location, and history of SAH (from other aneurysms). The authors reported an overall mortality rate of 65% when aneurysms ruptured. Like the first report, the second arm of the ISUIA study has also been criticized for its nonrandomized design (or intervention selection bias) and short follow-up (<5 years) in over half of the patients studied (8). Continued follow-up is ongoing.

Juvela and colleagues (9,10) followed 142 patients with 181 unruptured aneurysms for a total of 1944 patient-years and reported a nearly constant rupture rate of 1.3% annually over three decades of follow-up. Interestingly, despite reporting that rupture rates increased with aneurysm size, 29 of the 33 aneurysms that they reported to have eventually ruptured were smaller than 10 mm (which may be accounted for by the fact that most aneurysms studied by this group were <10 mm). Importantly, they also reported a 52% mortality rate associated with rupture of previously unruptured aneurysms. Unlike the original ISUIA report, this study had no surgical selection bias because surgery was never offered to patients with unruptured aneurysms during the period studied. Possible weaknesses of the study are its small sample size and the lack of truly incidental aneurysms; most subjects studied (131) had a history of SAH from a different aneurysm.

Tsutsumi and colleagues (11) studied the natural history of aneurysms in 62 patients presenting for causes other than SAH. They also confirmed that aneurysm size is an important consideration in the natural history of aneurysms. They reported a 5- and 10-year cumulative rate of rupture of 4.5% and 19%, respectively, for aneurysms <10 mm and 33.5% and 55.9%, respectively, for aneurysms >10 mm. Moreover, the authors reported that 100% of those experiencing SAH had bad outcomes, with an 86% mortality rate. Although the profile of the studied population (elderly, Asian, and comorbid cerebrovascular disease) may not resemble all populations, this study highlighted that SAH rates and mortality may be considerably higher in select populations, even without a history of SAH, in stark contrast to the findings of the ISUIA (11). This is an important point to bear in mind in designing clinical trials for therapeutics because population selection may be key to identifying potential therapeutics. Considering these epidemiological studies, it is probably safe to assume that rupture rates are approximately 1–2% per year.

If a patient survives the initial insult of aneurysmal rupture, the major cause of morbidity and mortality becomes vasospasm. Two types of vasospasm are recognized: angiographic and clinical. Whereas angiographic vasospasm occurs in approximately two-thirds of all patients (i.e., angiographic evidence of arterial narrowing), clinical vasospasm occurs in approximately one-third of all patients (i.e., vasospasm that results in a clinical deficit). Note that most clinical trials study clinical (or symptomatic) vasospasm. Because of this high rate of vasospasm, it is important to briefly review the factors that may affect the risk of vasospasm. Most studies estimate vasospasm rates to be approximately 15–30% (12–15), with rates varying widely depending on certain factors. The most common cited predictor of vasospasm remains the amount of blood seen on head computed tomography (CT) in the subarachnoid space. In 1980, Fisher and colleagues (16) demonstrated that greater than 1 mm of blood in the basal cisterns was associated with a nearly 96% rate of clinical vasospasm in the patients studied. Subsequent prospective studies confirmed that the amount of cisternal blood was an important predictor of vasospasm (14,17,18). Other risk factors identified by multivariate analysis include age <50–60 years old, better World Federation of Neurological Surgeons (WFNS) grade measured on admission, and hyperglycemia in the intensive care unit (12,14).

Despite the limitations of each study, several significant factors emerge as important considerations in evaluating the natural history of aneurysms, SAH, and vasospasm. The prevalence of unruptured aneurysms is likely approximately 1%, suggesting that approximately 3 million people in the United States harbor unruptured aneurysms. These studies also suggest that the likely rupture rate across all aneurysms is ~1–2% per year and that certain aneurysms are likely to be at greater risk of rupture, including those with increasing diameters and aneurysms in certain anatomical distributions. Based on these estimates and incidence data, approximately 30,000 cases of aneurysmal SAH occur each year in the United States. With an estimated 60-day mortality rate of 30–50%, morbidity rate of 30% (in survivors), and incidence of vasospasm of 20–30%, we can estimate that approximately 15,000 people a year will die of aneurysmal SAH, 5000 people will become debilitated by an aneurysmal SAH, and between 5000 and 10,000 people will experience vasospasm as consequence of aneurysmal SAH (11). Because of the relatively high incidence and theoretically preventable morbidity and mortality in some patients (if rebleeding and the consequences of vasospasm can be eliminated), a great deal of attention has focused on developing appropriate treatments for aneurysmal SAH. Unfortunately, despite a number of phase III clinical trials, the number of proven efficacious treatments is minimal.

DIAGNOSIS, DEFINITION OF SUBPOPULATIONS, AND OUTCOMES ASSESSMENT

History and Presentation

Outside of trauma, the classic history and presentation for SAH is characterized as a sudden onset of severe and often unremitting headache. Patients often describe it as the "worst headache of my life." Associated symptoms include nausea, photophobia, and meningismus. Despite this classic history, presentations can vary widely, as illustrated by multiple clinical grading scales, including the Hunt-Hess score (19) and that developed by the WFNS (20). Patients can present with as little as a mild headache with no neurological deficits to as severe as comatose with flexure or extensor posturing (Table 2.1). Note that the severity of the clinical presentation does not necessarily correlate with the amount of subarachnoid blood.

Often (up to 70%), patients experience a mild headache or neck stiffness due to a small bleed in the days or weeks leading up to the significant bleed, a sign called the "sentinel leak." If diagnosed correctly, these sentinel leaks can be used as a guide and indication for early intervention to prevent further more catastrophic bleeds in the future.

Diagnosis

Noncontrast head CT is the primary means of diagnosing SAH. Such a scan is warranted in any patient presenting with an unusually severe headache or new neurological deficits to rule out an intracranial hemorrhage. Noncontrast head CT has a reported sensitivity of 95% under ideal conditions. CT will not only confirm SAH, but in some cases the distribution of blood can be suggestive of the site of a presumed aneurysm or other underlying pathology. Furthermore, in some cases, the distribution of blood may be suggestive of a perimesencephalic SAH, a specific subgroup of SAH that is important to identify because these patients are at extremely low risk for vasospasm and recurrent hemorrhage (21) (see later discussion).

In patients in whom there is still a high suspicion of SAH despite a negative head CT, based on history, lumbar puncture and analysis of cerebrospinal fluid (CSF) is warranted, representing the gold standard for diagnosing SAH. CSF is examined for either gross blood or xanthochromia (yellow color), representing blood breakdown products in the CSF supernatant. To be sure that blood in CSF represents SAH and not a traumatic spinal tap, serial vials of CSF are collected. In SAH, the red blood cell concentration

TABLE 2.1   Clinical grading of patients with SAH

Grade	WFNS (1988)	Hunt-Hess (Hunt and Hess, 1968)
1	GCS 15	Asymptomatic, mild headache, or slight nuchal rigidity
2	GCS 13–14 without deficits	Cranial nerve palsy, moderate to severe headache, nuchal rigidity
3	GCS 13–14 with either aphasia, hemiparesis, or hemiplegia	Mild focal deficit, lethargy/confusion
4	GCS 7–12, with or without characterizable deficits	Moderate to severe hemiparesis, decerebrate rigidity, stupor
5	GCS 3–6, with or without characterizable deficits	Deep coma, moribund appearance, decerebrate rigidity

is constant across vials, whereas in traumatic spinal taps, the concentration of red blood cells decreases in serial vials.

Definition of Subpopulations

As implied previously, several different processes can give rise to SAH. Although trauma and intracranial aneurysms represent the two largest groups, other processes include AVMs, underlying coagulopathies (intrinsic or pharmacologic), vascular disease, and malignancy. In some cases, the underlying pathology is never identified. Appropriate management requires defining the etiology so that the underlying pathology can be addressed and a better appraisal of secondary complications (such as vasospasm) can be formulated.

In the case of trauma, the etiology is clear by history and no further workup is necessary, except for follow-up CT 24–28 hours after initial insult to ensure that bleeding has stopped. In other cases, the head CT used to confirm SAH reveals or suggests an intracranial lesion that may be responsible for the SAH, such as an AVM or tumor. Although it is tempting to automatically attribute the SAH to these abnormalities, formal assessment of the intracranial vasculature is necessary. Whereas cerebral angiography represents the gold standard, magnetic resonance angiography and computed tomographic angiography (CTA) are alternative means of assessment. Innovations in CTA have made this modality particularly well received, replacing cerebral angiography in some hospitals (22,23).

The most critical question that angiography answers is whether the SAH is aneurysmal or nonaneurysmal. Nonaneurysmal hemorrhages include those arising from AVMs, other vascular malformation, tumors, and a subset called perimesencephalic SAH. Perimesencephalic SAH represents SAH with a particular distribution (mainly or only in the cisterns around the midbrain) in which no aneurysm or other

underlying anatomical abnormality can be identified, despite repeated angiography approximately 1–2 weeks after the bleeding event (24). It is important to identify nonaneurysmal SAH because patients with these bleeds are at low risk for vasospasm. On the other hand, patients with aneurysmal SAH are at significant risk of vasospasm ($\sim$20% overall and up to $\sim$90% in some subgroups, as described previously) and at significant risk for secondary injury due to rebleeding from the aneurysm that has not yet been secured.

Diagnosing Patients with Vasospasm

Two types of vasospasm exist: angiographic and clinical. A diagnosis of clinical vasospasm is based on identifying a deterioration in neurological status in the presence of definite evidence of arterial narrowing after other causes (e.g., hydrocephalus, seizures) have been excluded. Angiographic vasospasm is based on objective evidence of arterial narrowing (independent of clinical status). The objective methods for detecting and diagnosing vasospasm include measuring vascular diameters with cerebral angiography (which is the gold standard), determining blood velocities using transcranial Doppler (TCD), and CTA (25,26). Each method offers advantages and drawbacks. Whereas cerebral angiography is the stated gold standard, the significant drawback is the associated morbidity associated with the procedure itself, including bleeding risks, embolism, stroke, procedure site (i.e., groin) hematomas, pseudoaneurysm induction, and repeated exposures to intravascular dye. This can become a particular burden in patients who may need repeated assessment of their cerebrovasculature and are repeatedly subjected to the risks of the procedure, sometimes on a daily basis. Many centers now regularly use TCD to detect vasospasm, using a mean velocity threshold of 140–200 cm/s as a threshold for a spastic vessel and using the Lindegaard ratio or the ratio of

TABLE 2.2   Outcomes assessment after SAH

Score	Glasgow Outcome Scale	Modified Rankin Scale
0	—	Asymptomatic
1	Death	No significant disability, only minor deficits
2	Persistent vegetative state	Slight disability, yet independent
3	Severe disability, conscious yet disabled	Moderate disability, yet independent
4	Moderate disability yet independent	Moderately severe disability, unable to walk without assistance, dependent
5	Good recovery with only minor deficits	Severe disability, requiring constant nursing care

middle cerebral artery-to-ipsilateral extracranial internal carotid artery velocity ratio (27). Although TCD offers a convenient and portable method for assessing vessel diameter indirectly, it is greatly limited by operator experience and the presence of appropriate cranial windows through which to detect intracranial velocities. Finally, CTA has been used to assess for vasospasm. Like cerebral angiography, it exposes the patient to dye but it obviates the need for direct arterial access for each assessment, thereby minimizing the associated risks. Although many centers regularly use CTA, formal evaluation is still pending in a blinded trial.

Outcomes Assessment

Because of the devastating consequences of SAH, multiple outcomes assessment scales have been used to characterize patient outcomes. These are critical for evaluating and quantifying outcomes, especially with respect to designing, conducting, and analyzing the outcomes of clinical trials. It is key to recognize that these outcomes scales are functional outcomes scales rather than measures of surrogate endpoints. Although many studies measure surrogate markers and endpoints of SAH (e.g., reversal of angiographic vasospasm), one must keep in mind that ultimately the important outcomes are that of the functional status of the patient. Achieving resolution of angiographic vasospasm without altering functional outcomes is not desirable. The

different scales used to quantify patient outcomes are listed in Table 2.2, including the Glasgow Outcome Scale (GOS) and the Rankin Scale. Note that the Hunt-Hess scale and the WFNS scale (Table 2.1) are intended for initial assessment and not originally intended for outcomes assessment. There has been increased interest in using neuropsychological assessment in clinical studies, because these tests can offer a more sensitive and specific means of determining how patients are affected by the disease process.

CURRENT MANAGEMENT OF ANEURYSMAL SAH

The management of aneurysmal SAH is multifaceted, aiming to avert rebleeding from the underlying aneurysm, prevent or minimize the functional consequences of vasospasm, prevent seizures, correct electrolyte abnormalities, and prevent brain edema and swelling. Neurosurgeons have deemed each aspect of this treatment paradigm essential to improve outcomes. Current standards for management of each of these components are reviewed briefly below, including the rationale for why each is so important.

Rebleeding

Natural history studies have shown that the greatest risk of rerupture of a ruptured aneurysm is within 48 hours of the initial event (28). In light of these risks and with the goal of preventing rebleeding, one of the first and most important parts of aneurysmal SAH management includes rapidly identifying and treating reversible causes of active bleeding, including coagulopathies and platelet disorders, using fresh frozen plasma, vitamin K, and platelets, respectively. In addition, aggressive management of high blood pressure is critical because increased blood pressure is theoretically a risk for aneurysmal rerupture. This necessitates admission to an intensive care unit, direct arterial pressure monitoring (with an arterial line), the use of antihypertensive medication drips (including nitroprusside and nicardipine, for example), and minimizing patient stimulation with so-called subarachnoid precautions. Subarachnoid precautions include limiting the number of visitors, minimizing room volume, minimizing coughing by using antitussives (e.g., codeine, dextromethorphan), minimizing strain secondary to pain (i.e., using adequate analgesia), prescribing stool softeners to prevent constipation and undue strain, and using sedation as necessary to minimize the patient's startle reflex and to prevent acute increases in blood pressure.

Once these interventions have been implemented, the neurosurgeon focuses on isolating the aneurysm from the systemic circulation. Although the timing of surgery trials indicated that earlier surgery was not necessarily associated with improved outcomes (due to increased surgical mortality) (29) for a combination of reasons, early surgery or endovascular coiling has been adopted as the preferred contemporary management plan to prevent early rebleeding. Many attribute the decrease in morbidity and mortality from ruptured aneurysms to the move toward early surgery (30). Despite the desire for early intervention, because of the delicate nature of ruptured aneurysms, most surgeons advocate not operating in the middle of the night; it is preferred to operate the next day when the surgeon, interventionalists, and nurses are well rested and an experienced team with ancillary support is available. Some have questioned whether the modality of intervention (i.e., method for isolating the aneurysm) affects the rate of symptomatic vasospasm and ischemic infarction; Dehdashti and colleagues (31) reported that these rates were comparable between treatment groups, as were overall clinical outcomes.

Although studies have investigated the use of antifibrinolytics early after aneurysmal SAH, the use of these agents is not yet standard or universally advocated, although more recent data with refined protocols of administration show promise for this intervention. The studies investigating these agents are described later under Novel, Investigational, and Failed Therapeutics.

In the case of nonaneurysmal SAH, like in the case of aneurysmal SAH, initial therapy is aimed at identifying reversible causes of bleeding. However, early surgery or intervention is not necessary or advocated. For example, in the case of AVMs, studies have shown no early increase in the rate of rerupture. Therefore, early intervention has not been advocated. In any case, regardless of pathology, the goal of surgical intervention is always directed toward treating the underlying pathology.

Vasospasm

Beyond rebleeding, vasospasm is one of the most feared consequences of aneurysmal SAH because of its high rate of morbidity and mortality (1). Consequently, developing pharmacological agents or interventions that target the underlying biochemical pathways mediating vasospasm has become the holy grail of aneurysmal SAH research. Unfortunately, vasospasm is likely the result of a complex and poorly understood cascade of multiple biochemical pathways operating in parallel. The current strategies for prevention and treatment of vasospasm are therefore limited. The most commonly used approaches to managing vasospasm are scheduled calcium channel antagonists (i.e., nimodipine) and the so-called triple H therapy (HHH therapy), composed of hypertension, hypervolemia, and hemodilution.

Calcium Channel Antagonists

Nimodipine is a calcium channel antagonist shown to reduce poor outcomes and DINDs after aneurysmal SAH (32). Although some have questioned the quality of the evidence for the role of nimodipine in preventing and protecting against the sequela of vasospasm, many argue that the risks (largely that of hypotension) are minimal in light of its potential benefit (32). The usual dose of nimodipine is 60 mg by mouth every 4 hours. In those who are particularly susceptible to hypotension, an alternative regimen of 30 mg by mouth every 2 hours can be used.

Some of the key studies leading to this management recommendation are summarized briefly. Calcium channel antagonists were initially introduced into the management (and anticipated prevention) of vasospasm with the view that these agents would prevent vascular constriction. One of the earliest studies to investigate the role of nimodipine in SAH was conducted by Tanaka and colleagues (33) in cats, in which the authors reported that pial vessels dilated in response to intravenous nimodipine after experimental SAH, suggesting the reversal of spasm. The first randomized, placebo-controlled, double-blind study of nimodipine in SAH was conducted in monkeys; the study by Espinosa and colleagues (34) was equivocal, showing no significant different in vessel caliber between the treated and placebo groups and DIND in 1 of 15 subjects in the placebo group and

in 0 of 15 subjects in the treated group. Although this study was important with regards to study design, because of its small number of subjects, the power, utility, and therefore implications of this study are limited.

To date, several prospective, randomized, placebo-controlled, double-blind, randomized trials have investigated the efficacy of calcium channel antagonists to reduce mortality and DINDs (35–38). Early studies consistently demonstrated a significant improvement in morbidity in patients treated with nimodipine (relative to placebo). Interestingly, despite differences in morbidity, the placebo and nimodipine-treated groups consistently showed no significant difference angiographically. This inconsistency between outcomes and angiographic evidence of vasospasm suggests that the protective effects of nimodipine may not be due to its vasodilating abilities but rather by some other neuroprotective effects (37,38). This has been hypothesized to be mediated by dilation of leptomeningeal collaterals and to cause decreased intracellular calcium accumulation and therefore decreased cell death (39). Despite some studies indicating a morbidity benefit, Haley and colleagues reported on behalf of the Cooperative Aneurysm Study that although patients treated with high-dose intravenous calcium antagonists had a reduction in symptomatic vasospasm, the two groups had similar outcomes at 3 months (35,36). Despite these negative findings, the authors concluded that calcium channel antagonists still reduce the risk of symptomatic cerebral vasospasm in a significant proportion of patients and therefore still have a role in treating and preventing vasospasm. The inconsistency across studies probably arises from a lack of power to detect a small but probably statistically and clinically significant difference between the two treatment groups. Therefore, as explained above, because of the minimal risk of this intervention, most conclude that the benefits of calcium channel antagonist therapy outweigh the risks.

HHH Therapy

HHH therapy is composed of hypertension, hypervolemia, and hemodilution. The goal of this therapy is to maintain cerebral perfusion even in the face of vasospasm. The rationale for hypertension and hypervolemia are similar: to increase vascular flow by increasing intravascular pressures and volumes to overcome the vascular narrowing and loss of autoregulation associated with vasospasm. The goal of hemodilution is also intended to increase blood flow by decreasing the viscosity of the blood, based on the Hagen-Poiseuille law. A combination of both crystalloid and colloid fluids are used, including normal saline and albumin, respectively. In some cases, pressors are indicated. The pressor of choice is usually phenylephrine because of its systemic peripheral vasoconstricting properties but lack of vasoconstricting effects on the cerebrovasculature. In general, the target systolic blood pressure is greater than 160 mm Hg. However, the systolic blood pressure target is titrated to reversal of neurological deficit and is based on clinical examination (up to a maximum systolic pressure of approximately 220 mm Hg). Because HHH therapy requires hypertension, the suspected ruptured aneurysm must be secured (either surgically or endovascularly) before its implementation. No randomized prospective trials of HHH therapy once vasospasm has been diagnosed have been reported due to ethical considerations; because it is one of the few therapies available for vasospasm once it has been detected, ethical considerations have prevented withholding such therapy in the face of pending morbidity.

Although HHH was originally devised as a means of treating vasospasm, a handful of studies have also investigated the utility of HHH therapy as a prophylactic measure (i.e., to prevent the onset of vasospasm). At least two prospective randomized trials of HHH therapy compared with normovolemia have suggested that HHH therapy offers no prophylactic benefit (7,40). The power and validity of these studies, however, are significantly limited by the small number of patients enrolled in each (32 and 82, respectively) and incomplete reporting of endpoints. Treggiari and colleagues (41) provided a systematic review of prospective comparative trials of HHH prophylaxis. They identified four trials, only two of which were randomized, namely the trials of Lennihan and colleagues (7) and Rosenwasser and coworkers (42). Across studies, they reported HHH prophylaxis was associated with a reduced risk of symptomatic vasospasm (relative risk [RR], 0.45; 95% confidence interval [CI], 0.32–0.65) and mortality (RR, 0.68; 95% CI, 0.53–0.87) but with no difference in DINDs (RR, 0.54; 95% CI, 0.20–1.49). Interestingly, the difference in mortality maintained statistical significance when only the two randomized trials were analyzed (RR, 0.4; 95% CI, 0.14–0.66), even though neither study found a statistical difference between groups when analyzed individually. This highlights the limitations of study power when small samples sizes are used. As noted by Treggiari and colleagues (41), meaningful comparisons between studies are difficult because of nonstandard definitions and reporting. In light of the possible benefits and lack of significant risks (if the aneurysm has been secured), most institutions prophylactically use some components of HHH therapy.

Intra-arterial Papaverine and Balloon Angioplasty

Intra-arterial infusion of papaverine has become yet another part of the arsenal of therapeutics at some institutions when vasospasm has been diagnosed. Kaku and colleagues (43) were the first to report the use of intra-arterial papaverine in 1992, reporting that 34 of 37 targeted vascular territories were successfully dilated and 8 of 10 patients showed neurological improvement. Kassell and colleagues (44) reported shortly afterward on 12 patients who were treated with intra-arterial papaverine. Eight demonstrated immediate angiographic improvement, and four demonstrated reversal of neurological deficits. Nonetheless, two of the treated patients deteriorated further after initial treatment due to recurrent vasospasm. The authors further noted that vasospasm in the posterior circulation and middle cerebral artery distribution appeared to be more responsive to intra-arterial papaverine.

These initial reports, like many initial reports, were not comparative studies but observational studies of the effects of a novel intervention. Despite over 10 years of reports and use, only two comparative (but nonrandomized) studies have been done to attempt to rigorously validate the use of this intervention. Elliot and colleagues (45), comparing intra-arterial papaverine with balloon angioplasty in a retrospective cohort study, found that patients who were treated with papaverine had a significantly higher rate of requiring retreatment for recurrent vasospasm (1% vs. 42%, $p < 0.001$). Moreover, the authors reported that changes in blood flow velocity induced by papaverine treatment were not sustainable beyond 24 hours, whereas dilation achieved by balloon angioplasty was sustainable. The reliability of this study is limited, however, because of its retrospective nature, its associated inherent bias, and its relatively small sample size for papaverine-treated patients. Polin and colleagues (46) used the prospectively collected database of the North American Trial of Tirilazad for Aneurysmal SAH to compare 31 patients treated with intra-arterial papaverine with matched patients (according to degree of vasospasm and Glasgow Coma Scale scores) who received only medical management. The authors found no statistical difference in the 3-month GOS scores between the groups.

These trials, although not prospective and not randomized, demonstrate the importance of selecting appropriate endpoints in evaluating the efficacy of a new treatment modality. Although papaverine indeed dilates spastic cerebral vessels, these studies suggest the dilation is not sustained and that it does not alter long-term outcomes. Studies need to be continued to evaluate the timing, concentration, and appropriate use of intra-arterial papaverine, because its role in treating vasospasm has not yet been rigorously or thoroughly investigated. Nevertheless, intra-arterial papaverine infusion remains part of standard treatment protocols due to numerous positive anecdotal reports. There has been interest in the benefits of intra-arterial verapamil (which preliminarily results in both angiographic and clinical improvement), although this has not yet been assessed in a prospective manner (47).

Numerous case reports document the perceived utility of angioplasty in the setting of vasospasm. Angioplasty consists of physically dilating constricted vessels with a balloon that has been placed within the lumen of the artery (via femoral cannulation). As described in the previous paragraph, the effects of angioplasty are believed to be more dependable and to achieve greater vasodilation than chemically dilating the vessel with papaverine. Zubkov and colleagues (48) were the first to report the use of angioplasty to treat spastic vessels; they reported on 33 patients in whom 104 of 105 attempted vessels were dilated with reported improvement in neurological symptoms. Despite a number of reports that angioplasty successfully dilates vessels, Polin and colleagues (49) investigated whether this intervention affected outcomes. Using a prospectively collected database, they compared 38 patients treated with angioplasty with 38 matched control subjects and found no significant difference in GOS at 3 months of follow-up. Despite its lack of observed clinical effect (which may have been masked by small sample size), most institutions continue to use angioplasty regularly because there are few other alternatives. Some have reported an interest in prophylactic balloon angioplasty (for which a pilot study was reported with excellent outcomes), but a formal clinical trial has not yet been reported (50).

Seizures

The routine use of anticonvulsant medications as seizure prophylaxis in immediate post-hemorrhagic management has become controversial. Approximately 20–25% of patients have seizures after SAH—most occur pre-hospitalization or in the immediate post-hemorrhage period (51–53). Although the rate of seizures is relatively high, less than 10% of patients go on to have epilepsy (51). Because of the high rate of early seizures, for several years, patients with aneurysmal subarachnoid hemorrhage were routinely loaded on anticonvulsant medication on presentation to the hospital. More recent studies suggest however that such routine prophylaxis results in worse outcomes, both with respective to functional status and cognitive status (53a, 53b). These results have prompted many institutions to stop using canticonvulsants routinely for seizure prophylaxis in patients with aneurysmal subarachnoid hemorrhage.

When seizure prophylaxis is used, it is difficult to predict the appropriate duration of therapy because no accurate predictors of late seizures have been identified (52). Importantly, early antiepileptic therapy does not affect the rate of at which SAH patients develop epilepsy. Further randomized, prospective, placebo-controlled studies will have to be conducted to better define the role of seizure prophylaxis after ancurysmal SAH. With regards to seizures, however, it is noteworthy that Lin and colleagues reported that onset seizures (i.e., seizures within 12 hours of

presentation) were a significant predictor of DINDs (51).

Electrolyte Abnormalities

Electrolyte abnormalities are not uncommon after aneurysmal SAH. Most importantly, hyponatremia, or low sodium concentration, is often reported. Although from a fluid–electrolyte standpoint it is critical to assess whether the hyponatremia is associated with hypervolemia, hypovolemia, or euvolemia, in aneurysmal SAH hyponatremia is most often attributed to cerebral salt wasting. Whereas hyponatremia is often associated with relative hypervolemia (or free fluid excess), the hyponatremia associated with cerebral salt wasting is believed to be due to a true loss of sodium. The interested reader should refer to the review by Palmer (54) for a thorough discussion of cerebral salt wasting. Because this hyponatremia is due to a lack of salt, therapy involves using hypertonic saline or salt tablets to replenish sodium stores. The selection of 1.8% or 3% saline depends on the sodium concentration, the predicted intravascular volume (based on weight), and the rate at which one wishes to correct the hyponatremia.

Some studies have investigated the possibility of promoting salt retention as a means of treating the hyponatremia of cerebral salt wasting rather than replenishing lost salt. Because the major route of salt loss is via the kidneys, studies have investigated using mineralocorticoids, such as fludrocortisone to promote salt retention. Mori and colleagues (55) randomized SAH patients into two groups, one receiving fludrocortisone and the other receiving placebo, demonstrating that fludrocortisone effectively reduced the need for sodium and water to prevent hyponatremia and achieve hypervolemia, respectively. A similar study was conducted with hydrocortisone, in which hydrocortisone therapy seemed to promote hypervolemia and significantly reduce the incidence of hyponatremia (56). Hydrocortisone offers the advantage of having a shorter half-life and being easier to titrate and thereby avoiding heart failure. Despite these encouraging studies for the role of mineralocorticoids in SAH, their use is not part of standard current management, although their use is increasing.

Cerebral Edema and Hydrocephalus

By virtue of having the brain enclosed within a fixed volume (the skull), significant pathological consequences can ensue from excessive edema or hydrocephalus. Hydrocephalus is most likely to occur if the hemorrhage communicates with the ventricular space. The treatment of choice in the case of hydrocephalus is ventriculostomy, which provides a means for both measuring intracranial pressure and draining CSF to reduce intracranial pressure. At times pharmacological agents, including hypertonic saline and mannitol, can also be used to reduce the volume of intracranial contents and therefore intracranial pressure. Finally, hyperventilation is often used as a temporary means of reducing intracranial volume (and pressure) by reducing PCO_2 and therefore cerebral blood flow and volume.

Perhaps the most controversial pharmacological intervention is the use of steroids in SAH. The benefits of steroid therapy in SAH remain unproven. Many would argue that the possible benefits are far outweighed by the added complications associated with high-dose steroid use, the most important of which is infection. Schurkamper and colleagues (57) reported in a retrospective study that regardless of neurological score, patients treated with larger doses of steroids had significantly decreased rates of hydrocephalus and significantly increased rates of favorable outcomes. This study was retrospective, and therefore it is unclear what kind of selection bias may have been introduced. However, it raises the question of whether

it would be appropriate to restudy the role of dexamethasone, or other steroids, in the acute phase of SAH.

NOVEL, INVESTIGATIONAL, AND FAILED THERAPEUTICS

Clinical trials for the management of SAH have addressed each point of SAH management. However, most investigations have focused on the prevention and effective treatment of vasospasm, because of its associated morbidity and mortality and the perceived opportunity to prevent it.

As described earlier, vasospasm is likely due to a complex cascade of parallel yet interacting biochemical pathways, likely including endothelium-derived factors (including nitric oxide and oxygen free radicals), vascular smooth muscle–derived factors (e.g., calcium channel activation and protein kinase C activation), proinflammatory mediators (e.g., histamine and bradykinin), and stress-induced gene activation (e.g., heat shock proteins and heme oxygenase-1) (58). Ultimately, these pathways result in vascular constriction, vascular smooth muscle proliferation, reduced perfusion, and neuronal injury.

Although clinical trials were originally designed to treat the symptoms and consequences of vasospasm (i.e., by increasing perfusion or by administering neuroprotective agents), more recently clinical trials have tried to address different aspects of the biochemical pathways that result in vasospasm (i.e., symptomatic treatment vs. treating the underlying disease). These latter trials have resulted in many intriguing yet nonsignificant results, suggesting none is addressing a final common pathway but only components of this complex cascade. Both types of studies are briefly reviewed here to highlight some of the failures of clinical trials in the field of SAH and to draw attention to some trials that may hold promise. Note that not every trial fits precisely into a specific category, largely because of so many persistent unknowns about SAH. Therefore, when appropriate, trials are cross-referenced between sections.

Preventing Rebleeding

The key to preventing rebleeding from a ruptured aneurysm is to isolate it from the circulation as soon as possible using either surgical or endovascular intervention. Some have recognized, however, that there is an opportunity for therapeutics between the time of hemorrhage and time of definitive treatment (surgical or endovascular); this is a critical time during which a devastating rerupture could occur.

The most thoroughly investigated agents to date have been antifibrinolytics, which are intended to prevent lysis of clots that are preventing a ruptured aneurysm from rebleeding. The theoretical disadvantage of such agents is that they may predispose patients to clotting after surgery and, in the face of vasospasm, make such patients more susceptible to DINDs and worse outcomes. Confirming this theory, both Fodstad (59) and Kassell and colleagues (29) reported on prospectively collected data on 105 patients and 672 patients, respectively, that patients treated with antifibrinolytics experienced significantly decreased rates of early rebleeding, increased rates of vasospasm and DINDs, and relatively no difference in overall outcomes and mortality. A meta-analysis by Roos and Vermeulen (60) supported these findings. Because of the increased rate of DINDs, most centers opt for early surgical intervention rather than administering antifibrinolytics. In light of data suggesting the efficacy of short-term administration without an increased risk of vasospasm and hydrocephalus (61,62), some centers have adopted their use (63).

Clot Evacuation

One of the earliest hypothesized strategies for prevention of vasospasm was to remove the subarachnoid blood as soon as possible to minimize the risk that the adjacent blood vessels would constrict via removal of the putative spasmogenic substrate. Lending credence to these early hypotheses, studies have shown that vasospasm is probably at least in part due to the release of oxyhemoglobin from erythrocytes in the subarachnoid space, triggering the endothelin pathway and leading to vascular constriction and smooth muscle proliferation. Therefore, theoretical evacuation of this blood should benefit the patient. Studies have looked at both surgical evacuation (while attempting to clip the aneurysm) and the use of fibrinolytics.

Surgical Evacuation

Surgical evacuation of the subarachnoid clot is done during open surgical procedures to clip the ruptured aneurysm. Primate studies, in which clot was either removed after 48, 72, or 96 hours or never removed, showed that the severity of vasospasm paralleled the duration of contact between the blood clot and the cerebral vessels (64). The authors therefore concluded that clot removal at early operation is likely to be useful only if it is performed within 48 hours of SAH. Studies in humans, as part of the Timing of Aneurysm Surgery trials, suggested that the timing of surgery does not make a significant difference in morbidity and mortality (65). A thorough discussion of the timing of aneurysm surgery is beyond the scope of this chapter. However, despite these studies, most neurosurgeons choose to operate on a ruptured aneurysm as soon as possible, both to achieve clot removal and to secure the aneurysm to prevent rebleeding.

Fibrinolytics

The role of fibrinolytics in the setting of SAH has also been the subject of multiple clinical trials, with the belief that these agents can help eliminate, at least in part, the offending agents that trigger the vasospasm pathway. Findlay and colleagues (66) originally reported on the relative safety of intrathecal administration of tissue plasminogen activator (tPA) in monkeys, reporting that intrathecal therapy appeared to reduce the degree of vasospasm, reduced the incidence of DINDs, and completely cleared the subarachnoid space of blood clot. In a follow-up study of timing of intrathecal administration of tPA, the authors concluded that treatment within 72 hours of SAH was effective in preventing vasospasm in primates (67). In their first report in humans, these same authors reported on intracisternal injection of tPA in 15 patients (68). They reported that all but one patient had complete or partial cisternal clot clearing (by CT) and that the one patient to not show clot clearance was the only one to experience symptomatic vasospasm. Eight of the 14 remaining patients had mild to moderate arterial narrowing despite treatment. Although interesting and intriguing, this study was limited because of its observational nature, the lack of a control group, and the lack of long-term outcome assessment. Mizoi and colleagues (69) were the first to report a prospective comparative study of tPA-treated patients versus no tPA treatment, concluding that intrathecal injection of 2 mg of tPA daily for 5 days (a total of 10 mg) is effective in preventing the development of vasospasm. The validity of this study, however, is severely limited by the selection criteria for the different treatment arms: To be treated with tPA, the patients' CTs had to be quantifiably different from the control group.

The first, and only, multicenter, randomized, blinded, placebo-controlled trial of the role of intracisternal fibrinolytic therapy in the setting of SAH was reported in 1995 by Findlay and colleagues (70). Studying 100 patients, they reported, contrary to earlier reports, that the overall incidence of angiographic vasospasm was similar between the two groups (74.4% in treated vs. 64.6% in placebo), although there was a trend toward lesser degrees of vasospasm. When only those patients with thick clots were considered, the authors reported a 56% relative risk reduction in the incidence of severe vasospasm in the treated group ($p < 0.05$). Other clinically important trends in the tPA-treated group (that did not reach statistical significance) included reduced delayed neurological worsening, a lower 14-day mortality rate, and improved 3-month outcome rate. Overall, bleeding complication rates did not differ between the two groups. The major limitation of this study, as the authors themselves suggested, is that they largely focused on radiographic endpoints rather than clinical endpoints. As the authors stated, "Although the fibrinolytic treatment used in this study may reduce angiographic vasospasm, its efficacy in preventing clinical vasospasm and its ischemic consequences require reexamination in a larger randomized trial" (70). Such a trial has not yet been reported.

Some investigators turned their attention toward the possibility of using fibrinolytics in the setting of endovascular embolization. Hamada and colleagues (70a) reported a prospective randomized trial of coil embolization followed by intrathecal urokinase infusion into the cisterna magna in 110 patients. This study improved on the former clinical trial in that they looked at outcomes 6 months after SAH. They reported symptomatic vasospasm in 9.4% of treated and 28.1% of untreated subjects ($p = 0.012$), improved outcomes in the treated groups (90.6%) compared with the untreated group (75.4%) ($p = 0.036$), but no difference in mortality between the groups (3.8% vs. 5.3%, respectively). The authors therefore concluded that intrathecal urokinase therapy was beneficial for patients, resulting in a lower rate of permanent neurological deficits, despite no difference in mortality. Although apparently at odds with the former report, significant differences exist in that the patients in the latter study never had a craniotomy and were followed for a longer period of time. This latest study highlights the need for further investigation of the role of fibrinolytics in SAH to prevent vasospasm and its sequelae.

Preventing Vascular Constriction

Several studies have demonstrated that nitric oxide plays a critical role in maintaining basal cerebrovascular tone (71,72). Nitric oxide has therefore been a natural point of interest for vasospasm investigations. Studies have looked at both basal nitric oxide levels and the effect of replenishing nitric oxide (by administering both nitric oxide precursors, such as L-arginine and pharmacological nitric oxide donors). Animal studies suggest that nitric oxide levels decrease after experimental SAH (73). This reduction in nitric oxide levels was associated with decreased frequency and amplitude of vasomotion (nitric oxide plays a critical role in regulating baseline cerebrovascular vasomotion) and was associated with histopathological evidence of cerebral ischemia (73,74). These blood flow alterations and histopathological changes can be alleviated by administration of L-arginine, a precursor of nitric oxide (73), or other nitric oxide donor complexes (74). Based on animal studies such as these, trials in humans have shown angiographic improvement of vasospasm with nitric oxide donors, such as nitropaste (75), and intraventricular/thecal injection of sodium nitroprusside (with reversal of associated weakness) (76,77). Despite these studies, the precise role of nitric oxide in vasospasm remains unclear, and therefore the use of nitric oxide donors or analogues are not yet a part of the normal management of vasospasm. Formal clinical trials need to be conducted, with the knowledge

of these precursor studies, to elaborate the efficacy of nitric oxide management in vasospasm and its effect on overall outcomes. No prospective randomized trials have been conducted using nitric oxide–related pharmaceuticals to date.

The roles of intra-arterial papaverine and nicardipine in preventing vasoconstriction were discussed previously under Current Management of Aneurysmal SAH. These two interventions are now accepted as standard of care at many institutions.

Vascular (Smooth Muscle) Activation

Despite the fact that the name "vasospasm" suggests that the underlying mechanism of vasoconstriction is due to muscle "spasm," it is now well accepted that smooth muscle constriction plays only a minor and early part in the pathophysiological mechanism of vasospasm. Several studies indicate that after 24–48 hours, vasospasm is largely mediated by vascular and smooth muscle activation, triggered by the release of endothelin and subsequent initiation of a vascular endothelial growth factor (VEGF)-mediated cascade, the activation of rho kinase II pathways (discussed later under Neuroprotection), and the presence of reactive oxygen species. These and other multiple pathways in parallel ultimately result in vasospasm that gives rise to DINDs and mortality. Because of the extensiveness of the vasospasm literature, only a selected number of potential targets are discussed as examples.

Therapies Targeted at the Endothelin/VEGF Cascade

Endothelin has become of primary interest in the discussion of pathophysiological mechanisms resulting in vasospasm. It is hypothesized that oxyhemoglobin released from lysed erythrocytes in the subarachnoid space causes vessel contraction. This process is potentiated by endothelin-1 (which is created from endothelin by endothelin-converting enzyme) via rho kinase and protein kinase C. The downstream effects of endothelin-1 are likely mediated by the release of VEGF, one of the most potent mediators of cerebral angiogenesis (78). (VEGF may therefore be another potential target for intervention.) Studies have implicated this pathway *in vitro* by measuring cerebrovascular smooth muscle contraction in the presence of varying amounts of endothelin-1 and inhibitors of the endothelin receptor and rho kinase (79–82). Consistent with these theories, studies have shown that plasma endothelin-1 concentrations correlate with the incidence of DINDs after SAH; Juvela (83) concluded that "an increased ET conversion rate in endothelium predicts ischemic symptoms."

Based on the animal trials similar to those described above and observational studies in humans, a phase II trial was reported using an endothelin receptor antagonist, TAK-044 (84). The multicentered, double-blind, placebo-controlled trial consisted of 420 subjects. Endpoints included DINDS within 3 months of first dose of medication, DINDS within 10 days of first dose, evidence of new cerebral infarct (on CT or postmortem), GOS at 3 months, and adverse events. The authors reported a lower incidence of DINDs at 3 months in the treatment group (29.5% vs. 36.6%; RR, 0.8; 95% CI, 0.61–1.06) with no other significantly different endpoints. These preliminary results did not, and should not, discourage the investigators because this investigation was intended to be a phase II trial; the authors intend to conduct a full phase III trial, with appropriate power (84).

Therapies Targeted at Reactive Oxygen Species

Because of the release of oxyhemoglobin by hemolyzed erythrocytes in the subarachnoid space, many hypothesize that there is an increased concentration of reactive oxygen species in the subarachnoid environment, thereby giving rise to vasospasm (85). Investigations have therefore focused on administering agents that would decrease the availability of these reactive oxygen species. One of the earliest animal studies to investigate the benefits of antioxidant therapy was reported in 1984. The authors (86) administered an antioxidant intrathecally in a canine model of SAH and measured basilar artery diameters angiographically. They reported that chronic spasm was significantly suppressed in a dose-dependent manner. The same authors conducted a similar study 12 years later in humans (86a). They reported a multicenter, placebo-controlled, double-blind, clinical trial to verify the beneficial effects of free radical scavenging on DINDs and overall outcomes (86). Studying 162 patients, the authors reported a 34.5% reduction in the incidence of DINDs in the treated group compared with placebo (35.5% vs. 54.2%, $p < 0.05$), a significantly improved GOS at 1 month ($p < 0.05$) and a marginal improvement in outcomes at 3 months, and a significantly reduced cumulative incidence of death ($p < 0.05$) (86).

These reactive oxygen species are likely responsible for lipid peroxidation, the target of tirilazad therapy. Tirilazad is discussed later under Neuroprotection.

Other investigators have reported on different strategies for reducing free radicals. Aladag and colleagues (87) reported such an experiment in an animal model of SAH in which they administered a superoxide dismutase mimetic to scavenge superoxide anions and peroxynitrite. The authors reported reduced basilar artery constriction in treated animals versus untreated animals.

Maintaining Perfusion

Although the investigations discussed thus far have focused on identifying and targeting the pathways resulting in vasospasm, many clinical trials still focus on developing modalities for treating the symptoms of vasospasm. These studies largely entail pursuing interventions that maintain and even increase cerebral perfusion in the face of vasospasm. The original intervention and the current standard of care for maintaining cerebral perfusion is HHH therapy. HHH therapy was discussed in detail previously under Current Management of Aneurysmal SAH. However, because of equivocal outcomes despite HHH therapy, many investigators continue to pursue other therapeutics to maintain perfusion during vasospasm and improve outcomes. These interventions largely focus on maintaining the patency of the cerebrovasculature using antiplatelet, or antithrombotic, therapy. In most cases, these therapies are initiated after exclusion of the aneurysm from systemic circulation by either surgical or endovascular intervention.

Antiplatelet Therapy

The use of antiplatelets to prevent DINDs and to improve outcomes after SAH has been motivated by the theoretical advantage of limiting platelet aggregation in an already constricted vessel and by the report that some antiplatelet agents may inhibit vasoconstriction mediated through oxyhemoglobin (one of the implicated pathways in vasospasm) (88). In an observational study, Juvela (89) reported that the relative risk of cerebral infarct after SAH (as determined by head CT) was 0.18 (95% CI, 0.04–0.84) in those patients who had been on aspirin before the SAH compared with those with no history of aspirin use, suggesting a benefit to antiplatelet therapy in the face of SAH. Five studies to date have investigated the benefits of antiplatelet therapy in a randomized controlled design (90–94). Only some of the studies were completely blinded (90,91,93). These studies unanimously report no difference

in long-term outcomes in patients treated with antiplatelets as opposed to placebo. With regards to DINDs, only three trials reported results, and only one (93) demonstrated a significant difference between treatment groups. A meta-analysis by Dorhout Mees and colleagues (95) suggested that the relative risk of DINDs in patients treated with antiplatelet therapy across trials was still significantly less than those not given antiplatelet therapy (RR, 0.65, 95% CI, 0.47–0.89). As with other trials of therapeutic interventions, these trials are admirable because they randomized patients and had appropriate control subjects. Unfortunately, for the most part, study samples were prohibitively small (sometimes as small as 11 patients), significantly restricting the power of these studies to truly identify differences between groups. Accordingly, Dorhout Mees and colleagues (95) concluded that a more thorough randomized trial is needed to assess the validity and usefulness of antiplatelet therapy in altering outcomes after SAH. A pilot phase I trial has already been reported for a larger clinical trial (90).

Neuroprotection

All the interventions described thus far have attempted to either target underlying vasospasm mechanisms or provide a means of overcoming the effects of vasospasm. An alternative is to administer a neuroprotectant agent, an agent that has not necessarily been specifically designed for vasospasm but has been shown in other neurological disease processes to be protective. The primary distinction is that the agents administered may protect the brain by supplementary pathways rather than necessarily targeting molecules and processes primarily involved in the pathophysiological pathways. Three prime examples of neuroprotectants that have been studied extensively are tirilazad, fasudil, and magnesium.

Tirilazad

Tirilazad is a potent inhibitor of oxygen free radical–induced, iron-catalyzed, lipid peroxidation in microvascular and nervous tissue that was designed from a theoretical standpoint, without any apropos evidence that it would interfere with the natural pathophysiology of ischemic stroke. It has been shown in animal models of ischemic stroke to be neuroprotective (96,97). Despite these animal studies suggesting possible improvement, randomized prospective studies suggested that tirilazad may in fact increase morbidity and mortality (98,99). Based on the theoretical benefits this drug could confer, several clinical trials were also conducted in patients with SAH to determine whether it may alter morbidity and mortality in patients who have vasospasm.

The first phase III clinical trial investigating the efficacy of tirilazad to reduce ischemic symptoms from vasospasm and improve outcomes was published by Kassell and colleagues in 1996 (100). The study serves as a model for a well-designed, randomized, double-blind, placebo-controlled, clinical trial with adequate enrollment (1023 patients) to ensure adequate power; it was conducted across 41 neurosurgical centers in Europe, Australia, and New Zealand. Although different doses of tirilazad were administered, only the group receiving 6 mg/kg/day demonstrated a reduced mortality ($p = 0.01$), an improved 3 month GOS ($p = 0.01$), and a trend toward decreased symptomatic vasospasm ($p = 0.048$). Interestingly, the benefits were predominately found in men. The authors therefore concluded that it improves overall outcomes in patients who have experienced an aneurysmal SAH (especially men).

Because of the decreased efficacy of tirilazad in women in the original clinical trial and the apparent dose–response relationship from the original study, follow-up trials were conducted, both internationally (819 patients) and within North America (832 patients), to assess the efficacy of high-dose tirilazad (15 mg/kg/day) in women (101,102). As before, the international study revealed a statistically significant mortality advantage among patients treated with tirilazad who were Hunt-Hess grade IV or V at time of admission compared with placebo-treated patients (24.6% vs. 43.4%, respectively, $p = 0.016$) (101). Interestingly, the outcomes advantage was not seen in the North American trial; although there was a significant reduction in symptomatic vasospasm, 3-month mortality was not different in this accompanying study. The authors attributed this difference to possibly relatively better therapy of placebo-treated patients in the North American trial with other effective therapies, such as HHH therapy (102).

Fasudil

Fasudil is discussed under the heading Neuroprotectants because, despite extensive investigations, its precise mechanism of altering outcomes in SAH patients remains unclear. Fasudil was originally designed as an intracellular calcium antagonist, with the intent of achieving vasodilation. The first clinical trial using fasudil was reported in 1992 from Japan in which 267 patients with Hunt-Hess grades I–IV were randomized to either receive fasudil or placebo (103). The groups were matched clinically and demographically. Fasudil reportedly reduced angiographic vasospasm by 23% (61% vs. 38%, $p = 0.0023$), ischemic CT lesions by 22% (38% vs. 16%, $p = 0.0013$), and symptomatic vasospasm by 15% (50% vs. 35%, $p = 0.0247$). More importantly, the authors reported that treated patients had improved rate of good GOS scores (26% vs. 12%, $p = 0.0152$) (103). This study promoted the widespread use of Fasudil in Japan to reduce the rate of vasospasm and its pathophysiological consequences.

Studies, however, suggest that the mechanism of action of fasudil may not be limited to calcium antagonism. Rather, a possibly more clinically significant role of fasudil in vasospasm may be its role in rho kinase II inhibition, thereby preventing the activation of pathways that promote vascular smooth muscle cell contraction and proliferation. Rho kinase is thought to play a critical role in endothelin-1 mediated vasoconstriction and proliferation. Fasudil has been shown *in vitro* to reduce neointimal formation after balloon injury to blood vessels (104). The complexity of action of fasudil is yet again emphasized by Kim and colleagues (105), who reported the benefits of fasudil are, at least in part, due to prevention of calponin degradation, a filament associated protein that is degraded in vasospasm. A full discussion of fasudil is beyond the scope of this chapter because of its complexity and unknown precise mechanism. However, the investigations surrounding this agent highlight that the precise mechanisms of an agent do not need to be known in advance and that the perceived mechanism of action is not necessarily the true mechanism of action. However, in all cases, the utility of the agent should be investigated thoroughly in animals before human trials.

Magnesium

Hypomagnesemia is frequent after SAH and has been noted to be associated with the severity of SAH. Moreover, hypomagnesemia between postbleed days 2 and 12 has been described as predictive of DINDs (106). Accordingly, there has been increasing interest in magnesium as a therapeutic for vasospasm. Magnesium may increase cerebral blood flow, reduce the contraction of cerebral arteries caused by various stimuli, and act as a nonspecific neuroprotectant (107). Although blinded and controlled animal studies do not demonstrate an angiographic advantage (in terms

of reducing vasospasm), there is persistent interest in this compound as a neuroprotectant.

A prospective, randomized, single-blind, clinical trial of high-dose magnesium therapy after aneurysmal SAH in 40 patients reported a trend (not statistically significant) in which a higher percentage of patients obtained GOS scores of 4 or 5 in those treated with magnesium compared with control subjects. As the authors suggested, a larger study is needed to evaluate this trend because the power of this study was limited by an *n* of 40 (108).

Nimodipine

Many have suggested that the beneficial effects of nimodipine and other calcium antagonists are by way of a nonspecific neuroprotective mechanism. This suggestion has been made in light of the fact that patients treated with nimodipine seem to fare better but do not necessarily have improved angiographic appearance of vasospasm. For a more complete discussion of the mechanisms of action of nimodipine, see the previous discussion under Current Management of Aneurysmal SAH.

Other Neuroprotective Agents

The search continues for other potential targets for neuroprotection. As illustrated by a study by Ivanova and colleagues (109) in which they identified increased levels of neurotoxic agents such as 3-aminopropanal in the CSF of patients after SAH, one can search indefinitely for potential targets for intervention. Ivanova and colleagues reported that 3-aminopropanal levels correlate with the degree of cerebral injury as measured by admission Hunt-Hess grade and that modifying this chemical level by administering N-2-mercaptopropionyl glycine altered 3-aminopropranal levels and altered infarct volume. As the vasospasm literature demonstrates, an endless number of potential targets can be identified. It is critical to assess in animal models, in advance, whether targeting the molecule of interest indeed affects predetermined endpoints. As is clear from the tirilazad studies, the development and testing of nonspecific neuroprotectant agents can be an involved process that may end in unclear equivocal results.

BIOLOGICAL MARKERS, SURROGATE ENDPOINTS, AND OUTCOMES

The issue of selecting appropriate endpoints for clinical trials has been discussed throughout the chapter thus far and in particular when highlighting the limitations of each study. Because of the importance of this issue, the major points are reiterated here. Studies of vasospasm after SAH have clearly demonstrated that the most important endpoint that can be assessed in any clinical trial is functional outcomes and mortality. Although vasospasm has been defined angiographically and physiologically (as measured by TCD), evaluating the efficacy of treatments using these criteria is inadequate. As illustrated well in the case of investigations of nimodipine and fasudil, for example, it is possible to alter outcomes without evidence of angiographic resolution of vasospasm. Conversely, angiographic reversal of vasospasm does not necessarily correlate with improved outcomes, as summarized in the section describing angioplasty as a therapy for vasospasm. Moreover, it is important to recognize that therapeutics that benefit one aspect of care can be detrimental and worsen outcomes in other respects, as illustrated by certain antifibrinolytic trials.

While measuring and reporting intermediate and surrogate endpoints of blood flow and velocities, vessel diameters, and serum concentrations of various intermediary molecules (as described for various

studies earlier in the chapter), it is absolutely necessary for critical appraisal of studies and for understanding the pathophysiological mechanisms resulting in vasospasm and DINDs. They should not be used as substitutes for valid endpoint measures for evaluating novel therapeutics.

WHY DO SO MANY CLINICAL TRIALS FAIL?

Much of the difficulty in addressing the management of SAH and the pathophysiological consequences of vasospasm arises from the multifactorial nature of the disease process. A seemingly endless number of factors and potential therapeutic targets have been associated with vasospasm, suggesting that vasospasm does not arise from activation of a single simple pathway. Rather, it is the result of a complex multifactorial cascade. It is therefore not surprising that clinical trials that target only a single part of this complex cascade fail or report only partial success, at best. One would expect greater success from clinical trials that plan to target multiple parts of this complex cascade simultaneously or a final common pathway, such that compensatory mechanisms do not outcompete the proposed therapeutic interventions. Similarly, the management of SAH and vasospasm is complicated by ostensibly competing interests. Although initially the goal of therapy is to prevent rebleeding (e.g., antifibrinolytics), the goal of later phase therapy is to ensure perfusion. Unfortunately, the interventions introduced during the early phase may result in poorer perfusion later in the disease course, as is evident by the increased rate of ischemic complications in patients treated with traditional regimens of antifibrinolytics (described previously in detail).

However, the failure of so many trials is not limited to the complexity of vasospasm but can also be attributed to several shortcomings in study design (as noted throughout the chapter). Most importantly, most studies have been underpowered. SAH is a complex disease process that does not occur in isolation; it is often accompanied by other neurological insults, including vasospasm and resulting cerebral ischemia, cerebral edema, electrolyte abnormalities, and seizures. Therefore, to account for inherent heterogeneity of samples, it is critical to use large sample sizes (on the order of that used by the Cooperative Aneurysm Study) to detect clinically significant changes associated with a specific therapeutic regimen.

The issue of heterogeneity is more complex when one considers that in general, all aneurysmal SAHs are considered a single disease entity, instead of evaluating the natural history of aneurysms with distinct locations and sizes separately. For example, although both supraclinoid and basilar tip aneurysms can result in SAH, it is conceivable and probable that the management of these hemorrhages may be different in some respects, because of differences in etiology, perfusion pressures, and surgical accessibility. Ideally, each type of SAH and aneurysm resulting in SAH should be characterized independently to determine the best management option because each one represents a distinct disease process. Patient demographics should also be considered. For example, studies of patients with intracerebral hematomas have shown that patients of different ages respond differently to different types of therapies (110). The question therefore remains, should people of all ages be treated equally and with the same interventions? Multivariate analysis is therefore critical for informed analysis and interpretation of study data. To accomplish this, as stated in the previous paragraph, larger samples must be studied.

Consideration of study design also dictates that we assess the quality of the literature currently. As with any novel therapeutic, the literature is buried in observational studies often without any control groups and case reports or anecdotes. Although these are critical for stimulating discussion and ideas, follow-up studies should be well thought out and designed instead of publishing further observational studies that merely duplicate prior results.

Finally, as discussed in the previous section, for clinical trials to be useful, it is critical that appropriate endpoints (and intermediate or surrogate endpoints) be measured so that the true efficacy of a particular therapeutic regimen can be evaluated. Studies have often measured surrogate endpoints (such as angiographic dilation) rather than outcomes, limiting their overall utility and contribution to our understanding of therapeutics for SAH.

This multiplicity of pathways for damage and outcomes may be a major contributing reason for the failure of phase III clinical trials. These limitations argue for better-designed and controlled studies with larger populations, such as the Surgical Treatment of Intracerebral Hemorrhage trial, which aimed to characterize optimal management strategies for patients with intracerebral hemorrhage. Increasing sample size (as determined by proper power analysis) will help overcome the limitations of sample heterogeneity (111).

EMERGENT CLINICAL TECHNOLOGIES AND METHODOLOGIES

As has been demonstrated in this chapter, SAH is a complicated problem, likely multifactorial in origin. Although emergent clinical technologies and methodologies are exciting and offer the opportunity to detect further changes that accompany clinical vasospasm, these advances promise to further complicate future clinical trials by adding more factors and outcomes to account for. Although more complicated, these advances may still help the field progress.

Perhaps the most important advance with respect to clinical trials of SAH is the increased attention given to more sensitive and specific measures of patient's functional status, both in terms of developing new scales and paying more attention to previously developed scales, such as the National Institute of Health Stroke Scale, the sickness impact profile, and the Barthel Index. Part of the failure of clinical trials to detect differences between treatment and control groups in the past has not only been the lack of trial power (as discussed throughout this chapter), but also the lack of a sensitive and specific measures of outcomes to truly differentiate between outcomes. As illustrated in Table 2.2 the outcome measures most often used are crude, possibly preventing the detection of small but clinically significant differences in functional outcomes. Developing more sensitive scales is critical for more precisely assessing how interventions affect outcomes.

The other major technological advances that impact management of SAH are in the field of imaging. With the development of new technologies such as CTA and new pulse sequences for magnetic resonance imaging, it will be increasingly tempting to use imaging to assess the efficacy of novel management interventions. By offering increased spatial resolution, multiple viewing perspectives, and the opportunity to assess tissue metabolism (e.g., magnetic resonance spectroscopy), to mention only a handful of imaging advances, these techniques will tempt investigators. Although it is clearly important to assess vascular dilation, tissue perfusion, and neuronal metabolism, one must keep in mind that the patient's functional outcome is still the most important endpoint that should be assessed in clinical trials. Although these advances will surely help us begin to unravel the mechanisms and effects of vasospasm,

these new imaging techniques will add significantly more to the amount of data to be analyzed without necessarily advancing our understanding at the same rate.

Advances in molecular biology will also surely play a role in creating management paradigms and designing new clinical trials in SAH. Keeping in mind the primary goals of SAH management, including preventing rebleeding, maintaining perfusion, and preventing vasospasm, it will be important to continually survey the literature for advances that may help support these three goals. One such opportunity is advances with factor VIIa, a hemostatic therapy that is intended to stimulate coagulation in individuals in whom the coagulation cascade is otherwise normal (113). Factor VIIa promotes local hemostasis at sites of vascular injury in both patients with and without coagulopathies (114,115). Although several studies have investigated the use of antifibrinolytics early after SAH (before aneurysmal obliteration), some have suggested investigating the possible role of factor VIIa in managing rebleeding. No randomized trials have yet to be reported with this therapy in patients with SAH to assess its value. Although being an attractive theory, it should be emphasized that this potential therapeutic has several risk factors, the most important of which are a theoretical increased risk of embolic cerebrovascular accidents due to hypercoagulability and possible increased risk of vasospasm. Other opportunities may include the development of rheological agents that decrease red blood cell adhesion and other general neuroprotective agents. Other points for intervention are mentioned throughout the chapter.

CONCLUSION

The management of SAH and vasospasm remains challenging, with few proven interventions that alter clinical outcomes.

Currently, the only regularly used intervention in North America with some evidence from clinical trials supporting their efficacy include early definitive therapy, seizure prophylaxis, electrolyte normalization, intracerebral pressure monitoring and correction, nimodipine, HHH therapy, balloon angioplasty, and intra-arterial papaverine therapy. As discussed, even some of these therapies remain controversial. Although numerous clinical trials have been conducted in search of interventions that will alter outcomes, most have shown little, no, or inconsistent benefits of proposed interventions, compared with placebo. The failure of these clinical trials is multifactorial, related to the underlying complexity of the mechanisms resulting in vasospasm, the failure of clinical trials to recruit enough participants to provide adequate power to assess the efficacy of interventions, and the lack of appropriate, sensitive, and specific measures of clinical endpoints. Even so, the morbidity and mortality of vasospasm is theoretically preventable. Therefore, it is critical that clinical trials continue to investigate means of preventing early rebleeding, maintaining perfusion, and preventing vascular constriction and subsequent neurological deterioration. With increased attention to details, outcomes, and study design as well as an improved understanding of the biochemical pathways resulting in vasospasm, effective interventions will surely be developed and tested in successful clinical trials in the future.

References

1. Kassell NF, Sasaki T, Colohan AR, Nazar G. Cerebral vasospasm following aneurysmal subarachnoid hemorrhage. Stroke 1985;16:562–572.
2. Housepian EM, Pool JL. A systematic analysis of intracranial aneurysms from the autopsy file of the Presbyterian Hospital, 1914 to 1956. J Neuropathol Exp Neurol 1958;17:409–423.
3. Inagawa T, Hirano A. Autopsy study of unruptured incidental intracranial aneurysms. Surg Neurol 1990;34:361–365.
4. McCormick WF, Nofzinger JD. Saccular intracranial aneurysms: an autopsy study. J Neurosurg 1965;22:155–159.
5. Winn HR, Jane JA Sr, Taylor J, Kaiser D, Britz GW. Prevalence of asymptomatic incidental aneurysms: review of 4568 arteriograms. J Neurosurg 2002;96:43–49.
6. Weir B. Unruptured intracranial aneurysms: a review. J Neurosurg 2002;96:3–42.
7. Lennihan L, Mayer SA, Fink ME, Beckford A, Paik MC, Zhang H, Wu YC, Klebanoff LM, Raps EC, Solomon RA. Effect of hypervolemic therapy on cerebral blood flow after subarachnoid hemorrhage: a randomized controlled trial. Stroke 2000;31:383–391.
8. Wiebers DO, Whisnant JP, Huston J 3rd, Meissner I, Brown RD Jr, Piepgras DG, Forbes GS, Thielen K, Nichols D, O'Fallon WM, Peacock J, Jaeger L, Kassell NF, Kongable-Beckman GL, Torner JC. Unruptured intracranial aneurysms: natural history, clinical outcome, and risks of surgical and endovascular treatment. Lancet 2003;362:103–110.
9. Juvela S, Porras M, Heiskanen O. Natural history of unruptured intracranial aneurysms: a long-term follow-up study. J Neurosurg 1993;79:174–182.
10. Juvela S, Porras M, Poussa K. Natural history of unruptured intracranial aneurysms: probability of and risk factors for aneurysm rupture. J Neurosurg 2002;93:379–387.
11. Tsutsumi K, Ueki K, Morita A, Kirino T. Risk of rupture from incidental cerebral aneurysms. J Neurosurg 2000;93:550–553.
12. Charpentier C, Audibert G, Guillemin F, Civit T, Ducrocq X, Bracard S, Hepner H, Picard L, Laxenaire MC. Multivariate analysis of predictors of cerebral vasospasm occurrence after aneurysmal subarachnoid hemorrhage. Stroke 1999;30:1402–1408.
13. Inagawa T. Cerebral vasospasm in elderly patients treated by early operation for ruptured intracranial aneurysms. Acta Neurochir (Wien) 1992;115:79–85.
14. Macdonald RL, Rosengart A, Huo D, Karrison T. Factors associated with the development of vasospasm after planned surgical treatment of aneurysmal subarachnoid hemorrhage. J Neurosurg 2003;99:644–652.
15. Solomon RA, Onesti ST, Klebanoff L. Relationship between the timing of aneurysm surgery and the development of delayed cerebral ischemia. J Neurosurg 1991;75:56–61.
16. Fisher CM, Kistler JP, Davis JM. Relation of cerebral vasospasm to subarachnoid hemorrhage visualized by computerized tomographic scanning. Neurosurgery 1980;6:1–9.

17. Kistler JP, Crowell RM, Davis KR, Heros R, Ojemann RG, Zervas T, Fisher CM. The relation of cerebral vasospasm to the extent and location of subarachnoid blood visualized by CT scan: a prospective study. Neurology 1983;33:424–436.

18. Qureshi AI, Sung GY, Razumovsky AY, Lane K, Straw RN, Ulatowski JA. Early identification of patients at risk for symptomatic vasospasm after aneurysmal subarachnoid hemorrhage. Crit Care Med 2000;28:984–990.

19. Hunt WE, Hess RM. Surgical risk as related to time of intervention in the repair of intracranial aneurysms. J Neurosurg 1968;28:14–20.

20. Report of World Federation of Neurological Surgeons Committee on a universal subarachnoid hemorrhage grading scale. J Neurosurg 1988; 68:985–986.

21. Canhao P, Ferro JM, Pinto AN, Melo TP, Campos JG. Perimesencephalic and nonperimesencephalic subarachnoid haemorrhages with negative angiograms. Acta Neurochir (Wien) 1995;132:14–19.

22. Boet R, Poon WS, Lam JM, Yu SC. The surgical treatment of intracranial aneurysms based on computer tomographic angiography alone—streamlining the acute management of symptomatic aneurysms. Acta Neurochir (Wien) 2003;145:101–105; discussion 105.

23. Kangasniemi M, Makela T, Koskinen S, Porras M, Poussa K, Hernesniemi J. Detection of intracranial aneurysms with two-dimensional and three-dimensional multislice helical computed tomographic angiography. Neurosurgery 2004; 54:336–340; discussion 340–331.

24. van Gijn J, van Dongen KJ, Vermeulen M, Hijdra A. Perimesencephalic hemorrhage: a non-aneurysmal and benign form of subarachnoid hemorrhage. Neurology 1985;35:493–497.

25. Anderson GB, Ashforth R, Steinke DE, Findlay JM. CT angiography for the detection of cerebral vasospasm in patients with acute subarachnoid hemorrhage. AJNR Am J Neuroradiol 2000;21: 1011–1015.

26. Otawara Y, Ogasawara K, Ogawa A, Sasaki M, Takahashi K. Evaluation of vasospasm after subarachnoid hemorrhage by use of multislice computed tomographic angiography. Neurosurgery 2002;51:939–942; discussion 942–933.

27. Vora YY, Suarez-Almazor M, Steinke DE, Martin ML, Findlay JM. Role of transcranial Doppler monitoring in the diagnosis of cerebral vasospasm after subarachnoid hemorrhage. Neurosurgery 1999;44:1237–1247; discussion 1247–1238.

28. Inagawa T, Kamiya K, Ogasawara H, Yano T. Rebleeding of ruptured intracranial aneurysms in the acute stage. Surg Neurol 1987;28:93–99.

29. Kassell NF, Torner JC, Adams HP Jr. Antifibrinolytic therapy in the acute period following aneurysmal subarachnoid hemorrhage.

30. Sen J, Belli A, Albon H, Morgan L, Petzold A, Kitchen N. Triple-H therapy in the management of aneurysmal subarachnoid haemorrhage. Lancet Neurol 2003;2:614–621.

31. Dehdashti AR, Mermillod B, Rufenacht DA, Reverdin A, de Tribolet N. Does treatment modality of intracranial ruptured aneurysms influence the incidence of cerebral vasospasm and clinical outcome? Cerebrovasc Dis 2004; 17:53–60.

32. Rinkel GJ, Feigin VL, Algra A, Vermeulen M, van Gijn J. Calcium antagonists for aneurysmal subarachnoid haemorrhage. Cochrane Database Syst Rev CD000277, 2002.

33. Tanaka K, Gotoh F, Muramatsu F, Fukuuchi Y, Okayasu H, Suzuki N, Kobari M. Effect of nimodipine, a calcium antagonist, on cerebral vasospasm after subarachnoid hemorrhage in cats. Arzneimittelforschung 1982;32: 1529–1534.

34. Espinosa F, Weir B, Overton T, Castor W, Grace M, Boisvert D. A randomized placebo-controlled double-blind trial of nimodipine after SAH in monkeys. Part 1. Clinical and radiological findings. J Neurosurg 1984;60:1167–1175.

35. Haley EC Jr, Kassell NF, Torner JC. A randomized controlled trial of high-dose intravenous nicardipine in aneurysmal subarachnoid hemorrhage. A report of the Cooperative Aneurysm Study. J Neurosurg 1993;78:537–547.

36. Haley EC Jr, Kassell NF, Torner JC. A randomized trial of nicardipine in subarachnoid hemorrhage: angiographic and transcranial Doppler ultrasound results. A report of the Cooperative Aneurysm Study. J Neurosurg 1993;78:548–553.

37. Petruk KC, West M, Mohr G, Weir BK, Benoit BG, Gentili F, Disney LB, Khan MI, Grace M, Holness RO, et al. Nimodipine treatment in poor-grade aneurysm patients. Results of a multicenter double-blind placebo-controlled trial. J Neurosurg 1988;68:505–517.

38. Philippon J, Grob R, Dagreou F, Guggiari M, Rivierez M, Viars P. Prevention of vasospasm in subarachnoid haemorrhage. A controlled study with nimodipine. Acta Neurochir (Wien) 1986;82:110–114.

39. Zornow MH, Prough DS. Neuroprotective properties of calcium-channel blockers. New Horiz 1996;4:107–114.

40. Egge A, Waterloo K, Sjoholm H, Solberg T, Ingebrigtsen T, Romner B. Prophylactic hyperdynamic postoperative fluid therapy after aneurysmal subarachnoid hemorrhage: a clinical, prospective, randomized, controlled study. Neurosurgery 2001;49:593–605.

41. Treggiari MM, Walder B, Suter PM, Romand JA. Systematic review of the prevention of delayed

ischemic neurological deficits with hypertension, hypervolemia, and hemodilution therapy following subarachnoid hemorrhage. J Neurosurg 2003;98:978–984.

42. Rosenwasser RH, Delgado TE, Buchheit WA, Freed MH. Control of hypertension and prophylaxis against vasospasm in cases of subarachnoid hemorrhage: a preliminary report. Neurosurgery 1983;12:658–661.

43. Kaku Y, Yonekawa Y, Tsukahara T, Kazekawa K. Superselective intra-arterial infusion of papaverine for the treatment of cerebral vasospasm after subarachnoid hemorrhage. J Neurosurg 1992; 77:842–847.

44. Kassell NF, Helm G, Simmons N, Phillips CD, Cail WS. Treatment of cerebral vasospasm with intra-arterial papaverine. J Neurosurg 1992; 77:848–852.

45. Elliott JP, Newell DW, Lam DJ, Eskridge JM, Douville CM, Le Roux PD, Lewis DH, Mayberg MR, Grady MS, Winn HR. Comparison of balloon angioplasty and papaverine infusion for the treatment of vasospasm following aneurysmal subarachnoid hemorrhage. J Neurosurg 1998; 88:277–284.

46. Polin RS, Hansen CA, German P, Chadduck JB, Kassell NF. Intra-arterially administered papaverine for the treatment of symptomatic cerebral vasospasm. Neurosurgery 1998;42:1256–1264.

47. Feng L, Fitzsimmons BF, Young WL, Berman MF, Lin E, Aagaard BD, Duong H, Pile-Spellman J. Intraarterially administered verapamil as adjunct therapy for cerebral vasospasm: safety and 2-year experience. AJNR Am J Neuroradiol 2002;23:1284–1290.

48. Zubkov YN, Nikiforov BM, Shustin VA. Balloon catheter technique for dilatation of constricted cerebral arteries after aneurysmal SAH. Acta Neurochir (Wien) 1984;70:65–79.

49. Polin RS, Coenen VA, Hansen CA, Shin P, Baskaya MK, Nanda A, Kassell NF. Efficacy of transluminal angioplasty for the management of symptomatic cerebral vasospasm following aneurysmal subarachnoid hemorrhage. J Neurosurg 2000;92:284–290.

50. Muizelaar JP, Zwienenberg M, Rudisill NA, Hecht ST. The prophylactic use of transluminal balloon angioplasty in patients with Fisher grade 3 subarachnoid hemorrhage: a pilot study. J Neurosurg 1999;91:51–58.

51. Lin CL, Dumont AS, Lieu AS, Yen CP, Hwang SL, Kwan AL, Kassell NF, Howng SL. Characterization of perioperative seizures and epilepsy following aneurysmal subarachnoid hemorrhage. J Neurosurg 2003;99:978–985.

52. Rhoney DH, Tipps LB, Murry KR, Basham MC, Michael DB, Coplin WM. Anticonvulsant prophylaxis and timing of seizures after aneurysmal subarachnoid hemorrhage. Neurology 2000;55: 258–265.

53. Sundaram MB, Chow F. Seizures associated with spontaneous subarachnoid hemorrhage. Can J Neurol Sci 1986;13:229–231.

53a. Rosengart Ak, Novakovic R, Huo D, Frank JI, Goldenberg FD, Baldwin ME, Macdonald RL. Impact of prophylactic anticonvulsive use on outcome in subarachnoid hemorrhage. Stroke 2004;35:250.

53b. Naidech AM, Kreiter KT, Janjua N, Ostapkovich N, Parra A, Commichau C, Connolly ES, Mayer SA, Fitzsimmons BM. Phenytoin exposure is associated with functional and cognitive disability after subarachnoid hemorrhage. Stroke 2005;26:583–587.

54. Palmer BF. Hyponatremia in patients with central nervous system disease: SIADH versus CSW. Trends Endocrinol Metab 2003;14:182–187.

55. Mori T, Katayama Y, Kawamata T, Hirayama T. Improved efficiency of hypervolemic therapy with inhibition of natriuresis by fludrocortisone in patients with aneurysmal subarachnoid hemorrhage. J Neurosurg 1999;91:947–952.

56. Moro N, Katayama Y, Kojima J, Mori T, Kawamata T. Prophylactic management of excessive natriuresis with hydrocortisone for efficient hypervolemic therapy after subarachnoid hemorrhage. Stroke 2003;34:2807–2811.

57. Schurkamper M, Medele R, Zausinger S, Schmid-Elsaesser R, Steiger HJ. Dexamethasone in the treatment of subarachnoid hemorrhage revisited: a comparative analysis of the effect of the total dose on complications and outcome. J Clin Neurosci 2004;11:20–24.

58. Bhardwaj A. SAH-induced cerebral vasospasm: unraveling molecular mechanisms of a complex disease. Stroke 2003;34:427–433.

59. Fodstad H. Antifibrinolytic treatment in subarachnoid haemorrhage: present state. Acta Neurochir (Wien) 1982;63:233–244.

60. Roos Y, Vermeulen M. Prevention of early rebleeding. J Neurosurg 2003;98:1148–1149; author reply 1149–1150.

61. Hillman J, Fridriksson S, Nilsson O, Yu Z, Saveland H, Jakobsson KE. Immediate administration of tranexamic acid and reduced incidence of early rebleeding after aneurysmal subarachnoid hemorrhage: a prospective randomized study. J Neurosurg 2002;97:771–778.

62. Iplikcioglu AC, Berkman MZ. The effect of short-term antifibrinolytic therapy on experimental vasospasm. Surg Neurol 2003;59:10–16; discussion 16–17.

63. Lanzino G, Wang H. Prevention of early rebleeding. J Neurosurg 2003;98:1146–1147; author reply 1148.

64. Handa Y, Weir BK, Nosko M, Mosewich R, Tsuji T, Grace M. The effect of timing of clot removal

on chronic vasospasm in a primate model. J Neurosurg 1987;67:558–564.

65. Kassell NF, Torner JC, Jane JA, Haley EC Jr, Adams HP. The International Cooperative Study on the Timing of Aneurysm Surgery. Part 2. Surgical results. J Neurosurg 1990;73:37–47.

66. Findlay JM, Weir BK, Steinke D, Tanabe T, Gordon P, Grace M. Effect of intrathecal thrombolytic therapy on subarachnoid clot and chronic vasospasm in a primate model of SAH. J Neurosurg 1998;69:723–735.

67. Findlay JM, Weir BK, Kanamaru K, Grace M, Baughman R. The effect of timing of intrathecal fibrinolytic therapy on cerebral vasospasm in a primate model of subarachnoid hemorrhage. Neurosurgery 1990;26:201–206.

68. Findlay JM, Weir BK, Kassell NF, Disney LB, Grace MG. Intracisternal recombinant tissue plasminogen activator after aneurysmal subarachnoid hemorrhage. J Neurosurg 1991;75: 181–188.

69. Mizoi K, Yoshimoto T, Takahashi A, Fujiwara S, Koshu K, Sugawara T. Prospective study on the prevention of cerebral vasospasm by intrathecal fibrinolytic therapy with tissue-type plasminogen activator. J Neurosurg 1993;78:430–437.

70. Findlay JM, Kassell NF, Weir BK, Haley EC Jr, Kongable G, Germanson T, Truskowski L, Alves WM, Holness RO, Knuckey NW, et al. A randomized trial of intraoperative, intracisternal tissue plasminogen activator for the prevention of vasospasm. Neurosurgery 1995; 37:168–176.

70a. Hamada J, Kai Y, Morioka M, Yano S, Mizuno T, Hirano T, Kazekawa K, Ushio Y. Effect on cerebral vasospasm of coil embolization followed by microcatheter intrathecal urokinase infusion into the cisterna magna: a prospective randomized study. Stroke 2003;34:2549–2554.

71. Brian JE Jr, Faraci FM, Heistad DD. Recent insights into the regulation of cerebral circulation. Clin Exp Pharmacol Physiol 1996;23:449–457.

72. Faraci FM, Brian JE Jr. Nitric oxide and the cerebral circulation. Stroke 1994;25:692–703.

73. Sun BL, Zhang SM, Xia ZL, Yang MF, Yuan H, Zhang J, Xiu RJ. L-arginine improves cerebral blood perfusion and vasomotion of microvessels following subarachnoid hemorrhage in rats. Clin Hemorheol Microcirc 2003;29:391–400.

74. Sonmez OF, Unal B, Inaloz S, Sahin B, Yilmaz M, Aydin A, Kaplan S. Therapeutic effects of intracarotid infusion of spermine/nitric oxide complex on cerebral vasospasm. Acta Neurochir (Wien) 2002;144:921–928; discussion 928.

75. Lesley WS, Lazo A, Chaloupka JC, Weigele JB. Successful treatment of cerebral vasospasm by use of transdermal nitroglycerin ointment (Nitropaste). AJNR Am J Neuroradiol 2003;24: 1234–1236.

76. Kumar R, Pathak A, Mathuriya SN, Khandelwal N. Intraventricular sodium nitroprusside therapy: a future promise for refractory subarachnoid hemorrhage-induced vasospasm. Neurol India 2003;51:197–202.

77. Pathak A, Mathuriya SN, Khandelwal N, Verma K. Intermittent low dose intrathecal sodium nitroprusside therapy for treatment of symptomatic aneurysmal SAH-induced vasospasm. Br J Neurosurg 2003;17:306–310.

78. Josko J. Cerebral angiogenesis and expression of VEGF after subarachnoid hemorrhage (SAH) in rats. Brain Res 2003;981:58–69.

79. Lan C, Das D, Wloskowicz A, Vollrath B. Endothelin-1 modulates hemoglobin-mediated signaling in cerebrovascular smooth muscle via RhoA/Rho kinase and protein kinase C. Am J Physiol Heart Circ Physiol 2004;286: H165–H173.

80. Vatter H, Zimmermann M, Weyrauch E, Lange BN, Setzer M, Raabe A, Seifert V. Cerebrovascular characterization of the novel nonpeptide endothelin-A receptor antagonist LU 208075. Clin Neuropharmacol 2003;26:73–83.

81. Zimmermann M, Jung CS, Vatter H, Raabe A, Seifert V. Effect of endothelin-converting enzyme inhibitors on big endothelin–1 induced contraction in isolated rat basilar artery. Acta Neurochir (Wien) 2002;144:1213–1219.

82. Zimmermann M, Jung CS, Vatter H, Raabe A, Seifert V. [D-Val22]big ET-1[16-38] inhibits endothelin-converting enzyme activity: a promising concept in the prevention of cerebral vasospasm. Neurosurg Rev 2003;26:125–132.

83. Juvela S. Plasma endothelin and big endothelin concentrations and serum endothelin-converting enzyme activity following aneurysmal subarachnoid hemorrhage. J Neurosurg 2002; 97:1287–1293.

84. Shaw MD, Vermeulen M, Murray GD, Pickard JD, Bell BA, Teasdale GM. Efficacy and safety of the endothelin, receptor antagonist TAK-044 in treating subarachnoid hemorrhage: a report by the Steering Committee on behalf of the UK/Netherlands/Eire TAK-044 Subarachnoid Haemorrhage Study Group. J Neurosurg 2000; 93:992–997.

85. Shishido T, Suzuki R, Qian L, Hirakawa K. The role of superoxide anions in the pathogenesis of cerebral vasospasm. Stroke 1994;25:864–868.

86. Asano T, Sasaki T, Koide T, Takakura K, Sano K. Experimental evaluation of the beneficial effect of an antioxidant on cerebral vasospasm. Neurol Res 1984;6:49–53.

86a. Asano T, Takakua K, Sano K, Kikuchi H, Nagai H, Saito I, Tamura A, Ochiai C, Sasaki T. Effects of a hydroxyl radical scavenger on delayed ischemic neurological deficits following aneurysmal subarachnoid hemorrhage: results of a multicenter, placebo-controlled double-blind trial. J Neurosurg 84;5:792–803.

87. Aladag MA, Turkoz Y, Sahna E, Parlakpinar H, Gul M. The attenuation of vasospasm by using a sod mimetic after experimental subarachnoidal haemorrhage in rats. Acta Neurochir (Wien) 2003;145:673–677.

88. Kawakami,M, Kodama N, Toda N. Suppression of the cerebral vasospastic actions of oxyhemoglobin by ascorbic acid. Neurosurgery 1991;28:33–39; discussion 39–40.

89. Juvela S. Aspirin and delayed cerebral ischemia after aneurysmal subarachnoid hemorrhage. J Neurosurg 1995;82:945–952.

90. Hop JW, Rinkel GJ, Algra A, Berkelbach van der Sprenkel JW, van Gijn J. Randomized pilot trial of postoperative aspirin in subarachnoid hemorrhage. Neurology 2000;54:872–878.

91. Mendelow AD, Stockdill G, Steers AJ, Hayes J, Gillingham FJ. Double-blind trial of aspirin in patient receiving tranexamic acid for subarachnoid hemorrhage. Acta Neurochir (Wien) 1982;62:195–202.

92. Shaw MD, Foy PM, Conway M, Pickard JD, Maloney P, Spillane JA, Chadwick DW. Dipyridamole and postoperative ischemic deficits in aneurysmal subarachnoid hemorrhage. J Neurosurg 1985;63:699–703.

93. Suzuki S, Sano K, Handa H, Asano T, Tamura A, Yonekawa Y, Ono H, Tachibana N, Hanaoka K. Clinical study of OKY-046, a thromboxane synthetase inhibitor, in prevention of cerebral vasospasms and delayed cerebral ischaemic symptoms after subarachnoid haemorrhage due to aneurysmal rupture: a randomized double-blind study. Neurol Res 1989;11:79–88.

94. Tokiyoshi K, Ohnishi T, Nii Y. Efficacy and toxicity of thromboxane synthetase inhibitor for cerebral vasospasm after subarachnoid hemorrhage. Surg Neurol 1991;36:112–118.

95. Dorhout Mees SM, Rinkel GJ, Hop JW, Algra A, van Gijn J. Antiplatelet therapy in aneurysmal subarachnoid hemorrhage: a systematic review. Stroke 2003;34:2285–2289.

96. Beck T, Bielenberg GW. The effects of two 21-aminosteroids on overt infarct size 48 hours after middle cerebral artery occlusion in the rat. Brain Res 1991;560:159–162.

97. Xue D, Slivka A, Buchan AM. Tirilazad reduces cortical infarction after transient but not permanent focal cerebral ischemia in rats. Stroke 1992;23:894–899.

98. The RANTTAS Investigators. A randomized trial of tirilazad mesylate in patients with acute stroke (RANTTAS). Stroke 1996;27:1453–1458.

99. Bath PM, Iddenden R, Bath FJ, Orgogozo JM. Tirilazad for acute ischaemic stroke. Cochrane Database Syst Rev CD002087:2001.

100. Kassell NF, Haley EC Jr, Apperson-Hansen C, Alves WM. Randomized, double-blind, vehicle-controlled trial of tirilazad mesylate in patients with aneurysmal subarachnoid hemorrhage: a cooperative study in Europe, Australia, and New Zealand. J Neurosurg 1996;84:221–228.

101. Lanzino G, Kassell NF. Double-blind, randomized, vehicle-controlled study of high-dose tirilazad mesylate in women with aneurysmal subarachnoid hemorrhage. Part II. A cooperative study in North America. J Neurosurg 1999;90:1018–1024.

102. Lanzino G, Kassell NF, Dorsch NW, Pasqualin A, Brandt L, Schmiedek P, Truskowski LL, Alves WM. Double-blind, randomized, vehicle-controlled study of high-dose tirilazad mesylate in women with aneurysmal subarachnoid hemorrhage. Part I. A cooperative study in Europe, Australia, New Zealand, and South Africa. J Neurosurg 1999;90:1011–1017.

103. Shibuya M, Suzuki Y, Sugita K, Saito I, Sasaki T, Takakura K, Nagata I, Kikuchi H, Takemae T, Hidaka H, et al. Effect of AT877 on cerebral vasospasm after aneurysmal subarachnoid hemorrhage. Results of a prospective placebo-controlled double-blind trial. J Neurosurg 1992;76:571–577.

104. Negoro N, Hoshiga M, Seto M, Kohbayashi E, Ii M, Fukui R, Shibata N, Nakakoji T, Nishiguchi F, Sasaki Y, Ishihara T, Ohsawa N. The kinase inhibitor fasudil (HA-1077) reduces intimal hyperplasia through inhibiting migration and enhancing cell loss of vascular smooth muscle cells. Biochem Biophys Res Commun 1999;262:211–215.

105. Kim I, Leinweber BD, Morgalla M, Butler WE, Seto M, Sasaki Y, Peterson JW, Morgan KG. Thin and thick filament regulation of contractility in experimental cerebral vasospasm. Neurosurgery 2000;46:440–446; discussion 446–447.

106. van den Bergh WM, Algra A, van der Sprenkel JW, Tulleken CA, Rinkel GJ. Hypomagnesemia after aneurysmal subarachnoid hemorrhage. Neurosurgery 2003;52:276–281.

107. Macdonald RL, Curry DJ, Aihara Y, Zhang ZD, Jahromi BS, Yassari R. Magnesium and experimental vasospasm. J Neurosurg 2004;100:106–110.

108. Veyna RS, Seyfried D, Burke DG, Zimmerman C, Mlynarek M, Nichols V, Marrocco A, Thomas AJ, Mitsias PD, Malik GM. Magnesium sulfate therapy after aneurysmal subarachnoid hemorrhage. J Neurosurg 2002;96:510–514.

109. Ivanova S, Batliwalla F, Mocco J, Kiss S, Huang J, Mack W, Coon A, Eaton JW, Al-Abed Y, Gregersen PK, Shohami E, Connolly ES Jr, Tracey KJ. Neuroprotection in cerebral ischemia

by neutralization of 3-aminopropanal. Proc Natl Acad Sci USA 2002;99:5579–5584.

110. Auer LM, Deinsberger W, Niederkorn K, Gell G, Kleinert R, Schneider G, Holzer P, Bone G, Mokry M, Korner E, Kleinert G, Hanusch S. Endoscopic surgery versus medical treatment for spontaneous intracerebral hemorrhage: a randomised study. J Neurosurg 1989;70:530–535.

111. Kanaya H, Kuroda K: Development in neurosurgical approaches to hypertensive intracerebral hemorrhage in Japan, in Kaufman H (ed): Intracerebral Hematomas. New York: Raven Press, 1992, pp 197–210.

112. Mendelow AD, Gregson BA, Fernandes HM, Murray GM, Teasdale GM, Hope DT, Karimi A, Shaw MD, Barer DH. Early surgery versus initial conservative treatment in patients with spontaneous supratentorial intracerebral haematomas in the International Surgical Trial in Intracerebral Haemorrhage (STICH): a randomised trial. Lancet 2005;365–387–397.

113. Mayer S. Ultra-early hemostatic therapy for intracerebral hemorrhage. Stroke 2003;34: 224–229.

114. Friederich PW, Henny CP, Messelink EJ, Geerdink MG, Keller T, Kurth KH, Buller HR, Levi M. Effect of recombinant activated factor VII on perioperative blood loss in patients undergoing retropubic prostatectomy: a double-blind placebo-controlled randomised trial. Lancet 2003;361:201–205.

115. Hedner U, Ingerslev J. Clinical use of recombinant FVIIa (rFVIIa). Trasnfus Sci 1998; 19:163–176.

3

Spontaneous Intracerebral Hemorrhage

Michael D. Hill

Intracerebral hemorrhage (ICH) is a devastating disease that has received scant attention. Therapy has traditionally been surgical or nonsurgical followed by supportive care, and perhaps as a result, in-hospital mortality is high. Recent developments in the therapy of ischemic stroke have led to more observational research on ICH and set the scene for the completion of a large randomized trial of surgical treatment of ICH and the first large study of medical treatment for ICH. Like ischemic stroke, therapies directed at ICH require rapid delivery and tremendous infrastructure.

The important clinical outcome for patients is freedom from disability rather than death. This presents a complex issue for clinical trialists because the measurement of disability is inexact. Commonly used and validated scales exist but are dependent not only on the type of stroke and degree of brain injury, but on psychosocial, educational, and socioeconomic factors. Further, measurement error compromises patients' assessment, and the increased variance results in reduced statistical power.

Surrogate outcomes in ICH are not readily apparent. However, expansion of hematoma volume in the first few hours after ICH onset is associated with poor outcome. Neuroimaging assessment of hematoma volume at defined time points may therefore be an important surrogate outcome. The emergence of clinical trials for ICH will build on advances in ischemic stroke management and observational epidemiological studies of ICH. The development of innovative surrogate outcomes will arise from such trials.

There are many questions to be answered in the acute management of ICH. How should hypertension be treated? Can hyperacute treatment impact outcome? Are there selected patients who should undergo surgical decompression of the hemorrhage? What are the best methods to prevent and treat deep venous thrombosis and pulmonary embolism among patients with ICH? What is the role of stroke unit care for ICH patients? Clinical trialists will need to bring innovation to solve the difficulties of testing new therapies in ICH, and trials will bring new therapies and more questions to the fore.

WHY ACUTE ICH CLINICAL TRIALS ARE DIFFICULT

ICH has been an orphan condition among stroke. Few randomized trials in ICH have been conducted, making comments on why such trials are difficult more of a theoretical exercise. However, many trials in ischemic stroke have been conducted and produced neutral results. Trialists in ICH can learn from the experience with ischemic stroke.

Traditionally, ICH patients have been managed by neurosurgeons, and selected patients have been offered surgical intervention. Only in the last year have clinical trials convincingly shown that surgery is generally ineffective at preventing death and reducing disability (1). Paradoxically, hemorrhage patients have been included in ischemic stroke neuroprotective trials, these decisions having been made with a view to marketing rather than science. This occurrence has been driven by the lack of infrastructure for rapid emergent imaging in some jurisdictions, so that sponsors have favored enrolling patients before imaging. None of these trials has shown a beneficial effect of the therapy, but this kind of decision making in trial design emphasizes that ICH patient management is in flux. Increasingly, ICH is a medical disease that should be managed on a stroke unit. Medical interventions, other than stroke unit care, are few. This is a burgeoning area for clinical trials. Many questions about acute management need to be answered. It is very likely that treatment for ICH will have to be hyperacute, in the same manner as acute ischemic stroke. Any interventions will need to be applied within the early hours after hemorrhage onset. This approach requires substantial infrastructure, including an acute stroke team and the 24-hour availability of immediate neuroimaging. Fortunately, such infrastructure has been and is being gradually developed for ischemic stroke, and new ICH trials will be able to piggyback on the acute stroke infrastructure.

ICH in human patients, like ischemic stroke patients, is highly heterogenous. To some extent, this is a neuroanatomical phenomenon because small hemorrhages in strategic locations can result in substantial disability. To learn from the lessons of ischemic stroke trials, it is very important

for ICH trialists to consider both the ICH subtype and the specific biological action of the therapy when developing new treatments.

The best outcomes measurements for ICH trials are unclear. Among stroke patients, disability is more important than death, yet early mortality is so high in ICH that it is a logical first outcome for large ICH trials. Any reduction in mortality with a new treatment must be carefully balanced against an increase in morbidity; the last thing physicians want to do is to save someone from death to result in a life of permanent severe disability.

ICH management is when evaluating the various clinical management strategies across the globe. Surgical therapy predominates in many countries, although this may change as results of the Surgical Trial in Intracerebral Haemorrhage (STICH) are interpreted (1a). This challenge of heterogeneity in management on a global basis may present substantial problems in the development of new therapies because large trials in ICH require a global reach to accrue patients in a reasonable time frame. In summary, multiple barriers exist in the performance of quality trials in ICH. These are infrastructure, heterogeneity of phenotype, and the establishment of the best outcome measure(s).

ANIMAL MODELS: RELEVANT ISSUES FOR CLINICAL DEVELOPMENT PROGRAMS

There are two fundamental animal models of ICH. The first uses injection of autologous blood directly into the brain. The second uses a small injection of bacterial collagenase into the striatum, dissolving the extracellular matrix and inducing hemorrhage (2,3). Neither model exactly mimics spontaneous vessel rupture in humans. A disadvantage of the injection models is tracking of blood back along the needle track into the subarachnoid space.

This can be overcome to some extent by the double-injection method (4). By contrast, bacterial collagenase can induce a significant inflammatory reaction (5).

A complete summary of results from animal models for the ICH is not appropriate in this chapter. The reader is referred elsewhere (6). The collagenase models have been reported in rats and pigs, whereas the injection model has been used in mice, rats, rabbits, cats, dogs, pigs, and primates. Both have contributed extensively to the understanding of the molecular events occurring after ICH within the limitations of the models.

A key issue in therapy development is to comprehensively examine the proposed interventions in animal models. One of the proposed failures of neuroprotective strategies in ischemic stroke has been the lack of a complete assessment of the neuroprotective drug in animal models. A few select positive results must be confirmed. For example, a new drug for ICH should be shown to be effective in animals models of ICH in at least two species, at time points for intervention that mimic the clinical situation, with dosages that mimic or can be achieved in humans, and using both surrogate and neurobehavioral outcomes. Animal assessment of neurobehavioral outcomes is critically important because it is these very types of outcomes that will be used in humans to assess disability. These same principles apply to surgical intervention or combined surgical plus pharmacological intervention for hemorrhage.

Many of these observations follow on from a clear understanding of causal inference. In medicine, events occur for multiple reasons. Drugs and surgery have pleiotropic effects, implying that simplistic understanding of even main effects can be wrong. A current example is the statin agents that block the key enzyme in cholesterol anabolism, hydroxymethylglutaryl-CoA reductase, and were originally developed for cholesterol lowering. These agents appear to be anti-inflammatory and endothelial stabilizing and are effective at reducing the burden of atherosclerotic disease apparently beyond their cholesterol-lowering effect.

Rothman and Greenland (7) defined causation for epidemiologists by its components. An outcome is determined when a sufficient cause, composed of one or (more commonly) many different variables, is present. Some variables are also necessary causes, but this is not usual in medicine. The same kind of paradigm can be brought to bear when evaluating new therapies for ICH at the experimental level. Further, the criteria for each variable component of the sufficient cause can be assessed against the causal criteria of Doll and Hill (8,9): strength of the association, consistency of effect, specificity of the effect, temporal association, biological gradient or dose response, and biological plausibility. A new pharmacological or surgical approach would ideally be a necessary cause, meaning that without the new agent, the improved outcome would not occur. It would be strongly associated with the outcome, show a consistency of effect across various experimental models and animal species, always or nearly always show a specificity to the outcome, be temporally associated with the outcome, show a dose–response relationship or a threshold effect, and have a clearly described and biologically plausible mechanism of action.

Many of these recommendations are common sense and follow common logic. Their applications to the clinical trials in stroke are discussed in the Stroke Therapy Academic Industry Roundtable conference proceedings (10).

EPIDEMIOLOGY AND NATURAL HISTORY

Incidence

ICH can be classified by location and by mechanism (Table 3.1). Deep ICH affects

TABLE 3.1   Subclassification of ICH

Mechanistic classification	Anatomical classification
Primary	
(spontaneous/idiopathic)	Lobar
• Hypertensive hemorrhage	
	Subcortical
• Amyloid angiopathy	• Ganglionic (putamen/caudate)
• Microarteriovenous malformation	
	• Thalamic
Secondary	• Cerebellar
• Arteriovenous malformation	
	• Brainstem
	Primary
• Tumor	intraventricular
• Coagulopathy	
• Other	

TABLE 3.2   Incidence per 100,000 population per year

Group	Incidence
Asians[a]	80–120
Black[b]	16
White[c]	7–15

[a]From Ueda K, Hasuo Y, Kiyohara Y, et al. Intracerebral hemorrhage in a Japanese community, Hisayama: incidence, changing pattern during long-term follow-up, and related factors. Stroke 1988;19: 48–52; and Yano K, Reed DM, MacLean CJ. Serum cholesterol and hemorrhagic stroke in the Honolulu Heart Program. Stroke 1987;18:311–324.

[b]From Broderick JP, Brott T, Tomsick T, et al. The risk of subarachnoid and intracerebral hemorrhages in blacks as compared with whites. N Engl J Med 1992;326:733–736.

[c]From Drury I, Whisnant JP, Garraway WM. Primary intracerebral hemorrhage: impact of CT on incidence. Neurology 1984;34:653–657.

the subcortical structures and includes basal ganglia, thalamic, brainstem, and cerebellar hemorrhage. Lobar hemorrhages are typically centered at the cortical–subcortical junction and affect the hemispheres, often extending out to the pia-arachnoid. Mechanistically, ICH can be primary or secondary. Primary ICH usually includes hypertensive ICH, amyloid angiopathy-related ICH, and idiopathic ICH. Secondary ICH is caused by an underlying medical or structural condition; anticoagulation and thrombolytic therapy are examples of medical conditions predisposing to ICH, whereas arteriovenous malformation and moya moya disease are examples of structural lesions associated with ICH.

Spontaneous ICH makes up 10–30% of all strokes, which varies by ethnic origin and is probably declining in age-adjusted incidence. A much higher proportion of Asians suffer ICH compared with white populations. The incidence of ICH is 7–15 per 100,000 among whites compared with 80–120 per 100,000 among Asians. African-Americans have an intermediate risk (Table 3.2).

The incidence of ICH in Canada has been reported from administrative data and increases with age. In 1991–1992 the incidence rate per 100,000 population in men varied from 4.3 in the below 55 age group to 114.2 in the over 85 age group. Among women, the rates were slightly lower, varying from 3.1 in the below 55 age group to 92.0 in the over 85 age group (11). More recent data suggest that over the last decade whereas the absolute numbers of ICH have remained stable, the age-adjusted incidence has fallen (12).

Declining incidences over the past five decades have been attributed to changes in risk profiles, in particular the reduction and improved control of blood pressure (13). Early mortality from ICH is higher than from ischemic stroke. In Toronto, 30-day mortality was 27.4% among 423 patients with primary ICH (14); this is analogous to experience elsewhere. Mortality may be higher among patients with secondary causes of ICH and with subcortical rather than cortical hemorrhage (Table 3.3). Nevertheless, declining mortality from ICH also has been observed overall with a parallel decline in 30-day mortality. The decline in in-hospital mortality is presumably due to improvements

TABLE 3.3 Mortality by ICH type, Toronto series

Primary ICH type	n (%)	M:F ratio	Mean age (years)	30-day mortality (%)
Lobar	154 (36.4)	1.1	66.7	29 (18.8)
Nonlobar	269 (63.6)	1.3	65.3	87 (32.3)
• Cerebellar	52 (12.3)	1.2	68.7	14 (26.9)
• Thalamic	73 (17.3)	1.1	67.0	21 (28.8)
• Putamenal/caudate	121 (28.6)	1.5	65.1	43 (35.5)
• Brainstem	21 (5.0)	2.0	52.5	8 (38.1)
• Unspecified	2 (0.5)			1 (50.0)
Total	423	1.2	65.9	116 (27.4)

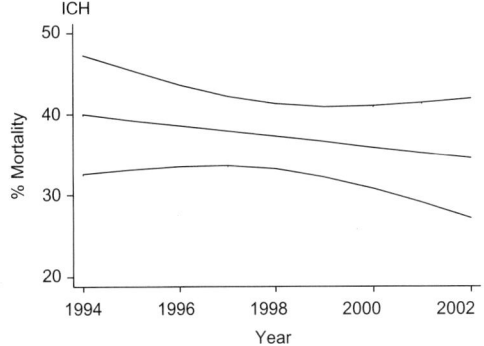

FIGURE 3.1 Declining in-hospital mortality for ICH patients in Calgary, Alberta, Canada. (From ref. 12.)

in the management of complications such as pneumonia and pulmonary thromboembolism, but this is unclear (Fig. 3.1).

Because prevalence includes both new cases and survivors and because ICH is not uniformly fatal, the prevalence of ICH is higher than the incidence. Among all strokes there are six to eight times as many survivors as there are new cases annually (15). At least one-third of survivors are significantly disabled with consequent high social and economic burden.

Risk Factors

Hypertension is the most important risk factor for ICH. However, most neurologists have seen cases of "cryptogenic" hemorrhage in a typical hypertensive location such as the putamen in the absence of any history hypertension. It is important

then to note that hypertension probably accounts for less than two-thirds of all ICH (14). There is a clear dose–response relationship between alcohol consumption and ICH, even after adjusting for other factors (16); alcohol consumption does not just increase the risk of ICH by its known hypertensive effect.

Cigarette smoking is associated with an increased incidence of subarachnoid hemorrhage due to rupture of intracranial aneurysms; however, the relationship with ICH is less clear (16). The Honolulu Heart Program suggested an increased risk of all intracranial hemorrhage but did not classify the hemorrhage type (17). Low serum cholesterol has been associated with an increased risk of ICH (18,19). This observation has been seen in both Asian and white populations. Confounding by diets rich in polyunsaturated fatty acids such as eicosapentaenoic acid (from fish oil), which acts to reduce platelet aggregation, has been suggested as a possible explanation. Importantly, clinical trials of lipid-lowering therapy have not shown an increased risk of ICH among treated patients (20).

Iatrogenic hemorrhage due to anticoagulant therapy is an increasingly important problem. The coumarins and heparins are drugs with a narrow therapeutic index. The incidence of ICH associated with warfarin use for atrial fibrillation is estimated to be 1% annually when the International Normalization Ratio is maintained between 2 and 3 (21). Intensity of

TABLE 3.4   Causes of ICH

Modifiable	Nonmodifiable
Hypertension	Cerebral amyloid angiopathy
Alcohol	Arteriovenous malformations (AVM, AVF)
Cigarette smoking	Hemorrhagic transformation of ischemic stroke
Serum cholesterol	Neoplasia
Anticoagulant therapy	Venous sinus thrombosis
Antiplatelet therapy	Moyamoya disease
Thrombolytic therapy	Trauma
Sympathomimetic drugs	Systemic disease: liver failure, leukemia
Postendarterectomy/stent	Vasculitis, hemorrhagic leukoencephalitis of Hurst
hyperperfusion syndrome	Other

anticoagulation, age, and so-called leuko-ariosis increase the risk of anticoagulant-related ICH (22); genotype may also be important (see later). Bleeding on unfractionated heparin is exponentially higher as the activated partial thromboplastin time rises. The risk of hemorrhage is probably lower with low-molecular-weight heparin.

Although antiplatelet therapy reduces the risk of thrombotic events, it also marginally increases the risk of ICH (23,24). Wolf (25) pointed out that although the risk was elevated in the Physicians Health Study and the Swedish Aspirin Low-Dose Trial, the risk of ICH was not elevated in the Canadian aspirin trial. The combination of acetylsalicylic acid plus clopidogrel resulted in an increased risk of ICH in the Clopidogrel in unstable Angina to prevent recurrent event (CURE) trial of unstable angina (note there was an absolute risk increase in major bleeding of 0.9%) (26); similarly, the risk of all types of hemorrhage was higher in the acetylsalicylic acid plus clopidogrel arm of the Management of a therothrombosis with clopidogrel in high-risk patients (MATCH) study of secondary stroke prevention.

The risks of ICH after systemic and local thrombolytic therapy are real. In acute ischemic stroke, the risks of hemorrhage are higher than with the treatment of myocardial infarction. Although the pooled results of the tissue plasminogen activator

stroke studies demonstrated an ICH rate of 6.0% (27), interim results from the Canadian alteplase for stroke effectiveness study (CASES) collaboration suggest a rate of 4.7% (95% confidence interval, 3.2–6.1). By contrast, the risk of ICH after thrombolysis for myocardial infarction is 1% (28). New and better thrombolytic drugs such as tenecteplase and desmoteplase may minimize these risks while maintaining efficacy.

Evidence confirms what has been commonly taught: that the sympathomimetic drug, phenylpropanolamine, a commonly taken appetite suppressant and nasal decongestant, is associated with an increased risk of ICH, particularly in women (29). This resulted in withdrawal of phenylpropanolamine from the North American market. Stroke and ICH have been reported with all other major decongestant medications (pseudoephedrine, oxymetazoline, phenylephrine) that are available over the counter (30). Equally, ICH is associated with illicit sympathomimetic agents such as amphetamine, methamphetamine, and cocaine.

Nonmodifiable Risk Factors

Congophilic/cerebral amyloid angiopathy (CAA) increases in incidence with age and results in vessel wall fragility and rupture. Deposition of amyloid material in the vessel wall can only be diagnosed pathologically, often at autopsy. It remains unknown what proportion of lobar hemor-rhages is due to hypertension and what proportion is due to CAA. Until such time as a biochemical or molecular imaging marker is identified, the epidemiology of this condition as it relates to the causation of ICH remains speculative. However, if CAA is proven, this implies a much greater risk of recurrence (18). Patients with recurrent lobar hemorrhage commonly have CAA. Similarly, patients with CAA may be more prone to ICH if they are anticoagulated with warfarin (31). Familial forms of congophilic angiopathy have been reported among Dutch (32), Icelandic (33), and German families (34). Several proteins and hence their genes are now implicated in CAA. These are cystatin-C (Icelandic type), gel-solin (35), presenilin-1 (36), presenilin-2 (38), apolipoprotein ε_2 and ε_4 (37), α_1-antichymo-trypsin (38), and probably the CAA protein itself. There is a relationship between Alzheimer's disease and CAA, particularly the heritable forms of Alzheimer's disease. Most sporadic CAA remains unexplained.

One potential new marker of CAA or hypertensive hemorrhage is the presence of microhemorrhage on gradient recall echo magnetic resonance imaging (MRI) sequences. MRI is sensitive to the presence of iron present as hemosiderin in macro-phages. A $T2^*$ sequence results in a "blooming" effect on MRI, which is seen as a dark black microhemorrhage. Micro-hemorrhages are associated with an increa-sed risk of subsequent clinical ICH (39).

When vascular malformations underlie the first presentation of ICH, they are most often unsuspected. However, if they are discovered before hemorrhage, then treat-ment may dramatically reduce the risks of recurrence. Cavernous malformations may require surgical excision, particularly if they are subcortically located because of a high annual event rate (10% a year) (40). Many cavernomas present with seizure, allowing surgical treatment before hemorrhage. Among arteriovenous malfor-mations and arteriovenous fistulas, flow

dynamics may be an important predictor of the risk of hemorrhage (41), along with clinical and angiographic grading.

DIAGNOSIS AND SUBPOPULATIONS

The diagnosis of ICH is based on brain imaging. Brain computed tomography (CT) has been the standard, but it is now clear that MRI, including gradient recall echo sequences, is as good as or better than CT (42). Gradient recall echo MRI is clearly more sensitive at detecting small hemor-rhages or areas of old hemorrhage because of its sensitivity to the presence of iron in macrophages. Clinical tools to make the diagnosis of hemorrhage lack sensitivity and specificity; the presentation of acute ischemic stroke and acute ICH can be identical.

In the immediate phase in the emergency department, the diagnosis of subtypes of hemorrhage is based on localization. Where is the epicenter of the hemorrhage? Divid-ing hemorrhages into hemispheric and brainstem/cerebellar locations and then further into lobar and deep subtypes provides the basis for descriptive nosol-ogy. Immediate prognostic information is obtained from imaging based on two factors: the size of the hemorrhage and the presence or absence of intraventricular extension. The volume of hematoma can be reliably estimated from CT or MRI using the ABC/2 rule, based on estimating the volume of an ellipsoid.

The prognostic information can be ranked using an ICH score developed by Tuhrim and colleagues. This is a validated model using four factors to predict out-come after supratentorial ICH. The factors are ICH volume, Glasgow Coma Scale, intraventricular extension of hemorrhage, and pulse pressure (43,44). Lesion volume has been shown elsewhere to be the major predictor of outcome (45,46). The simple formula for calculating lesion volume, the

ABC/2 rule, has been validated based on CT appearance (47) as follows:

A = the largest diameter of the hemorrhage on the CT slice with the larger area of ICH (units = cm)
B = the largest perpendicular diameter on the same slice as A (units = cm)
C = the number of 1-cm slices containing hemorrhage:

$$Volume = (A \times B \times C) \div 2$$

Note that all units are in centimeters. If, as is increasingly common, the variable C is based on 5-mm slice thickness, C must be divided by 2.

In posterior fossa hemorrhage (cerebellar, brainstem) the clinical state of the patient, age, and the presence of hydrocephalus may be the most important predictors of outcome (48). The decision to operate does not appear to influence outcome (49).

Other risk factors that have been explored include pulse pressure, pulsatility index (50), and the onset of clinical and electrographic seizures after ICH. Wide pulse pressure predicts ICH after thrombolytic treatment of myocardial infarction (51). Seizures, both clinical and electrographic, are common after ICH and predictive of poorer immediate outcome (52).

Achieving an adequate scientifically valid description of disease populations, especially subpopulations, for which effective treatment may be demonstrated is essential for clinical development. For example, it is well understood that the inclusion of patients with excessive good prognosis and excessive poor prognosis may add little to our ability to demonstrate a treatment effect because of ceiling or floor effects. This is particularly true when disability is the relevant outcome. This is a statistical concept in trial design; the power to show an effect depends on the expected outcomes and the minimal clinically important difference. Choosing the

relevant population for a clinical trial in ICH then depends on whether mortality or disability is the primary endpoint.

In hemispheric ICH, the key variables have been defined by Tuhrim. How to incorporate posterior fossa ICH prognosis is less well determined. Because the ICH score has been internally and externally validated, it would be quite reasonable to develop a stratified randomization using this score as key stratification variable.

CURRENT DISEASE MANAGEMENT

ICH has been traditionally managed as a surgical disease. Removal of a hematoma has been thought to be lifesaving or disability preventing in selected circumstances. However, this approach has never been consistently shown to be true in randomized trials (53). Most trials were small and subject to critique. A meta-analysis of these trials shows that no benefit, and also no substantial harm, accrues to surgically treated patients (54). The general issue has now largely been laid to rest with the completion of the STICH trial. The results have not yet been published but have been presented and suggest that overall there is no morbidity or mortality benefit to surgical treatment of spontaneous ICH.

Nevertheless, it is likely that some patients benefit from surgery and some are harmed. Hyperacute surgical evacuation of the hematoma is dominated by early rebleeding (55). Patients who do not bleed do exceptionally well. If the bleeding could be stopped medically, the issue of surgical treatment may need to be reexamined.

The treatment of ICH then becomes a medical issue, much like ischemic stroke. Stroke unit care is the ideal method to care for stroke patients of all types, and this type of treatment should optimally be available for all victims of ICH. Stroke unit care probably helps by reducing in-hospital complications and promoting early rehabilitation, but the relative importance of the components of stroke unit care is not known, particularly in the management of ICH.

The best acute and subacute management of many facets of ICH care remain unresolved. For example, most patients have acute hypertension, and its management remains unclear. The International Hypertension Society has identified the management of hypertension in acute stroke as a priority area for research (56). General guidelines exist, but these are based on best estimates of experts rather than solid evidence. A major theoretical concern about the existence of penumbral tissue surrounding a hematoma was researched (57). MR studies have suggested that patients with ICH may have hemispheric hypoperfusion and diaschisis but probably do not have substantial perihematomal ischemia (58,59). One study using single positron emission CT suggested that perihematomal ischemia may exist (60), but this has not been confirmed with positron emission tomography (61). Moreover, subacute lowering of blood pressure has not been shown to induce perihematomal ischemia in small to medium-sized hematomas (62). Theoretically, there is good justification for a randomized clinical trial of aggressive blood pressure lowering in the acute phase.

ISSUES IN CLINICAL PROTOCOL DEVELOPMENT

As much as possible, human protocols should be modeled on what works in relevant animal models with little regard for marketing concerns (63). Dosing regimens should be decided based on careful dose escalation studies at phases I and II in humans. The lessons in dosing from neuroprotection trials for ischemic stroke will apply to ICH: Under-dosing in humans will lead to trial failure. Similarly, the optimal duration of treatment should be determined based on animal models.

All these decisions will always be governed to some extent by practical considerations. The key concept for trial development is rigor in basing all decisions on dose, dosing regimen, and dosing duration on careful preclinical animal and/or phase I/II human studies. Without this kind of effort, a positive trial at phase III may be as much luck as good design.

OUTCOMES

The importance of clinical outcomes compared with surrogate outcomes cannot be overstated. Patients and regulators will only accept new treatments if they improve patient experience and result in improved function. Patients have no concern for the mechanism of action of the treatment. Principally, the same applies to government regulators. Neuroscience patients often require multiple clinical endpoints to adequately describe damage and outcome. However, the trend toward the use of global functional outcome assessments in stroke trials reflects this bottom-line approach. Am I going to be better with this treatment, doctor? And, more specifically, in stroke medicine, can I get back to my life again?

The Glasgow Outcome Scale (GOS), although originally developed for assessing outcome after head injury, lends itself well to stroke trials. An improved eight-point compared with the previous five-point version, using a structured interview, allows for finer discrimination in outcome and better interrater variability (64–66). Similarly, the modified Rankin Scale (mRS) score is a seven-point scale shown to have better reliability with the use of a structured interview (67,68) and is a routine scale for assessment of outcome in ischemic stroke. Both are similar, and although they are strongly focused on neurological disability, they also reflect other multiple components of outcome (69,70). Indeed, the mRS is well correlated with stroke-specific, quality-of-life measures.

Use of surrogate clinical outcome measures creates problems of complexity in describing the safety and efficacy of novel drugs. However, surrogate endpoints may provide an economy in trial design by allowing earlier outcome assessment and a reduction in outcome variance and by allowing smaller sample sizes at early stages in drug development. A good surrogate endpoint is very strongly predictive of the clinical outcome. Few such endpoints exist in medicine. A good example is target vessel recanalization after thrombolysis for ischemic stroke. Recanalization occurs often after thrombolytic treatment, but a substantial portion of the time the patient does not improve. Recanalization as a surrogate clinical outcome is good but certainly not perfect.

In ICH, few surrogate outcomes are available. Much research is ongoing on potential biomarkers of brain injury. Proteins such as S100, neuron-specific enolase, and others are released into the blood after injury and are potential estimates of tissue damage (71). To date, none has shown the diagnostic sensitivity and specificity necessary to be useful diagnostically. However, most have shown good correlation with the volume of injury, as measured by neuroimaging. None has been used in outcome analyses to date.

Brain imaging may be the most likely candidate as a surrogate outcome in ICH. Because the volume of ICH is such a dominant predictor of early mortality and late disability among survivors, treatments to reduce the growth of hemorrhage may result in improved clinical outcomes.

EFFICACY EVALUATION

Outcome measures in stroke trials have varied over time. The focus on neurological outcome has led to the National Institute of Health Stroke Scale (NIHSS) score, the Scandinavian Stroke Scale, the Canadian Neurological Scale score, and others. However, the clinical phenotype may result from small or large lesions and lesions in various neuroanatomical locations leading to difficulties in selecting the study population. Functional scores include the Barthel Index of activities of daily living. Global assessments include the mRS score and the Glasgow Outcome Scale. These latter scales unintentionally incorporate socioeconomic and other factors into the scale, resulting in a functional disability assessment that is heavily weighted to neurological disability. Tilley and colleagues (72) combined four scales (NIHSS, mRS, modified Barthel Index, GOS) using a statistical technique to produce a global outcome assessment. The major theoretical advantage of this approach is that it represents an assessment of multiple domains of outcome and provides slightly increased statistical power compared with any of the individual scales. Perhaps the most telling observation that was observed in the National Institute of Neurological Disorders and Stroke tPA Stroke Trial, where this new global outcome statistic was used, was the fact that discordant results between individual scales and the global outcome statistic did not occur.

The largest ICH trial, the STICH trial, used a global outcome measure as the primary outcome. Similarly, two phase II trials of ICH examining activated factor VII used the eight-point Glasgow outcome score. This likely represents an increasing emphasis in stroke trials toward convergence of the primary outcome measure on global scales assessing functional disability weighted to neurological deficits. ICH trials will do well to follow this lead.

An important issue is when after ICH to measure the primary outcome. Ischemic stroke trials have focused on the 90-day outcome because most ischemic stroke patients make the most gains in recovery during that time. ICH may be different, in part, because recovery often takes longer and the initial insult is often more severe. A 6-month timeline would allow more patients to achieve maximal recovery.

Given the dominance of the mRS and eight-point GOS in assessing stroke outcomes, attempts to examine the shift of patients along disability categories has been proposed as a way of increasing the power of a study and therefore allowing smaller sample sizes. If a "shift to the left" can be shown such that more patients are achieving lower levels of disability with treatment than control subjects, this provides evidence of treatment effect. As discussed above, clearly any reduction in mortality must also be associated with a reduction in disability all along the spectrum of the scale. This approach has not been taken in ischemic stroke, principally because patients have articulated the desire to be alive and normal and consider significant disability to be a fate worse than death. Indeed, with thrombolytic ischemic stroke treatment, the potential for rapid and complete or near complete recovery is real.

However, this approach of "shift to the left" may be well suited to ICH. There is no current treatment that allows for very rapid and complete recovery. ICH patients recover slowly and incompletely. A "shift to the left" in the disability scales may well indicate an important treatment effect. This approach needs to be assessed as a post-hoc outcome with current clinical trial data. Moreover, and more importantly, it must be shown empirically that this kind of outcome is of value to patients.

GOLD STANDARD MEASURES

As discussed previously, the global disability scales—mRS and GOS—are the current gold standards for stroke outcome assessment. With the use of the structured

interview in each, both are well validated and provide good interrater reliability. Most importantly, they provide a patient-centered neurological outcome assessment.

Other important measures of outcome ("silver standards") include neurological disability scales such as the NIHSS score, the Scandinavian Stroke Scale score, middle cerebral artery score, Canadian Neurological Scale score, and others. Neurological disability scales quantitate the neurological examination and allow the measurement of changes in neurological status over hours to days or longer. Neurological disability does not necessarily correlate well with the global functional outcome and so is more a measure of biological activity of the treatment. Neuropsychological batteries, depression ratings, and quality-of-life indexes are alternate measures addressing important domains of outcome after all types of stroke. Neuropsychological outcome is partly dependent on lesion location and is perhaps most relevant in assessing the loss of the ability to work among younger patients. Imaging can only really be considered a surrogate outcome, principally because imaging outcomes correlate only roughly with clinical scores and scales.

The final important consideration in outcomes assessment is economic. Although generally if a new intervention results in improved clinical outcomes, the cost may be ignored, the cost-effectiveness of any new intervention is a relevant consideration for health policy and management. Highly cost-effective interventions are the most likely to be widely endorsed across multiple health jurisdictions and by multiple payers.

SURROGATE ENDPOINTS

There are no well-accepted surrogate endpoints for ICH. The fact that approximately 30% of ICH patients suffer expansion of the hematoma in the first few hours after stroke onset has led to the idea that brain imaging assessment could be a surrogate outcome. Prevention of hematoma enlargement is a logical surrogate because ICH volume is so tightly correlated with outcome.

There are no current biochemical surrogate markers. Serum levels of metalloproteinase-9 have been shown to be associated with thrombolysis-related hemorrhage (73). Diagnostic markers of ICH have yet to be uncovered.

SAFETY EVALUATION

The major adverse events after spontaneous ICH are neurological death, aspiration pneumonia, deep venous thrombosis/pulmonary thromboembolism, and seizures. All new treatments must take into account the expected high rate of these events. For example, efforts to reduce hematoma expansion using the procoagulant molecule, factor VIIa, might result in an increased incidence of deep venous thrombosis and pulmonary embolism, but this needs to be assessed relative to improved clinical outcomes. Aspiration pneumonia can be prevented with careful attention to swallowing, as on a stroke unit, but any new treatment that suppresses the immune system may result in increased morbidity for pneumonia.

The best method is to evaluate safety, first in phase I dose-escalation studies in humans and then by randomized assessment in phase II or phase III studies. Comparison with a concurrent control, with blinded treatment allocation and blinded outcome assessment, remains the gold standard method to determine the true rate of adverse events.

References

1. Mendelow AD; Investigators and the Steering Committee. The International Surgical Trial in Intracerebral Haemorrhage (ISTICH). Acta Neurochir 2003;86(suppl):441–443.

1a. Mendelow AD, Gregson BA, Fernandes HM, Murray GD, Teasdale GM, Hope TD, Karimi A, Donald M, Shaw M, and Barer DH for the STICH investigators. Early surgery versus initial conservative treatment in patients with spontaneous supratentorial intracerebral hematomas in the International Surgical Trial in intracerebral hemorrhage (STICH): a randomized trial. Lancet 2005;365:387–97.

2. Rosenberg GA, Mun-Bryce S, Wesley M, Kornfeld M. Collagenase-induced intracerebral hemorrhage in rats. Stroke. 1990; 21: 801–807.

3. Rosenberg GA, Estrada E, Kelley RO, Kornfeld M. Bacterial collagenase disrupts extracellular matrix and opens blood–brain barrier in rat. Neurosci Lett. 1993;160: 117–119.

4. Belayev L, Saul I, Curbelo K, Busto R, Belayev A, Zhang Y, Riyamongkol P, Zhao W, Ginsberg MD. Experimental intracerebral hemorrhage in the mouse: histological, behavioral, and hemodynamic characterization of a double-injection model. Stroke 2003;34:2221–2227.

5. Del Bigio MR, Yan HJ, Buist R, Peeling J. Experimental intracerebral hemorrhage in rats. Magnetic resonance imaging and histopathological correlates. Stroke 1996;27:2312–2319.

6. Andaluz N, Zuccarello M, Wagner KR. Experimental animal models of intracerebral hemorrhage. Neurosurg Clin North Am 2002;13: 385–393.

7. Rothman KJ, Greenland S. Causation and causal inference. In: Rothman KJ, Greenland S, eds. Modern Epidemiology, 2nd ed. Philadelphia: Lippincott Raven Publishers, 1998.

8. Doll R, Hill AB. Smoking and carcinoma of the lung: preliminary report. Br Med J 1950;2: 739–748.

9. Doll R, Hill AB. A study of the aetiology of carcinoma of the lung. Br Med J 1952;2:1271–1286.

10. Fisher M; Stroke Therapy Academic Industry Roundtable. Recommendations for advancing development of acute stroke therapies: Stroke Therapy Academic Industry Roundtable 3. Stroke 2003;34:1539–1546.

11. Mayo NE, Neville D, Kirkland S, Ostbye T, Mustard CA, Reeder B, Joffres M, Brauer G, Levy AR. Hospitalization and case-fatality rates for stroke in Canada from 1982 through 1991. The Canadian Collaborative Study Group of Stroke Hospitalizations. Stroke 1996;27: 1215–1220.

12. Field TS, Green TL, Roy K, Pedersen J, Hill MD. Trends in stroke occurrence in Calgary. Can J Neurol Sci 2004;31:387–393.

13. Furlan AJ, Whisnant JP, Elveback LR. The decreasing incidence of primary intracerebral hemorrhage: a population study. Ann Neurol 1979;5:367–373.

14. Hill MD, Silver FL, Austin PC, Tu JV. Rate of stroke recurrence in patients with primary intracerebral hemorrhage. Stroke 2000;31:123–127.

15. www.heartandstroke.ca, Accessed April 2001.

16. Abbott RD, Yin Y,Reed DM, Yano K. Risk of stroke in male cigarette smokers. N Engl J Med 1986;315:717–720.

17. Donahue RP, Abbott RD, Reed DM, Yano K. Alcohol and hemorrhagic stroke: the Honolulu Heart Program. JAMA 1986;255:2311–2314.

18. Jacobs D, Blackburn H, Higgins M, Reed D, Iso H, McMillan G, et al. Report of the conference on low blood cholesterol: mortality associations. Circulation 1992;86:1046–1060.

19. Tanaka H, Ueda Y, Hayashi M, Date C, Baba T, Yamashita H, Shoji H, Tanaka Y, Owada K, Detels R. Risk factors for cerebral hemorrhage and cerebral infarction in a Japanese rural community. Stroke 1982;13:62–73.

20. Heart Protection Study Collaborative Group. MRC/BHF Heart Protection Study of cholesterol lowering with simvastatin in 20,536 high-risk individuals: a randomised placebo-controlled trial. Lancet 2002;360:7–22.

21. Hart RG. What causes intracerebral hemorrhage during warfarin therapy? Neurology 2000;55:907–908.

22. Smith EE, Rosand J, Knudsen KA, Hylek EM, Greenberg SM. Leukoariosis is associated with warfarin-related hemorrhage following ischemic stroke. Neurology 2002;59:193–197.

23. Hart RG, Halperin JL, McBride R, Benavente O, Man-Son-Hing M, Kronmal RA. Aspirin for the primary prevention of stroke and other major vascular events: meta-analysis and hypotheses. Arch Neurol 2000;57:326–332.

24. He J, Whelton PK, Vu B, Klag MJ. Aspirin and risk of hemorrhagic stroke: a meta-analysis of randomized controlled trials. JAMA 1998;280:1930–1935.

25. Wolf PA. Epidemiology of intracerebral hemorrhage. In: Kase CS, Caplan LR, eds. Intracerebral Hemorrhage. Butterworth-Heinemann, 1994.

26. www.theheart.org. Accessed April 2001.

27. Hacke W, Donnan G, Fieschi C, Kaste M, von Kummer R, Broderick JP, Brott T, Frankel M, Grotta JC, Haley EC Jr, Kwiatkowski T, Levine SR, Lewandowski C, Lu M, Lyden P, Marler JR, Patel S, Tilley BC, Albers G, Bluhmki E, Wilhelm M, Hamilton S; ATLANTIS Trials Investigators; ECASS Trials Investigators; NINDS rt-PA Study Group Investigators. Association of outcome with early stroke treatment: pooled analysis of ATLANTIS, ECASS, and NINDS rt-PA stroke trials. Lancet 2004;363:768–774.

28. GUSTO Investigators. An international randomized trial comparing four thrombolytic strategies for acute myocardial infarction. N Engl J Med 1993;329:673–682.

29. Kernan WN, Viscoli CM, Brass LM, Broderick JP, Brott T, Feldmann E, Morgenstern LB, Wilterdink JL, Horwitz RI. Phenylpropanolamine and the risk of hemorrhagic stroke. N Engl J Med. 2000;343:1826–1832.

30. Cantu, C, Arauz A, Murillo-Bonilla L, et al. Stroke associated with sympathomimetics contained in over-the-counter cough and cold drugs. Stroke 2003;34:1667–1673.

31. Rosand J, Hylek EM, O'Donnell HC, Greenberg SM. Warfarin-associated hemorrhage and cerebral amyloid angiopathy: a genetic and pathologic study. Neurology 2000;55:947–951.

32. Van Broeckhoven C, Haan J, Bakker E, Hardy JA, Van Hul W, Wehnert A, Begter-Van Der Vlis M, Roos RAC. Amyloid β protein precursor gene and hereditary cerebral hemorrhage and amyloidosis (Dutch). Science 1990;248:1120–1122.

33. Gudmundsson G, Hallgrimsson J, Jonasson TA, Bjarnasson O. Hereditary cerebral hemorrhage with amyloidosis. Brain 1972;95:387–404.

34. Nochlin D, Bird TD, Nemens EJ, Ball MJ, Sumi SM. Amyloid angiopathy in a Volga German family with Alzheimer's disease and a presenilin-2 mutation. Ann Neurol 1998;43:131–135.

35. Kiuru S, Salonen O, Haltia M. Gelsolin-related spinal and cerebral amyloid angiopathy. Ann Neurol 1999;45:305–311.

36. Yasuda M, Maeda S, Kawamata T, Tamaoka A, Yamamoto Y, Kuroda S, Maeda K, Tanaka C. Novel presenilin-1 mutation with widespread cortical amyloid deposition but limited cerebral amyloid angiopathy. J Neurol Neurosurg Psychiatry 2000;68:220–223.

37. O'Donnell HC, Rosand J, Knudsen KA, Furie KL, Segal AZ, Chiu RI, Ikeda D, Greenberg SM. Apolipoprotein E genotype and the risk of recurrent lobar intracerebral hemorrhage. N Engl J Med 2000;342:240–245.

38. Durany N, Ravid R, Riederer P, Cruz-Sanchez FF. Increased frequency of the alpha-1-antichymotrypsin T allele in cerebral amyloid angiopathy. Neuropathology 2000;20:184–189.

39. Lee SH, Bae HJ, Kwon SJ, Kim H, Kim YH, Yoon BW, Roh JK. Cerebral microbleeds are regionally associated with intracerebral hemorrhage. Neurology 2004;62:72–76.

40. Porter PJ, Willinsky RA, Harper W, Wallace MC. Cerebral cavernous malformations: natural history and prognosis after clinical deterioration with or without hemorrhage. J Neurosurg 1997;87:190–197.

41. Norris JS, Valiante TA, Wallace MC, Willinsky RA, Montanera WJ, terBrugge KG, Tymianski M. A simple relationship between radiological arteriovenous malformation hemodynamics and clinical presentation: a prospective, blinded analysis of 31 cases. J Neurosurg 1999;90:673–679.

42. Fiebach JB, Schellinger PD, Gass A, Kucinski T, Siebler M, Villringer A, Olkers P, Hirsch JG, Heiland S, Wilde P, Jansen O, Rother J, Hacke W,

Sartor K; Kompetenznetzwerk Schlaganfall B5. Stroke magnetic resonance imaging is accurate in hyperacute intracerebral hemorrhage: a multicenter study on the validity of stroke imaging. Stroke 2004;35:502–506.

43. Tuhrim S, Dambrosia JM, Price TR, Mohr JP, Wolf PA, Hier DB, Kase CS. Intracerebral hemorrhage: external validation and extension of a model for prediction of 30-day survival. Ann Neurol 1991;29:658–683.

44. Tuhrim S, Horowitz DR, Sacher M, Godbold JH. Validation and comparison of models predicting survival following intracerebral hemorrhage. Crit Care Med 1995;23:950–954.

45. Franke CL, van Swieten JC, Algra A, van Gijn J. Prognostic factors in patients with intracerebral haematoma. J Neurol Neurosurg Psych 1992; 55:653–657.

46. Helweg-Larsen S, Sommer W, Strange P, Lester J, Boysen G. Prognosis for patients treated conservatively for spontaneous intracerebral hematomas. Stroke 1984;15:1045–1048.

47. Kothari RU, Brott T, Broderick JP, Barsan WG, Sauerbeck LR, Zuccarello M, Khoury J. The ABCs of measuring intracerebral hemorrhage volumes. Stroke 1996;27:1304–1305.

48. St. Louis EK, Wijdicks EFM, Li Hongzhe, Atkinson JD. Predictors of poor outcome in patients with a spontaneous cerebellar hematoma. Can J Neurol Sci 2000;27:32–36.

49. Hill MD, Silver FL. Epidemiologic predictors of outcome after cerebellar hemorrhage. J Stroke Cerebrovasc Dis 2001;10:118–121.

50. Marti-Fabregas J, Belvis R, Guardia E, Cocho D, Munoz J, Marruecos L, Marti-Vilalta JL. Prognostic value of Pulsatility Index in acute intracerebral hemorrhage. Neurology 2003;61:1051–1056.

51. Selker HP, Beshansky JR, Schmid CH, Griffith JL, Longstreth WT Jr, O'Connor CM, Caplan LR, Massey EW, D'Agostino RB, Laks MM, et al. Presenting pulse pressure predicts thrombolytic therapy-related intracranial hemorrhage. Thrombolytic Predictive Instrument (TPI) Project results. Circulation 1994;90:1657–1661.

52. Vespa PM, O'Phelan K, Shah M, Mirabelli J, Starkman S, Kidwell C, Saver J, Nuwer MR, Frazee JG, McArthur DA, Martin NA. Acute seizures after intracerebral hemorrhage: a factor in progressive midline shift and outcome. Neurology 2003;60:1441–1446.

53. Teernstra OP, Evers SM, Lodder J, Leffers P, Franke CL, Blaauw G; Multicenter randomized controlled trial (SICHPA). Stereotactic treatment of intracerebral hematoma by means of a plasminogen activator: a multicenter randomized controlled trial (SICHPA). Stroke 2003;34:968–974.

54. Fernandes HM, Gregson B, Siddique S, Mendelow AD. Surgery in intracerebral

hemorrhage. The uncertainty continues. Stroke 2000; 31:2511–2516.

55. Morgenstern LB, Frankowski RF, Shedden P, Pasteur W, Grotta JC. Surgical treatment for intracerebral hemorrhage (STICH): a single-center, randomized clinical trial. Neurology 1998; 51:1359–1363.

56. Bath P, Chalmers J, Powers W, Beilin L, Davis S, Lenfant C, Mancia G, Neal B, Whitworth J, Zanchetti A; International Society of Hypertension Writing Group. International Society of Hypertension (ISH): statement on the management of blood pressure in acute stroke. J Hypertens 2003;21:665–672.

57. Schellinger PD, Fiebach JB, Hoffmann K, Becker K, Orakcioglu B, Kollmar R, Juttler E, Schramm P, Schwab S, Sartor K, Hacke W. Stroke MRI in intracerebral hemorrhage: is there a perihemorrhagic penumbra? Stroke 2003;34:1674–1679.

58. Kidwell CS, Saver JL, Mattiello J, Warach S, Liebeskind DS, Starkman S, Vespa PM, Villablanca JP, Martin NA, Frazee J, Alger JR. Diffusion-perfusion MR evaluation of perihematomal injury in hyperacute intracerebral hemorrhage. Neurology 2001;57:1611–1617.

59. Warach S. Is there a perihematomal ischemic penumbra? More questions and an overlooked clue. Stroke 2003;34:1680.

60. Siddique MS, Fernandes HM, Wooldridge TD, Fenwick JD, Slomka P, Mendelow AD. Reversible ischemia around intracerebral hemorrhage: a single-photon emission computerized tomography study. J Neurosurg 2002;96:736–741.

61. Zazulia AR, Diringer MN, Videen TO, Adams RE, Yundt K, Aiyagari V, Grubb RL Jr, Powers WJ. Hypoperfusion without ischemia surrounding acute intracerebral hemorrhage. J Cerebr Blood Flow Metab 2001;21:804–810.

62. Powers WJ, Zazulia AR, Videen TO, Adams RE, Yundt KD, Aiyagari V, Grubb RL Jr, Diringer MN. Autoregulation of cerebral blood flow surrounding acute (6 to 22 hours) intracerebral hemorrhage. Neurology 2001;57:18–24.

63. Stroke Therapy Academic Industry Roundtable II (STAIR-II). Recommendations for clinical trial evaluation of acute stroke therapies. Stroke 2001;32:1598–1606.

64. Wilson JT, Pettigrew LE, Teasdale GM. Structured interviews for the Glasgow Outcome Scale and the extended Glasgow Outcome Scale: guidelines for their use. J Neurotrauma 1998;15:573–585.

65. Wilson JT, Edwards P, Fiddes H, Stewart E, Teasdale GM. Reliability of postal questionnaires for the Glasgow Outcome Scale. J Neurotrauma 2002;19:999–1005.

66. Pettigrew LE, Wilson JT, Teasdale GM. Reliability of ratings on the Glasgow Outcome Scales from

in-person and telephone structured interviews. J Head Trauma Rehabil 2003;18:252–258.

67. Wilson JT, Hareendran A, Grant M, Baird T, Schulz UG, Muir KW, Bone I. Improving the assessment of outcomes in stroke: use of a structured interview to assign grades on the modified Rankin Scale. Stroke 2002;33:2243–2246.

68. Newcommon NJ, Green TL, Haley E, Affleck L, Cooke T, Hill MD. Improving the assessment of outcomes in stroke: use of a structured interview to assign grades on the modified Rankin Scale. Stroke 2003;34:377–378.

69. van Swieten JC, Koudstaal PJ, Visser MC, Schouten HJ, van Gijn J. Interobserver agreement for the assessment of handicap in stroke patients. Stroke 1988;19:604–607.

70. de Haan R, Limburg M, Bossuyt P, van der Meulen J, Aaronson N. The clinical meaning of Rankin "handicap" grades after stroke. Stroke 1995;26:2027–2030.

71. Hill MD, Bayer N, Lawrence M, Jaeschke R, Jackowski G. Biochemical markers in acute ischemic stroke. Can Med Assoc J 2000;162:1139–1140.

72. Tilley BC, Marler J, Geller NL, Lu M, Legler J, Brott T, Lyden P, Grotta J. Use of a global test for multiple outcomes in stroke trials with application to the National Institute of Neurological Disorders and Stroke t-PA Stroke Trial. Stroke 1996;27:2136–2142.

73. Montaner J, Molina CA, Monasterio J, Abilleira S, Arenillas JF, Ribo M, Quintana M, Alvarez-Sabin J. Matrix metalloproteinase-9 pretreatment level predicts intracranial hemorrhagic complications after thrombolysis in human stroke. Circulation 2003;107:598–603.

4

Traumatic Brain Injury

Wayne M. Alves and Lawrence F. Marshall

Advances in understanding of the mechanisms of damage and the pathophysiology of traumatic brain injury (TBI) during the 1990s, especially the neurochemical cascade associated with secondary brain injuries, were offset by a disappointing legacy from phase 3 clinical development programs. Drugs and non-pharmacologic treatments tested without success include high-dose corticosteroids, free radical scavengers, calcium channel blockers, glutamate *N*-methyl-D-aspartate (NMDA)—receptor antagonists, and induced hypothermia. The TBI experience has been one of considerable disappointment as most of the more promising novel neuroprotectants failed to live up to their preclinical expectations and failed phase 3 clinical trials became the norm (1–4). Apparent safety problems or early evidence of clinical futility cut short very expensive clinical development projects. Interaction of novel experimental drug therapies with conventional treatment and management practices may have clouded the interpretations of results of phase 3 efficacy trials. As a result, there are no proven therapies available to minimize the effects of severe or moderate TBI. Consequently, best-management practices continue to focus on supportive care and intensive physiologic monitoring. It has been an exasperating experience.

Because TBI is conventionally treated earlier after injury than many other neuroemergencies, the opportunity exists to begin early therapy. The stakes following TBI are high because survivors' cognitive and behavioral disorders are profound. Permanent deficits are a major source of disability that unfortunately may lead to institutionalization. The recovering TBI survivor places an immense burden on families and stresses on rehabilitation and social service resources that are available but not designed to handle the demand that exists during the long recovery period.

HISTORICAL CONTEXT OF TRAUMATIC BRAIN INJURY CLINICAL TRIALS

In the 1990s, the *decade of the brain* brought with it a considerable amount of enthusiasm and optimism regarding TBI and the prospects for novel treatments in the form of new and innovative drugs targeted to specific pathophysiological processes (2,3,5). The clinical epidemiology of TBI was fairly well defined, and it was clear that although severe TBI did not have a high incidence compared with other major diseases, it was followed by potentially devastating consequences. It was clear that there was no singular pattern of recovery and morbidity was probably underestimated. Numerous complicating factors created patient management and treatment challenges requiring intensive care unit (ICU) settings, leading to the observed high costs of the disease (6–8). In addition there were few effective treatments (1,9).

Patients with TBI are clearly among the *sickest of the sick* in emergency medicine, and the supportive care and intensive physiological monitoring required involves numerous clinical tests and evaluations. At the start of the 1990s, it was understood that brain injury was complicated by a neurochemical cascade that produced devastating secondary insults or damage to brain tissue, eventually producing final clinical outcomes (10–13). Phospholipase activation and the subsequent arachidonic acid cascade and oxygen radical—mediated lipid peroxidation were the major pathophysiological mechanisms implicated as the final common pathway to neuronal cell death. Of particular interest was the role of these biochemical processes in neuronal cell membrane dysfunction and subsequent cell death and, by extension, permanent neurological impairments (14,15). Becker (16) elegantly postulated that the major operating hypothesis for the 1990s in the context of TBI (but equally applicable to other neuroemergencies) was that

…following injury to the brain, many cells are rendered dysfunctional, but are not mechanically disrupted. If the milieu of those cells is favorable, they may recover; if it is unfavorable, they may die. The challenge, then, is to define the milieu in which an injured cell finds itself and define the milieu ideal for recovery.

The sense of enthusiasm and optimism that came with the *decade of the brain* also stemmed, in part, from the rich drug development pipeline available for neuroemergencies (11–15,17–22). An unprecedented number of promising novel drugs, biologics, and devices for preventing or ameliorating the pernicious

consequences of neuroemergencies were aggressively put into clinical development testing in the 1990s. With notable exceptions, before then there were relatively few examples of modern multicenter randomized clinical trials in TBI (1,4,9).

The most explored therapeutic hypothesis for TBI during the 1990s was the role of the excitatory neurotransmitter glutamate in the secondary cell damage of TBI and ischemic stroke. Excessive glutamate release and activation of various glutamate receptors (NMDA, kainate, and α-amino-3-hydroxy-5—methylisoxazole-4-proprionic acid [AMPA]) led to strong membrane depolarization, excitation of neurons, further release of glutamate, and a resultant dramatic increase in Ca^{2+} influx and eventually cell death. Advances in neuroscience, clinical management, and clinical drug development also created excitement about the potential for novel antioxidants (especially 21-aminosteroids) (10), calcium channel blockers (13,14), NMDA antagonists (12), and the role of reducing metabolic demand using drugs or other treatment interventions (e.g., intentional hypothermia treatment) to mitigate the destructive processes of brain damage characteristic of neuroemergencies (11).

Testing of many of the novel neuroprotectants rested on a common underlying rationale. In a matter of hours or days, independent, albeit overlapping, pathological processes led to irreversible damage or death of neuronal cells. Early treatment in the form of drugs that can interrupt these processes must be developed and, if successful, should yield improved cerebral metabolism and function with better clinical outcomes. Despite this reasoning, virtually all compounds for traumatic brain injury taken forward in phase 3 clinical trials have failed to demonstrate efficacy. Table 4.1 summarizes the legacy of failure and disappointment of the 1990s.

TABLE 4.1 Legacy of Clinical Trials in Traumatic Brain Injury and Stroke in the 1990s

Negative or failed clinical trials
  Barbiturates
  Antioxidants
  Free radical "scavengers"
  Calcium channel blockers
  Anti—GP IIb/IIIa
  Hypothermia
  ICAM-1 antibody
  Adenosine deaminase inhibitor
  Bradykinin antagonist
  Phosphatidylcholine precursor
  Fibroblast growth factor
  Glutamate NMDA-antagonists
  Glycine NMDA-antagonists
  AMPA antagonists
  GABA agonists
  Maxi-K channel opener
  NO blocker
  Phenytoin precursor
  Serotonin agonists
  Sodium channel blocker
  r-Pro-urokinase
  Fibrinogen-cleaving enzyme

Uncertain legacy of clinical trials
  Streptokinase
  Ancrod
  Thrombin inhibitor
  Heparinoids
  Antiplatelets
  PAF antagonists
  Granulocyte adhesion blocker

Legacy of success
  rt-PA in acute ischemic stroke

AMPA, α-amino-3-hydroxy-5-methylisoxazole-4-proprionic acid; GABA, ≫-aminobutyric acid; NO, nitrous oxide; NMDA, N-methyl-D-aspartate; platelet activating factor (PAF); recombinant tissue plasminogen activator (rt-PA).

DRUG DEVELOPMENT PROGRAMS FOR TRAUMATIC BRAIN INJURY

Underlying Logic of Traumatic Brain Injury Trials

Ideally, TBI interventions rest on the observation that overlapping pathological processes that were initiated at the time of injury to brain may take several hours to several days to cause irreversible neuronal cell damage or death. Early treatments aimed at interrupting or quenching these processes should minimize or prevent their delayed consequences (16). If these processes are prevented from achieving their full course, we can expect improvement of brain metabolic or physiological function, which will result in improved clinical outcomes (11). Finally, the most reasonable way to pursue these hypotheses is to test them with novel or already approved drugs with demonstrated mechanisms of action that can interrupt these deleterious processes.

Emphasis in clinical drug development in TBI has been on secondary mechanisms of brain injury, with special interest on drugs affording global or focal brain protection from ischemic and hypoxic injury (2,3,5). In practice, most TBI clinical trials have focused only on a conventional drug testing paradigm: Give the treatment, ensure that it is safe (i.e., not detrimental and with manageable complications), and determine if the outcome is better than that of patients who did not receive the drug (1). In the case of neuroemergencies, this paradigm is fairly straightforward and acceptable if the drug is clearly effective. If it is not, the reasons for the apparent lack of efficacy may not be easily identified.

Several logical steps in the drug development process lead to a determination of the efficacy and effectiveness of a novel drug or new chemical entity (NCE). These comments are necessarily brief and more detailed discussions are available (1,23,24). First, screening and identification of an NCE worthy of clinical testing in humans is accomplished by using the drug in experimental (animal) models of the disease for which it is to be used clinically. Because animal models cannot fully replicate with exact fidelity the entire spectrum of a neuroemergency, such as TBI, this initial step may be problematic. The fact that preclinical TBI models may only be useful in a limited number of clinical situations is an obvious limitation, but our urgent need for effective TBI drugs or other treatments often leads to a new promising drug to be generalized to clinical conditions for which it was not originally tested. In many cases the drug's mechanism of action remains only putative rather than proven. The paradigm of serial animal-then-human testing has proven risky in TBI drug development.

Determining that a novel drug treatment for human use is efficacious generally requires fulfillment of several steps (1,23, 24). The initial safety of the compound must be demonstrated at least to proof of the ability to effectively manage the drug's side-effects profile. We have usually presumed, sometimes incorrectly, that drugs are delivered to their intended sites of action and have the expected pharmacologic effects, thereby producing desired neurophysiological responses. Our ability to clearly identify a neurophysiological response to a drug depends in part on the fidelity of the animal model to the clinical situation. Despite the extensive preclinical testing that precedes the human testing of many new compounds, this ideal may not be realized in practice. In general, demonstration of efficacy *proof of fact* requires a *measurable* improvement in the outcome of the disease process (25).

Classes of Novel Drugs Tested in Traumatic Brain Injury

Numerous TBI clinical trials can be performed to assess all of the off-the-shelf (i.e., already approved) drugs with presumed therapeutic value in neuro-emergency (2). The unprecedented pipeline of novel NCEs available in the 1990s gave renewed hope of finding efficacious and approvable drugs (3,5). Hypotheses ascribed the excitatory neurotransmitter glutamate a major role in secondary cell damage after acute ischemic stroke and TBI. A massive release of glutamate following cerebral ischemia and the subsequent activation of various glutamate receptors (e.g., NMDA, kainite, and AMPA receptors) leads to strong membrane depolarization, excitation, and further release of glutamate, resulting in a dramatic increase of Ca^{2+}-influx into neurons through various Ca^{2+} channels, and finally in cell death. Numerous strategies for therapeutic intervention were investigated in the 1990s and a variety of pharmacological approaches investigated, including glutamate-receptor antagonists (e.g., blockers for NMDA and AMPA receptors such as CNS-1102 and D-CPPene), Ca^{2+}-channel antagonists, lipid peroxidation inhibitors, and other principles (5,11, 13–15,18–22,26). All of the compounds taken forward into phase 3 clinical trials failed to show convincing evidence of efficacy.

The immediate damage due to the consequences of primary human TBI cannot be prevented or treated. Greater opportunity for intervention lies in preventing or treating the consequences of secondary insults associated with hypoxic—ischemic damage of extracranial etiology (27,28) and brain swelling due to venous congestion or brain edema. Complex biochemical cascades and axonal and neuronal derangements that follow for some time after primary injury lead to delayed secondary deterioration and poor outcome. In a considerable number of patients with TBI this cascade might be prevented by effective pharmacological intervention.

TRAUMATIC BRAIN INJURY DRUGS AND TREATMENTS

During the 1990s there was an enormous increase in the amount of drug development activity testing novel drugs for TBI, while in parallel aggressive surgical trials and needed evaluations of neurosurgical devices also took place. Table 1 in the Introduction to this volume lists many of the neuroprotectants in clinical development that has made this dramatic

TABLE 4.2   Commonality of Traumatic Brain Injury and Ischemic Stroke Damage

Ischemic damage contributes to traumatic brain injury (TBI) outcome

Preclinical evidence that trauma heightens brain's vulnerability to ischemia

Convergence of concepts of cellular and molecular mechanisms of stroke and TBI

Parallel triggering of excitotoxicity and apoptosis

Augmented lactate production and accumulation

Release of excitatory amino acids (EAAs) and loss of homeostasis

Glutamate- and aspartate-induced Ca^{2+} increases and other ionic fluxes

Role of inflammatory mediators (cytokines)

Effects of neurotrophins

effort possible. Most of these drugs were developed initially for TBI and ischemic stroke, and they represented an unprecedented effort to identify and develop NCEs for these populations. Table 4.2 lists some of the commonality of TBI and stroke pathology. As we have mentioned, there have been no definitive results from phase 3 efficacy trials providing noncontroversial proof of fact justifying the general use of the tested drugs in TBI.

TBI targets have emphasized calcium influx, glutamate release, and interruptions of cerebral blood flow. Numerous cellular-based approaches to ischemic TBI include antioxidants (e.g., tirilazad mesylate), intracellular adhesion molecule (ICAM) inhibitors (e.g., enlimomab), lactate inhibitors (e.g., CPS-211), and nitric oxide inhibitors (lubeluzole). Receptor-based approaches include NMDA/glycine antagonists, calcium channel blockers, and apoptosis inhibitors to influence targeting gene regulation mediating cell recovery or apoptosis.

We cannot describe the development status of each compound in detail in this chapter. However, some general comments are made and the reader should refer to numerous reviews that already exist in the TBI literature (5,11,13–15,18–22,26).

EARLY CLINICAL TRIALS OF DRUGS AND BIOLOGICS FOR TRAUMATIC BRAIN INJURY

Barbiturate Trials

In the 1980s, evidence from nonrandomized clinical studies indicated a potential benefit to managing unresponsive intracranial pressure (ICP) after TBI by ICP normalization with high-dose barbiturate therapies (29–31). Pentobarbital was tested and results were inconclusive in part because there appeared to be a greater than anticipated efficacy of conventional therapy. As such, no definitive statement can be made regarding the efficacy of high-dose barbiturates in the management of TBI. A lengthier discussion is available elsewhere (4).

High-Dose Corticosteroids Trials

Numerous small, randomized clinical trials of high-dose corticosteroids were conducted from the late 1970s through the mid 1980s. At the time, these studies comprised the largest series using modern clinical trials methodology to test a single clinical hypothesis in TBI (32–38). The high-dose steroids trials were limited to subjects with severe TBI on admission, as defined by the admission Glasgow coma scale (GCS) or subsequent changes in the GCS indicating early neurologic deterioration before treatment. There is insufficient information in the published reports to determine whether the studies were adequately power to detect clinically feasible differences between the treated and placebo groups. In retrospect, most of the studies were (or might have been) stopped early for safety considerations. It is not likely that that the apparent safety issues were manageable or limited to specific patient subpopulations. Yet, it might be thought that a definitive answer regarding *efficacy* of high-dose steroids may not be drawn from the studies reported to date.

On the other hand, Braakman and colleagues (32) terminated their trial on the grounds of futility when it was determined through sequential analysis procedures that the likelihood of a positive study for the 1-month survival endpoint could not be achieved after 161 subjects were enrolled and matured to the outcome assessment time. The 1-month survival endpoint may have been too premature an assessment to evaluate efficacy in a compound intended for use in TBI. The recent failure of the corticosteroid randomisation after significant head injury (CRASH) trial to demonstrate a favorable effects of glucocorticoids in a large series of patients covering the entire spectrum of head injury should put to rest the utility of these agents in TBI (39).

Tromethamine Trials

Carbon dioxide (CO_2)—induced cerebral vasoconstriction leading to decreased intracranial pressure was the underlying rationale for the use of sustained hyperventilation therapy in the management of neuroemergency patients (40,41). Of utmost concern in patients undergoing prolonged hyperventilation was the possible initiation of a counteracting destructive tissue acidosis. The lactate buffer tromethamine (THAM) was proposed to counteract the depletion of bicarbonate buffer in the cerebrospinal fluid (CSF) due to the short-lived deleterious effect of hyperventilation on CSF pH and subsequent arteriolar diameter. This hypothesis was tested in a number of studies but the major *proof of fact* trial was terminated for safety considerations on the basis of a sequential analysis of the data (40). Although the primary efficacy endpoint was the 12-month Glasgow outcome scale (GOS), at 3 months the *hyperventilation alone* group showed a decreased favorable outcome rate compared to the *normal ventilation* group. Age and admission motor scores were used as covariates and treatment by covariate interaction terms indicated that the effect was pronounced in the subgroup with GCS motor scores of 4 or 5, and so the study was terminated. When the subjects matured to the final analysis endpoint and the data were later analyzed, the decrease in favorable outcome was no longer evident at 12 months after injury. Wolf and colleagues (41) reported no significant findings regarding efficacy endpoints, but did replicate the finding of a decrease in the percent of time that ICP was greater than 20 mm Hg in the first 48 hours after injury. In addition, fewer subjects in the THAM group required barbiturate coma therapy. They concluded that the THAM subjects were worse at 3 months after injury, but there were no differences between the THAM and control groups at 6 or 12 months after injury (41).

Ion Channel Blocker Trials

Until the mid 1990s, the largest phase 3 multicenter clinical trials in TBI were the studies of the calcium channel blocker, nimodipine (42–45). Teasdale and the British/Finnish Group (9) reported a study involving 351 patients treated for 7 days after injury, initially at 1 mg/hr and then at 2 mg/hr if the patient's blood pressure tolerated it. The study was a randomized, double-blind, placebo-controlled trial and used the 6-month GOS as the primary efficacy endpoint. No clear evidence of efficacy emerged from the phase 3 nimodipine clinical trials.

Antioxidant (Free-Radical Scavenger) Trials

The oxygen radical superoxide anion has long been implicated in the pathophysiology of acute neuroemergencies (10,46–51). The most common presumed pathway is ischemic injury and/or ischemic—reperfusion injury, making this class of compounds appealing for ischemic stroke and TBI applications. Control of sustained high ICP and amelioration of vicious damage due to the secondary insults of hypoxia, hypotension, and increased intracranial hypertension due to the destructive effects of oxygen free—radicals formed the rationale to test poly-ethylene glycol-conjugated superoxide dismutase (PEG-SOD) in a double-blind, placebo-controlled phase 2 and 3 trials although the early phase effect sizes were modest (52–54). Completed phase 3 study findings were neutral (52–54).

Perhaps the most ambitious clinical development effort in the 1990s was the Upjohn Company's (subsequently Pharmacia & Upjohn) efforts on behalf of tirilazad mesylate, a novel antioxidant that had proved enormously promising in preclinical development studies (49,50,56). Tirilazad mesylate is a potent antioxidant with numerous putative mechanisms of action. Although tirilazad has been approved in a number of European countries, the development program was abandoned on the basis of a combination of scientific and regulatory reasons and business decisions.

Glutamate Antagonists

The most active area of clinical development in neuroemergency indications in the 1990s involved glutamate antagonists including the NMDA receptor—site antagonists, glycine-site antagonists, and AMPA/kainate-site antagonists (2,5,11,12, 57). Development of most NMDA antagonists was suspended or terminated, and full information is not as yet publicly available to determine all of the factors leading to these decisions. The well-known psychotomimetic effects of NMDA receptor—site antagonists were a major reason for slowing or termination of clinical development of some drugs. In addition, apparent safety or risk—benefit calculations are implicated, but the information available is only anecdotal or based on brief presentations in various scientific venues.

CDP-choline Trial

Decreases in cell-membrane phospholipids leading to failure of the sodium-potassium pump and eventually causing cytotoxic edema was the underlying mechanism of cellular damage used to justify the use of CDP-choline, a precursor in the synthesis of glycerophospholipids. CDP-choline was tested at a single center in Spain in a group of 216 patients with a history of TBI and GCS scores of 5 to 10 on admission (58). This study was a single-blind, standard-care controlled trial, and patients with open trauma and severe systemic disease were excluded from eligibility. The 3-month GOS was the primary efficacy endpoint, and other endpoints were posttraumatic symptoms (headaches, dizziness, memory problems), motor dysfunction, IQ, personality changes, and total ICU days. More deaths were observed in the CDP-choline group (18 [15.7%]) than in the conventional care control group (10 [9.9%]). Only survivors were compared in the final efficacy analysis, and the CDP-choline group did not appear to differ in favorable outcome (83.5% vs. 81.2% in the controls), 66.9% were considered *good recovery* compared with 50.5% in the controls (58). Although the CDP-choline group was consistently better than control in the other endpoints, most differences were not statistically significant.

WHY HAVE TRAUMATIC BRAIN INJURY CLINICAL TRIALS PROVED DIFFICULT?

Numerous reasons can easily be offered to explain the failure of so many randomized clinical trials in TBI. These include flaws in study design; inability of the drug to cross the blood—brain barrier; the clinical populations included in the trials; the adequacy of the drugs tested or the dosing regimens used; and even whether the

TABLE 4.3 Possible Reasons for Failure to
Achieve Success

Failure of therapeutic hypothesis (drugs do not work)
Failed trials due to design failures
Methodological flaws in phase 1 or 2 development
Suboptimal design or execution of phase 3 trials
Drug not delivered to target site due to failure to cross
  blood–brain barrier
Therapeutic window was not adequately defined
Dose or dose regimen was not adequately defined
Phase 3 populations not adequately defined
Multiplicity of damage and outcomes
Poor study design or insufficient power to detect true
  drug effect
Premature termination that does not withstand real
  scrutiny

TABLE 4.4 Characteristics of Traumatic Brain
Injury

Mechanisms of damage are limited
  Diffuse axonal injury
  Cerebral ischemia
  Brain swelling/edema
  Hemorrhage
  Traumatic subarachnoid hemorrhage
  Hydrocephalus
  Neurotransmitter failures
  Toxic substances cross blood–brain barrier
  Infection/inflammation
  Posttraumatic seizure
  Brain atrophy
**Each instance may be complex with a combination
  of diagnoses. Inputs may be complicated.**

underlying therapeutic hypotheses were correct. Table 4.3 lists some of the more common reasons.

Patients with TBI are seen in emergency and ICU settings that involve numerous drug and other therapeutic interventions and extensive examination points (16). In the case of TBI, we do not have a long history with regard to proven treatment interventions, efficacious drugs, or, until more recently, well-defined treatment protocols (1,9). Compared with many diseases, severe TBIs are relatively rare, thereby creating the need for many investigative sites and an inherent variability in patient inputs (i.e., prognostic factors) and endpoints in modern TBI clinical trials. Decision-making and clinical practice conventions in the treatment of TBI remains idiosyncratic and this mindset must at least be acknowledged in evaluating physician screening practices relevant for TBI clinical trials. Because of these factors, case report forms (CRFs) associated with TBI clinical trials are usually quite large. There are few gold standards (diagnostic, prognostic, or clinical endpoints) and subjective definitions and perceptions are still very important parts of the TBI knowledge base available to guide clinical development programs.

Now that the decade of the brain has passed, and we have the experience of numerous modern clinical trials, we can begin to better address the question of what we will need for success in clinical drug development in TBI. In many cases, the only answers will come from success in several areas of needed research.

We need to better understand the evolution of a TBI therapeutic hypothesis and seek to define an optimal clinical development process for TBI drugs (24). It has been evident in the case of TBI clinical trials that we need to improve our classifications of patients and improve endpoint measures. Although the mechanisms of damage in TBI are limited, they seldom occur in isolation. Typically, several pathophysiological mechanisms are combined in each patient. This multiplicity of damage and outcome of TBI, described in Table 4.4, may also be a contributing reason for the failure of phase 3 clinical trials.

We need to improve translational research from the laboratory to clinic and consider the potential value of leveraged *in vivo* models for evaluating novel TBI drugs. Surrogate endpoints are not a solution in their own right (59,60). On the other hand, complete and sensitive development of such measures could greatly improve the TBI clinical trials process. For example, magnetic resonance imaging (MRI) techniques of diffusion-weighted imaging and perfusion imaging following stroke have recently offered great promise as surrogate measures but have not demonstrated their value in obtaining approval for novel drugs. Finally, adoption of innovative clinical trials methodology in TBI trials is needed (1,7,23).

CURRENT ENVIRONMENT OF TRAUMATIC BRAIN INJURY DRUG DEVELOPMENT

Given the expected or potential complications and outcomes of TBI, what treatments do we have to offer today? Despite the number of novel drugs under clinical development thus far, it is clear that providing proof of fact of efficacious drugs for TBI will be hard to achieve. This resistance to proof of fact stems, at least in part, from the multiplicity of damage and outcomes characteristic of TBI. Improved TBI trials methodologies are needed for dealing with this multiplicity. The fundamental challenge is to answer the question: How can we *fairly* demonstrate safety, efficacy, and effectiveness of novel drugs and treatments for TBI? Achieving the proverbial *level playing field* has thus far proved elusive.

KEY ISSUES IN TRAUMATIC BRAIN INJURY CLINICAL TRIALS DESIGN AND ANALYSIS

Multiplicity of Damage and Outcome

TBI offers the best illustration of the underlying complexity of neuroemergency damage and outcome. Damage from trauma occurs to the brain, its blood vessels, and its coverings (i.e., the skull). As in other neuroemergencies, there are numerous ways of classifying the damage or underlying pathology that has occurred. One could simply use the nonmutually exclusive set of terms such as *diffuse axonal injury* (DAI) or *focal brain injury* (e.g., contusions or hematoma), or the

secondary mechanisms of injury such as hypoxia/ischemia, cerebral edema, or brain swelling.

Anatomical Classification

Improving our classifications of neuro-emergency populations to more adequately reflect the underlying clinical pathology is essential for future success in neuro-emergency drug development (61–63). These improved classifications must take into account a number of issues. Although each instance of TBI is likely to be heterogeneous and complex, the mechanisms of damage themselves are quite limited (1,4). TBI is heterogeneous in terms of the limited mechanisms of brain injury, underlying pathology, and the severity of injury as measured in physiological terms, (e.g., by the GCS). On the other hand, it is typical to observe complex instances of these limited mechanisms (i.e., they occur in *combination* in patients [see Table 4.4]). The relevant questions are as follows:

What injuries to the brain have occurred?
What metabolic derangements are the results?
What individual and whole-brain impairments resulted from these injuries?

In sum, the amount and complexity of brain injury is typically not made explicit in clinical trials.

An alternative way of classifying TBI is to describe the physiological response to trauma by use of the GCS, which grades coma severity, allows description of outcome in comatose patients, and reduces dependence on specific anatomical—clinical correlation for assessing coma (64). Table 4.5 presents the categories of the GCS and the scoring algorithm used in TBI. However, this alone is not sufficient and ways to improve classification of patients with TBI are being developed (27).

Improved TBI description would offer considerable value for clinical trials (56,65). We might better understand the influence of clinical management and specific therapies; improve our understanding of how tissue damage disrupts function and disorganizes the integration among neurons that are responsible for behavior; improve our understanding of pathophysiology of the diseases of *head injury*; improve interpretation of results from different centers (or studies); and, finally, improve the design, conduct, and analysis of data from clinical trials in TBI.

Potential for Pathological Outcomes

There always is a possibility that treatment-or-drug-related pathological outcomes may occur, which are a concern in neuro-emergency clinical trials. Theoretically, a new intervention or drug that can prevent death may at best result in no better than survival in a *persistent vegetative state* (PVS). It is also conceivable that patients who might have had a *severe disability* (SD) outcome could be pushed into the PVS or dead outcome categories. Of even more concern is an intervention in which perhaps a *good recovery* or *moderate disability*

TABLE 4.5 Glasgow Coma Scale (GCS) (66)

Eye opening (E)	4 Spontaneous
	3 To voice
	2 To pain
	1 No response
Verbal response (V)	5 Oriented conversation
	4 Confused, disoriented
	3 Inappropriate words
	2 Incomprehensible sounds
	1 No response
Best motor response (M)	6 Obeys commands
	5 Localizes
	4 Withdrawals (flexion)
	3 Abnormal flexion (posturing)
	2 Extension (posturing)
	1 No response

Total GCS = Eye score + Verbal score + Motor score

Source: Teasdale and Jennett 1974.

TABLE 4.6 Glasgow Outcome Scale (67)

Outcome	Original	Extended	Descriptors
Good recovery	GR	GR-upper	Includes full recovery without signs or symptoms
		GR-lower	Capacity to resume normal occupational and social activities
			May still have minor physical deficits or complaints
			Reasons other than physical or mental deficits may prevent resumption of some former activities
Moderate disability	MD	MD-upper	Independence in activities of daily living and mobility
		MD-lower	Some previous activities, either at work or social life, are no longer possible because of physical or mental deficit
			Resumption of activities at a lower level or resumption of most activities may be possible, in spite of deficits
			May be able to return to work "Independent but disabled"
Severe disability	SD	SD-upper	Due to many posttraumatic complaints or deficits resumption of former life and work are not possible
		SD-lower	Needs assistance of another person for some activities every day
			May range from continuous total dependency to need for assistant with only one activity, such as dressing
			Dependency due to physical and/or mental impairments "Conscious but dependent"
Persistent vegetative state	PVS	PVS	Sleep/awake, nonsentient
Dead	D	D	

Source: Jennett B, Snoek J, Bond MR, Brooks N. Disability after severe head injury: observations on the use of the Glasgow Outcome Scale. JNNP 1981;44:285–293.

would wind up with a SD, PVS, or dead outcome.

IMPROVED CLASSIFICATIONS OF TRAUMATIC BRAIN INJURY POPULATIONS

The problem we face in TBI clinical trials is that the terms used to describe head injuries are not mutually exclusive and there is little consensus regarding which terms should take precedence when more that one is needed to describe the patient. It is reasonable to assume that we need to go beyond reporting outcome based solely on *predominant pathological process.* Unanswered as yet is, for the purpose of *phenotyping* patients for inclusion or exclusion in a TBI clinical trial, whether we can describe or determine the following:

• The total amount or complexity of tissue damage so as to better understand the significance of injury severity for treatment and outcome correlation.
• The lesions that cause low GCS scores.
• If patients who have the same GCS score caused by different lesions have the same or different outcomes.
• How the degree and time course of recovery in specific impaired functions (i.e., motor, sensory, cognitive, behavioral) are influenced by the total damage and complexity of injuries.

In light of this, the following might be productive areas of methodology research for TBI clinical trials.

• Anatomic descriptors of the total amount and complexity of brain injuries;
• Improved physiological severity scores;
• Methods to relate anatomical and physiological severity to outcomes;
• Patient classification based on the type of brain injury as determined by acute anatomical *and* physiological severity measures;

• Ways of classifying patients into homogeneous injury-severity groups to improve sensitivity of study designs to evaluate drug interventions;
• Development and validation of neurological impairment indexes.

DAMAGE AND OUTCOME MEASURES

Glasgow Coma Scale

The GCS grades coma severity without depending on specific anatomical—clinical correlations and allows us to describe the outcomes of neuroemergency in terms of initial inputs. It is well understood that TBI severity is a continuum, although practical classifications into mild, moderate, or severe injury are useful. In the case of GCS scores of 3 to 5, mortality is high and success has generally been low, whereas for GCS scores of 6 to 8, aggressive triage and neurosurgical management may be the most important determinant of outcome. The situation is less clear in the case of GCS scores of 9 to 12 (moderate neuroemergency) and 13 to 15 (mild neuroemergency) where the sequelae may be best handled by other disciplines. Clinical trials have varied in terms of the range of GCS scores included, with GCS scores of 4 to 8 now typical.

The value of the GCS in grading coma severity is undeniable, but the underlying pathology, or anatomical injury severity, accounts for some (as much as 50%) of the variability in TBI outcome. This has profound implications for present neuroemergency clinical trials, since it is imperative that at least statistical *balance* in pathology be achieved through the study randomization procedures. Small, apparently insignificant differences between treatment groups play havoc with our ability to detect or measure treatment effects. In the worse case, imbalance in clinical trials may make it appear that a drug is harmful.

Glasgow Outcome Scale

We saw earlier that TBI simultaneously heterogeneous and complex. Primary and secondary brain injuries, complicating factors (e.g., hemorrhage or extracranial injuries) and competing risk factors (or co-morbidities) influence outcomes (27,31,61, 63,64). Since there is no singular and easily describable recovery pattern, typically measurements of recovery from TBI have been global clinical rating scales like the GOS. In part, these measures have value in maintaining acceptable levels of validity and reliability because the multifactorial nature of neuroemergency outcome may require more sophisticated and extensive quality-of-life measures (68,69).

Finally, one could use the original five GOS categories or the eight categories of the extended GOS. The eight-category GOS is still in a research phase, although its reliability appears to be at least as good as five-category GOS.

Computed Tomography Classification of Traumatic Brain Injury Populations

Classifying patients with TBI into meaningful *injury severity groups* is difficult and few validated classification schemes exist. In addition, an experienced central reader for computed tomography (CT)/magnetic resonance imaging (MRI) is necessary to prevent significant misclassification error. It is desirable to be able to identify the natural pathological groupings of TBI patients so more sensitive TBI clinical trials can be designed. The primary hurdle to doing so is that the terms of neuroemergency are not mutually exclusive and it is not always clear which should take precedence. Further, how do we factor both anatomical and physiological severity descriptors into a *single* classification system that is clinically meaningful, correlates with clinical outcomes, and is useful in subsetting patients for inclusion in clinical trials? TBI clinical trials

require a fairly simple way of describing the pathology in the population of patients enrolled as subjects. The most useful CT-based clinical classification has been provided by the National Traumatic Coma Data Bank investigation team, which is described in Table 4.7 (62,63). This group proposes that a minimal requirement for TBI clinical trials is intracranial diagnoses based on CT scan findings, with some modification of the original classification.

Prognostic Factors

Factors that have major prognostic value for TBI outcomes are hypotension (shock), hypoxia, age, and to a lesser extent, gender (56). In addition, isolated severe TBI is less frequent than multiple trauma, and *extracranial* injuries may interact in synergy with brain injuries to produce a greater than expected rate of unfavorable outcomes (27,28). Consequently, minimal inputs for TBI clinical trials design including injury severity (usually the GCS), age, gender, preemergency department evidence of

TABLE 4.7   Traumatic Coma Data Bank Computerized Tomography Injury Classification System

Category	Definition
Diffuse injury I	No visible pathology (normal computed tomography [CT])
Diffuse injury II	Midline shift 0–5 mm, with no high or mixed density Lesions >25 cc
Diffuse injury III (swelling) absent	Midline shift 0–5 mm with no high or mixed density Lesions >25 cc *plus* cisterns compressed or
Diffuse injury IV (shift)	Midline shift >5 mm with no high or mixed density Lesions >25 cc
V	Any lesion surgically evacuated
VI	Surgical mass lesion >25 cc not evacuated

Source: Marshall LF, Marshall SB, Klauber MR, et al. A new classification of head injury based on computerized tomography. J Neurosurg 1991;75: S14–S20.

TABLE 4.8 Outcome Assessment in Traumatic Brain Injury

Global outcome
Glasgow Outcome Scale (GOS)
Disability Rating Scale (DRS)

TCDB neuropsychological test battery
Galveston Orientation and Amnesia Test (GOAT)
Controlled Word Association (oral fluency)
Symbol Digit Modalities Test (sustained attention)
Grooved Pegboard (fine motor dexterity)
Rey-Osterrieth Complex Figure (visuoconstruction and memory)
Neurobehavioral Functioning Inventory (behavior/ quality of life)

hypoxia/hypotension, and possibly temperature on admission. How to factor these inputs into the trial's preplanned data analyses is less straightforward and likely to be controversial (1,68,70–72). What needs to be determined is whether some of these factors is sufficiently powerful to *swamp* the effects of others. This would allow some parsimony in the number of prognostic factors to be taken into account in designing TBI trials and in planning and conducting data analyses.

Outcomes in Traumatic Brain Injury Clinical Trials

Improved measurement of clinical and quality of life endpoints is needed in TBI populations. Several recent attempts to provide reasonable outcomes assessment schedules are laudable and are steps in the right direction (73–75). The problem we face in TBI may be stated as follows: Patients who apparently have the *same* brain injury have *different* outcomes, while patients with apparently *different* injuries have the *same* outcomes. It is likely that this paradox reflects how well we describe and measure the pathology and the severity of damage to the brain and the relative imprecision of our endpoint measurements. Our choice of endpoints depends on the expected outcomes associated with TBI.

TBI outcomes used as endpoints in current clinical trials include mortality and various descriptors of mobility and morbidity. What counts as relevant outcomes from the point of view of TBI clinical trials? A *crude chain* of expected TBI outcomes is given by the sequence listed below.

- Death
- Severe or moderate neurophysical impairments
- Personality changes and/or neurobehavioral impairments
- Social and behavioral problems
- Lack of employability
- Reduction in quality of life

Where in this survival chain do we draw the line in testing new TBI drugs or treatments? Perhaps drugs should be evaluated in terms of their ability to reduce physical and mental impairments (i.e., change in structure and function with known association to later clinical outcomes). This standard would place a considerable obligation to determine the likely measurable effects of a new drug or treatment and whether the endpoints selected (surrogate or long-term clinical) have the requisite level of precision and sensitivity.

Time Course of Recovery

A clear definition of relevant clinical outcomes is an essential requirement of good clinical trials design practices. In most TBI clinical trials, the 6-month GOS has been the primary study endpoint. It is important to consider the expected (if not actual) time to recovery in understanding when it is best assess drug efficacy because the selection of the time to assess outcome can dramatically alter the conclusions of a clinical trial (68). Although the 6-month GOS has been the outcome of choice, a 12–month score may be needed. Choi and colleagues (70), however, recently considered the transition of outcome states based on the GOS from 3 to 6 months after injury. They concluded that the 3-month outcome endpoint could potentially be a satisfactory primary endpoint for TBI clinical trials.

Respecifying the Glasgow Outcome Scale

There are numerous ways to specify the GOS as the primary endpoint. Most typical has been to distinguish between *favorable outcome* (GCS = GR [good recovery] + MD [moderate disability]) and *unfavorable outcome* (GCS = SD [severe disability] + PVS [persistent vegetative state] + D [death]). This is the most conservative because it is a binary outcome and, in general, results in the largest sample size requirements. One could also base outcome on three categories of the GOS–GR versus MD versus unfavorable outcome as defined previously. Similarly a four-GOS categories endpoint would be GR versus MD versus SD versus PVS plus D. The extended GOS (see Table 4.6) offers additional opportunities for defining positive outcomes of novel treatments. From a societal standpoint the social burden for the low end of the severe TBI population (GCS, 3–5) is substantial due to the complete dependency of these patients. To be able to move patients from the GCS score of 3 to 5 group to the GCS score of 6 to 8 (*high end severe*), who are indistinguishable from patients with moderate TBI in terms of cognitive outcomes, would be a positive result. Work by Murray and colleagues (76) on the *sliding dichotomy* offers a novel statistical methodology to exploit these shifts in outcome categories. In this approach, the definition of a *good outcome* for a specific patient is tailored to the baseline prognosis on enrolment into the trial. For a patient with a very severe injury, survival alone may be regarded as a good outcome. Each group of patients defined by their expected outcome on the basis of a baseline prognostic model would be dichotomized into *good* or *bad* outcomes and then usual testing by treatment group would be

performed. The value of this approach or its acceptance by regulatory agencies remains to be seen.

We may wish to gain credit for two *shifts* in patient population. First, *credit* could be given for moving patients from unfavorable to favorable outcome category, and second for moving patients from moderate disability to the good recovery category. For example, in considering sample size requirements for the three-category option (GR, MD, SD+PVS+D) where the total equals 10% and the mean score statistic is used (a chi-squared statistic with 1 df) would require that more than seven of ten patients end up in the good recovery (vs. MD) category before the required sample sizes would be marginally less than the sample sizes for the binary endpoint ($n = 916$ vs. $n = 992$).

Measurement of a putative mechanism of drug action is often considered a desirable secondary endpoint. Including so-called mechanistic endpoints as secondary endpoints in TBI clinical trial protocols is a common practice, but is often done without compelling evidence that the proposed *measure* is a valid and reliable measure of mechanism of action. For example, if traumatic subarachnoid hemorrhage (tSAH) reflects pathology that can be expected to lead to delayed ischemic deficits in patients with TBI, and therefore a poor outcome, we may need to be able to document that clinical ischemic deficits actually do occur and that the drug actually prevents or reduces their incidence and effects.

SHAPE OF OUTCOME DISTRIBUTION

Current approaches to the management of TBI assume that patients who sustain brain injury can expect a good prognosis if treatment is immediate and aggressive. Also assumed is that identification of the outcomes that are due to the direct and secondary effects of brain injury is possible and therefore can serve as indices of drug efficacy. Unfortunately the underlying shapes of TBI outcome distributions for a variety of outcome measures are largely unknown. We expect the distribution of the GOS will be *J-shaped* and should take appropriate steps in statistical planning and data analysis, but this knowledge is unavailable for other outcome measures, especially quality-of-life measures. Ideally, we want an outcome measure that at least fully ranks all outcomes from best to worst.

Not knowing the shape of TBI outcome distributions is significant for the question of *when* to measure outcome (i.e., the time-oriented aspect of the outcome distribution). The distribution covers the subacute phase (at which point surrogate measures may be more appropriate), through rehabilitation, and then recovery. Depending on lost-to—follow-up patterns and the patient mix (i.e., inputs) even apparently subtle differences in pathology between treatment groups can be reflected as substantial differences in outcome rates (1,4,23).

PHARMACOKINETIC— PHARMACODYNAMIC CORRELATIONS

Pharmacokinetic—pharmacodynamic (PK-PD) modeling can be used to identify the optimal plasma concentration range for a given endpoint. Bayer proposed a *randomized exposure controlled trials* (RECT) study design for its repinotan program. Such a design offered to provide the best risk—benefit ratio and best chance of success by providing a focused patient selection. However, in this study design, caution must be taken to ensure that the resultant patient population actually represents the underlying treatment population of interest. Furthermore, diagnostic tests may need to be developed to ensure that the highest possible proportion of patients achieve the optimal concentration range.

CENSORING TREATMENT POPULATIONS

Censoring excessively good or poor prognoses offers a constructive approach to finding a level playing field in TBI clinical trials. For example, through patient selection and exclusion criteria, patients with moderate to severe TBI would be included in the trial if they meet the following criteria:

- Patients who are *not obeying commands* at the time of entry (GCS best motor score 5 or less) *and* have an abnormal CT scan showing intracranial pathology compatible with TBI;
- In the case of diffuse injury category II, a 5-cm lesion or greater is necessary for inclusion.
- If, at the time of entry, the patient's response (GCS) cannot be evaluated because of sedation/paralysis, there *must* be a GCS motor score of 5 or less previously recorded, together with abnormal CT scan findings.

The impact of these and other optimal inclusion and exclusion criteria (described in Table 4.9) is shown in Table 4.10 The net effect of censoring excessively good and poor prognoses allows the more severe end of disability outcomes to increase in relative size. This provides a distribution of patients with a greater likelihood of shifting toward good recovery if the drug being tested actually works.

CONCLUSION

Success in TBI clinical trials has been elusive and has led to a reticence in initiating TBI drug development programs. TBI remains a critical area of unmet medical need, and recent thinking about TBI trials design and outcomes measurement offers promise in providing the *level playing field* desired. Validation of the

TABLE 4.9 Key Inclusion and Exclusion Criteria for Traumatic Brain Injury Clinical Trials

Inclusion

Glasgow Coma Scale (GCS) motor score of 5 or less and abnormal computed tomography showing intracranial pathology compatible with traumatic brain injury (TBI)

At least one reactive pupil after resuscitation

Treatment within <specify> hours of injury

Stable 4.condition (systolic blood pressure [SBP] = 90 mm Hg) before treatment started

Exclusion

Gunshot wound or penetrating injury

Pure epidural hematoma

Pure depressed skull fracture

Prognostic features associated with extremely high mortality

GCS = 3 after resuscitation

GCS = 4 with bilateral nonreacting pupils

Persistent severe hemodynamic instability after resuscitation

Surgical subdural hematoma in combination with document shock (SBP <90 mm Hg and/or pH = 7.1 on initial arterial blood gas

Patients likely to die within 24 hours (investigator opinion)

Source: Developed for a global TBI clinical trial by a team including representatives of ABIC and EBIC.

TABLE 4.10 Expected Glasgow Outcome Scale Distribution Using Proposed Inclusion/Exclusion Criteria (77)

Glasgow Outcome Scale	Expected	EBIC Survey
Good recovery	25%	31%
Moderate disability	35%	20%
Severe disability	25%	16%
Vegetative *plus* dead	15%	33%

EBIC, Source: Teasdale GM, Pettigrew LE, Wilson JT, Murray G, Jennett B. Analyzing outcome of treatment of severe head injury: a review and update on advancing the use of the Glasgow Outcome Scale. J Neurotrauma 1998;15:587–597.

---

concepts and approaches presented in this chapter must be provided in the context of clinical development programs for the TBI indication.

TABLE 4.11 Potential Strategies for Traumatic Brain Injury Clinical Trials Design and Analysis

Focus on concentration—outcome relationships.

Avoid reparative strategies.

Censor excessively good or poor prognoses.

Expand outcome categories and control interrater reliability through structured interviews.

Perform shift analyses over a range of ordinal outcome categories.

Identify early surrogate endpoints for later clinical outcomes (e.g., changes in intracranial pressure or neurological worsening).

References

1. Alves WA, Eisenberg HM. Head injury trials: past and present. In: Narayan RK, Wilberger JE, Jr., Povlishock JT, eds. Neurotrauma. New York: McGraw Hill, 1996: 947–967.

2. Bullock R. Experimental drug therapies for head injury. In: Narayan RK, Wilberger JE, Jr., Povlishock JT, eds. Neurotrauma. New York: McGraw Hill, 1996: 375–391.

3. Duhaime AC. Conventional drug therapies for head injury. In: Narayan RK, Wilberger JE, Jr., Povlishock JT, eds. Neurotrauma. New York: McGraw Hill, 1996:365–374.

4. Eisenberg HM. Head and spinal cord injury. In: Porter RJ, Schoenberg BS, eds. Controlled clinical trials in neurological disease. Boston/ Dordrecht/London: Kluwar Academic Publishers, 1990:171–183.

5. Faden AI. Pharmacological treatment approaches for brain and spinal cord trauma. In: Narayan RK, Wilberger JE, Jr., Povlishock JT, eds. Neurotrauma. New York: McGraw Hill, 1996:1479–1490.

6. Bennett BR, Jacobs LM, Schwartz RJ. Incidence, costs, and DRG-based reimbursement for traumatic brain injured patients: a 3-year experience. J Trauma 1989;29:556–565.

7. Cheung ME. Economics of head injury trials. In: Narayan RK, Wilberger JE, Jr., Povlishock JT, eds. Neurotrauma. New York: McGraw Hill, 1996:969–975.

8. Max W, MacKenzie EJ, Rice DP. Head injuries: costs and consequences. J Head Trauma Rehabil 1991;6:76–91.

9. Haines SJ. Randomized clinical trials in neurosurgery. Neurosurg 1983;12:259–264.

10. Hall ED, Braughler JM. Free radicals in CNS injury. In: Waxman SG, ed. Molecular and Cellular Approaches to the Treatment of Neurological Disease. New York: Raven Press Ltd, 1993:81–105.

11. Hovda DA: Metabolic dysfunction. In: Narayan RK, Wilberger JE, Jr., Povlishock JT, eds. Neurotrauma. New York: McGraw Hill, 1996:1459–1478.

12. Smith DH, McIntosh TK. Traumatic brain injury and excitatory amino acids. In: Narayan RK, Wilberger JE, Jr., Povlishock JT, eds. Neurotrauma. New York: McGraw Hill, 1996: 1445–1458.

13. Young W: Death by calcium: A way of life. In: Narayan RK, Wilberger JE, Jr., Povlishock JT, eds. Neurotrauma. New York: McGraw Hill, 1996:1421–1431.

14. Taylor CP, Meldrum BS. Na+ channels as targets for neuroprotective drugs. Trends in Pharmacol Sci 1995;16:309–316.

15. Yakovlev AG, Faden AI. Molecular strategies in CNS injury. J Neurotrauma 1995;12:767–777.

16. Becker D Common themes in head injury. In: Becker DP, Gudeman SK, eds. Textbook of Head Injury. Philadelphia: WB Saunders, 1989:1–22.

17. Gaab MR, Trost HA, Alcantara A, Karimi-Nejad A, Moskopp D, Schultheiss R, Bock WJ, Piek J, Klinge H, Scheil F, et al. "Ultrahigh" dexamethasone in acute brain injury. Results from a prospective randomized double-blind multicenter trial (GUDHIS). German Ultrahigh Dexamethasone Head Injury Study Grou Zentralbl Neurochir 1994;55:135–143.

18. Mattson M, Scheff W. Endogenous neuroprotection factors and traumatic brain injury: Mechanisms of action and implications for therapy. J Neurotrauma 1994;11:3–33.

19. McIntosh TK. Novel pharmacologic therapies in the treatment of experimental truamatic brain injury: a review. J Neurotrauma 1993; 10:215–261.

20. McIntosh TK. Neurochemical sequelae of traumatic brain injury: therapeutic implications. Cerebrovasc Brain Metab Rev 1994; 6:109–162.

21. Ott L, McClain CJ, Gillespe M, et al. Cytokines and metabolic dysfunction afte severe head injury. J Neurotrauma 1994;11:447–508.

22. Siesj<auo> BK. Basic mechanisms of traumatic brain damage. Ann Emerg Med 1993;22:959–972.

23. Contant CF Jr. Clinical trial design. In: Narayan RK, Wilberger JE, Jr., Povlishock JT, eds. Neurotrauma. New York: McGraw Hill, 1996:923–945.

24. Steiner J. Clinical Development: Strategic, Pre-Clinical, Clinical, and Regulatory Strategies. Buffalo Grove: Interpharm Press, Inc., 1997.

25. Clifton GL, Hayes RL, Levin HS, et al. Outcome measures for clinical trials involving traumatically brain-injured patients: report of a conference. Neurosurg 1992;31:975–978.

26. Stein SC, Spettell C. The head injury severity scale (HISS): a practical classification of closed-head injury. Brain Injury 1995;9:437–444.

27. Gennarelli TA, Champion HR, Sacco WJ, et al. Mortality of patients with head injury and extracranial injury treated in trauma centers. J Trauma 1989;29:1193–1202.

28. Kohi YM, Mendelow AD, Teasdale GM, Allardice GM. Extracranial insults and outcome in patients with acute head injury–relationship to the Glasgow Coma Scale. Injury 1984;16:25–29.

29. Eisenberg HM, Frankowski RF, Contant CF, et al. High-dose barbiturate control of elevated intracranial pressue in patients with severe head injury. J Neurosurg 1988;69:15–23.

30. Nordby HK, Nesbakken R. The effect of high dose barbiturate decompression after severe head injury: a controlled clinical trial. Acta Neurochirur 1984;72:157–166.

31. Rea GL, Rockswold GL. Barbiturate therapy in uncontrolled intracranial hypertension. Neurosurg 1983;12:401–404.

32. Braakman R, Schouten HJA, Blaauw-van Dishoeck M, et al. Megadose steroids in severe head injury: results of a prospective double-blind clinical trial. J Neurosurg 1983; 58:326–330.

33. Cooper PR, Moody S, Clark WK, et al. Dexamethasone and severe head injury: a prospective double-blind study. J Neurosurg 1979;51: 307–316.

34. Dearden NM, Gibson JS, Chir B, et al. Effect of high-dose dexamethasone on outcome from severe head injury. J Neurosurg 1986;64:81–88.

35. Giannotta SL, Weiss MH, Apuzzo MLJ, Martin E. High dose glucocorticoids in the management of severe head injury. Neurosurg 1984; 15:497–501.

36. Gudeman SK, Miller JD, Becker DP. Failure of high-dose steroid therapy to influence intracranial pressure in patients with severe head injury. J Neurosurg 1979;51:301–306.

37. Pitts LH, Kaktis JV. Effect of megadose steroids on ICP in traumatic coma. In: Shulman K, Marmarou A, Miller JD (eds.) Intracranial Pressure, IV. Berlin/Heidelberg/New York: Springer-Verlag, 1980:638–642.

38. Saul TG, Ducker TB, Salcman M, Carro E: Steroids in severe head injury: A prospective randomized clinical trial. J Neurosurg 1981; 54:596–600.

39. Roberts I, Yates D, Sandercock P, et al. Effect of intravenous corticosteroids on death within 14 days in 10008 adults with clinically significant head injury (MRC CRASH trial): randomised placebo-controlled trial. Lancet 2004; 364: 1321–1328.

40. Muizelaar JP, Marmarou A, Ward JD, et al. Adverse effects of prolonged hyperventilation in patients with severe head injury: a randomized clinical trial. J Neurosurg 1991;75:731–739.

41. Wolf AL, Levi L, Marmarou A, et al. Effect of THAM upon outcome in severe head injury: a randomized prospective clinical trial. J Neurosurg 1993;78:54–59.

42. Bailey I, Bell A, Gray J, Gullan R, Heiskanan O, Marks PV, Marsh H, Mendelow DA, Murray G, Ohman J, et al. A trial of the effect of nimodipine on outcome after head injury. Acta Neurochir (Wien) 1991;110:97–105.

43. British/Finnish Co-operative Head Injury Trial Group. The effect of nimodipine on outcome after head injury: a prospective randomised control trial. Acta Neurochir 1990;51(suppl): 315–316.

44. European Study Group on Nimodipine in Severe Head Injury. A multicenter trial of the efficacy of nimodipine on outcome after severe head injury. J Neurosurg 1994;80:797–804.

45. Teasdale G, Bailey I, Bell A, Gray J, Gullan R, Heiskanan O, Marks PV, Marsh H, Mendelow DA, Murray G, et al. A randomized trial of nimodipine in severe head injury: HIT I. British/Finnish Co-operative Head Injury Trial Group. J Neurotrauma 1992;9(suppl 2): S545–S550.

46. Braughler JM, Hall ED. Central nervous system trauma and stroke, I: biochemical consideration for oxygen free radical formation and lipid peroxidation. Free Radic Biol Med 1989;6:289–301.

47. Hall ED. Lipid antioxidants in acute central nervous system injury. Ann Emerg Med 1993;22: 1022–1027.

48. Hall ED. The role of oxygen radicals in traumatic injury: clinical implications. J Emerg Med 1993;11:31–36.

49. Hall ED, Andrus PK, Smith SL, et al. Neuroprotective efficacy of microvascularly-localized versus brain-penetrating antioxidants. Acta Neurochir 1996;66(suppl):107–113.

50. Hall ED. Inhibition of lipid peroxidation in central nervous system trauma and ischemia. J Neurol Sci 1995;134(suppl):79–83.

51. McCord J. Oxygen-derived free radicals in post-ischemic injury. N Engl J Med 1985;312: 159–163.

52. Muizelaar JP. Cerebral ischemia-reperfusion injury after severe head injury admits possible treatment with polyethyleneglycol-superoxide dismutase. Ann Emerg Med 1993;22:1014–1036.

53. Muizelaar JP. Clinical trials with Dismutec (pegorgotein; polyethylene glycol-conjugated superoxide dismutase; PEG-SOD) in the treatment of severe closed head injury. In: Armstrong D, ed. Free Radicals in Diagnostic Medicine. New York: Plenum Press, 1994:389–400.

54. Muizelaar JP, Marmarou A, Young HF, et al. Improving the outcome of severe head injury with the oxygen radical scavenger polyethylene glycol-conjugated superoxide dismutase: a phase II trial. J Neurosurg 1993;78:375–382.

55. Marshall LF, Maas AI, Marshall SB, Bricolo A, Fearnside M, Iannotti F, Klauber MR, Lagarrigue J, Lobato R, Persson L, Pickard JD, Piek J, Servadei F, Wellis GN, Morris GF, Means ED, Musch B. A multicenter trial on the efficacy of using tirilazad mesylate in cases of head injury. J Neurosurg 1998;89:519–525.

56. Marshall LF, Marshall SB. Pitfalls and advances from the international tirilazad trial in moderate and severe head injury. J Neurotrauma 1995; 12:929–932.

57. Marmarou A, Nichols J, Burgess J, Newell D, Troha J, Burnham D, Pitts L. Effects of the bradykinin antagonist Bradycor (deltibant, CP-1027) in severe traumatic brain injury: results of a multi-center, randomized, placebo-controlled trial. American Brain Injury Consortium Study Group. J Neurotrauma 1999;16:431–444.

58. Maldonado VC, Calatayud P<aae>rez JB, Escario JA. Effects of CDP-choline on the recovery of patients with head injury. J Neurol Sci 1991; 103(suppl):S15–S18.

59. Prentice RL. Surrogate endpoints in clinical trials: definition and operational criteria. Stat Med 1989;8:431–440.

60. Wagner AK, Bayir H, Ren D, Puccio A, Zafonte RD, Kochanek PM. Relationships between cerebrospinal fluid markers of excitotoxicity, ischemia, and oxidative damage after severe TBI: the impact of gender, age, and hypothermia. J Neurotrauma 2004;21:125–136.

61. Marion DW, Garlier PM. Problems with initial Glasgow Coma Scale assessment caused by prehospital treatment of patients with head injuries: results of a national survey. J Trauma 1994;36:89–95.

62. Marshall, LF, Marshall SB, Klauber MR, et al. The diagnosis of head injury requires a classification based on computed axial tomography. J Neurotrauma 1992;9(suppl):S287–S292.

63. Marshall LF, Marshall SB, Klauber MR, et al. A new classification of head injury based on computerized tomography. J Neurosurg 1991; 75(suppl):S14–S20.

64. Teasdale GM. Head injury. J Neurol Neurosurg Psychiatry 1995;58:526–539.

65. Marmarou A. Conduct of head injury trials in the United States: the American Brain Injury Consortium (ABIC). Acta Neurochir 1996;66 (suppl):118–121.

66. Teasdale G, Jennett B. Assessment of coma and impaired c consciousness: a practical scale. Lancet 1974;2:81–84.

67. Jennett B, Bond M. Assessment of outcome after severe brain damage. Lancet 1975;1:480–484.

68. Choi SC, Barnes TY. Predicting outcome in the head-injured patient. In: Narayan RK, Wilberger JE, Jr., Povlishock JT, eds. Neurotrauma. New York: McGraw Hill, 1996:779–792.

69. Levin HS. Neurobehavioral outcome of closed head injury: implications for clinical trials. J Neurotrauma 1995;12:601–610.

70. Choi SC, Barnes TY, Bullock R, et al. Temporal profile of outcomes in severe head injury. J Neurosurg 1994;81:169–173.

71. Klauber MR, Marshall LF, Luerssen TG, et al. Determinants of head injury mortality: importance of the low risk patient. Neurosurg 1989;24:31–36.

72. Temkin NR, Holubkov R, Machamer JE, et al. Classification and regression trees (CART) for prediction of function at 1 year following head trauma. J Neurosurg 1995;82:764–771.

73. Pettigrew LE, Wilson JT, Teasdale GM. Assessing disability after head injury: improved use of the Glasgow Outcome Scale. J Neurosurg 1998; 89:939–943.

74. Pettigrew LE, Wilson JT, Teasdale GM. Reliability of ratings on the Glasgow Outcome Scales from in-person and telephone structured interviews. J Head Trauma Rehabil 2003;18:252–258.

75. Wilson JT, Pettigrew LE, Teasdale GM. Structured interviews for the Glasgow Outcome Scale and the extended Glasgow Outcome Scale: guidelines for their use. J Neurotrauma 1998; 15:573–585.

76. Murray GD, Barer D, Choi S, et al. Design and analysis of phase III trials with ordered outcome scales: the concept of the sliding dichotomy. J Neurotrauma 2005;22:511–517.

77. Teasdale GM, Pettigrew LE, Wilson JT, Murray G, Jennett B. Analyzing outcome of treatment of severe head injury: a review and update on advancing the use of the Glasgow Outcome Scale. J Neurotrauma 1998; 15:587–597.

5

Acute Seizures
and Status Epilepticus

Susan Herman

Acute seizures are neurological emergencies with high morbidity and mortality. Most epileptic seizures are stopped within a few seconds to minutes by intrinsic seizure terminating mechanisms (1). Individual seizures are usually followed by a refractory period, during which additional seizures do not occur. Failure of the normal seizure-terminating mechanisms or the postictal refractory period leads to seizure emergencies. Prolonged or repetitive seizures can lead to neuronal damage and clinical sequelae such as focal neurological deficits and cognitive impairment. Acute seizure emergencies fall into several subgroups based on the duration of seizures and the patient population affected. Status epilepticus (SE), continuous seizures lasting more than 30 minutes or serial seizures over 30 minutes without return to normal consciousness between seizures, has a mortality of up to 30% and high morbidity (2). Seizures may also occur in flurries or clusters with return to baseline consciousness between seizures. Such acute repetitive seizures (ARS) frequently require emergency room visits or hospitalization and may progress to SE (3,4). Prolonged seizures secondary to fever are particularly common in infants and young children. Although febrile SE has relatively low morbidity and mortality, it may increase the risk for subsequent epilepsy (5,6) Seizures and SE in neonates are often secondary to an underlying neurological disorder and may cause or exacerbate neuronal damage (7).

Clinical trials to address acute seizures can have one of several purposes: (a) rapid termination of prolonged or repetitive seizures before SE occurs, (b) termination of SE itself, (c) amelioration of neuronal damage induced by prolonged seizures (neuroprotection), or (d) prevention of long-term sequelae, namely cognitive dysfunction and epilepsy (antiepileptogenesis). Randomized clinical trials (8,9) provide some guidelines for aims a and b, but available treatments for seizure emergencies and SE are effective in terminating seizures in only about 60% of patients. Treatment options for patients not responding to initial therapy are controversial. No trials have yet addressed neuroprotection or antiepileptogenesis after acute seizures and SE in humans. This chapter reviews the epidemiology, pathophysiology, current management, and trial design methodology for acute prolonged or repetitive seizures and SE.

DEFINITIONS

SE is traditionally defined as continuous seizures lasting more than 30 minutes or recurrent epileptic seizures without full recovery of consciousness between seizures (2,10–12), but this definition is controversial. Pathophysiologically, SE occurs when seizures progress to a point where they are unlikely to remit spontaneously (13). Because this point is almost impossible to determine in an individual patient, specific seizure durations are used to define SE. Position papers and studies of SE have used short time windows of 10 or even 5 minutes of continuous seizure activity (8,9,14). Several factors support this shift to a shorter time requirement before treatment is initiated (14). Isolated seizures rarely last more than 2 minutes (1). More than half of seizures lasting more than 10 minutes ultimately progress to SE, using the traditional 30-minute criterion (15). Similarly, new-onset seizures in children appear to fall into two distributions, one with a mean of 3.6 minutes (76% of cases) and the other with a mean of 31 minutes (24%) (16). A subset of patients is therefore predisposed to prolonged seizures, and the longer a seizure lasts, the less likely it is to stop within the next few minutes. Finally, the preponderance of animal data suggests that neuronal damage depends on the duration of continuous seizure activity (17). There appears to be a threshold duration below which neuronal damage does not occur, but this threshold has not been precisely determined in humans and may vary from individual to individual. In clinical practice, seizures should be treated aggressively if they last longer than 5 minutes.

The defined duration of acute seizures or SE for a particular trial depends on the purpose of the trial. A study aiming to prevent progression to SE would use a very short time window of 5 minutes, because early treatment is most effective in terminating seizures. A study attempting to show improvement in outcome, on the other hand, might require a longer duration of seizures. The 30-day mortality rate for seizures lasting 10 to 29 minutes is 2.6%, much lower than the 19% rate in SE lasting 30 or more minutes (15). Trials designed to show a difference in mortality rates secondary to SE should focus on patients meeting the standard criteria of a 30-minute duration to reduce the size of the required study sample.

There are as many types of acute seizures and SE as there are seizure types (Table 5.1). The international classification of epileptic seizures (12), which classifies seizures by site of onset in the brain, is the most appropriate classification of SE for use in clinical trial design. The types of seizures that should be included in clinical

TABLE 5.1 Classification of acute seizures and status epilepticus by seizure type (12)

GCSE
   Generalized onset
      Primary generalized tonic-clonic SE
      Clonic SE
      Tonic SE
      Myoclonic SE
   Partial onset
      Secondarily generalized tonic-clonic SE
NCSE
   Generalized onset
      Generalized absence SE
         Typical
         Atypical
      Subtle or electrographic GCSE
   Partial onset
      Simple partial SE
      Epilepsia partialis continua
      Complex partial SE
      Electrographic partial SE
Age-related SE
   Neonatal SE
   Febrile SE
Nonepileptic SE (psychogenic or pseudo-status
  epilepticus)
ARS
   Generalized onset
      Primary generalized tonic-clonic
      Tonic
      Clonic
      Myoclonic
      Absence (typical or atypical)
   Partial onset
      Simple partial
      Complex partial
      Secondarily generalized tonic-clonic

trials of SE are discussed more fully in the section on diagnosis and subpopulations. SE is usually grouped into convulsive SE (clinically obvious motor manifestations) and nonconvulsive SE (alteration of consciousness, electroencephalogram [EEG] usually required for diagnosis).

Generalized convulsive SE (GCSE) is the most common and most dangerous type of SE. Seizures may either be generalized at onset or partial onset with secondary generalization. In GCSE, the patient is unresponsive with generalized tonic followed by clonic motor activity, and the EEG shows continuous or recurrent bilateral ictal discharges. As SE persists, clinical signs become more subtle and may even disappear. Subtle GCSE is seen in late stages of SE and is characterized by coma with subtle twitches of the extremities or nystagmoid movements of the eyes (18). The EEG shows bilateral ictal discharges.

Nonconvulsive SE (NCSE) can be either partial or generalized, and an EEG is usually required for accurate diagnosis and classification. Partial SE may be either simple partial (no impairment of consciousness) or complex partial (consciousness impaired). Generalized NCSE, or absence SE, is characterized by alteration of consciousness, often described as an "epileptic twilight state," and generalized ictal discharges on EEG (19).

Some types of SE are seen only at particular ages. Febrile SE occurs in children between the ages of 6 months and 5 years in association with a high fever but no other acute neurological disease. Neonatal SE is defined as prolonged seizures in infants less than 30 days conceptional age. Seizure duration greater than 30 minutes is typically required to make a diagnosis of febrile or neonatal SE.

Some prolonged or repetitive seizures do not meet standard criteria for SE because they last for less than 30 minutes or occur serially but with return to normal consciousness between seizures. Such seizures also require rapid and effective treatment to prevent progression to SE. ARS (also called serial, cluster, recurrent, or repetitive seizures) are a change in seizure pattern in patients with epilepsy, with increase in severity, frequency, or duration of seizures (3). An episode of ARS may last for minutes to hours but rarely for more than 1 day. More than half of seizures lasting more than 10 minutes may progress to SE (15).

ANIMAL MODELS: RELEVANT ISSUES FOR CLINICAL DEVELOPMENT PROGRAMS

Optimal animal models of acute seizures and SE should reproduce the behavioral manifestations, EEG characteristics, neuropathological consequences, and therapy responsiveness of human SE. A variety of *in vivo* experimental models of SE has been developed by systemic or focal administration of chemical convulsants or by focal or generalized electrical stimulation of the brain (Table 5.2) (20). *In vitro* models of sustained seizure activity using slice preparations also exist. Available animal models cover a range of seizure types, species, and ages, reflecting the variety of SE types in humans. The most appropriate animal model for a study of acute seizures or SE depends on the question to be answered. Some models are most appropriate for studies of seizure-terminating drugs, others for neuropathological changes, and still others for chronic behavioral and epileptogenic effects of SE. The most commonly used animal models are reviewed below; an extensive discussion is beyond the scope of this chapter.

Systemic parenteral administration of a chemical convulsant is a simple and rapid method to induce acute seizures and SE. Chemical convulsants either increase neuronal excitation or decrease neuronal inhibition. Several agents produce prolonged seizures that approximate human SE. Because such agents remain

TABLE 5.2   Animal models of SE

Chemical convulsant	Mechanism	Seizure type
Generalized		
NMDA	Glutamate agonist	Generalized convulsions
Domoic acid	Glutamate agonist	Generalized spike wave
Kainic acid	Glutamate agonist	Limbic and secondarily generalized seizures
Quisqualic acid	Glutamate agonist	Generalized seizures
Pilocarpine	Acetylcholine agonist	Limbic motor seizures
Pilocarpine + lithium	Acetylcholine agonist	Limbic and secondarily generalized seizures, prolonged, high mortality
Soman	Acetylcholine agonist (cholinesterase inhibitor)	Generalized convulsions
Bicuculline	GABA antagonist	Generalized myoclonus, flexor spasms
Pentylenetetrazol	GABA antagonist	Generalized convulsions
Flurothyl	Sodium channel opening	Short repeated generalized seizures
4-Deoxypyridoxine	Pyridoxal phosphate antagonist	Generalized seizures
Allyglycine	Glutamic acid decarboxylase inhibitor	Prolonged recurrent convulsions
Hyperthermia		Febrile convulsions
Focal		
Alumina gel–4-deoxypyridoxine	Pyridoxal phosphate antagonist	Focal and secondarily generalized seizures
Bicuculline into area tempestas	GABA antagonist	Secondarily generalized convulsions
Cobalt-homocysteine		Secondarily generalized convulsions
Domoic acid	Glutamate agonist	Generalized seizures
Kainic acid	Glutamate agonist	Limbic and secondarily generalized seizures
Electrogenic models		
Amygdala		Limbic and secondarily generalized seizures
Perforant pathway		Limbic and secondarily generalized seizures

Na, sodium.

in the animal once SE is induced, however, results may be confounded by continued presence of the inducing agent. Systemic chemical convulsants have been used in mice, rats, cats, and primates to study SE neuropathology, systemic consequences (21,22), cognitive and behavioral sequelae (23), and the development of epilepsy.

Kainic acid and pilocarpine (either systemic or injected into hippocampus) induce severe acute limbic motor seizures and secondarily generalized seizures that can last for hours (24,25). After acute seizures, there is a latent or "silent" period and then development of spontaneous recurrent limbic seizures. Both agents generate lesions in the hippocampus, including loss of GABAergic

(γ-aminobutyric acid) interneurons in the dentate hilus, pyramidal cell death in hippocampal regions CA3 and CA1, and mossy fiber sprouting in the dentate gyrus. Widespread damage is seen in the hippocampus, contralateral hippocampus, thalamus, limbic cortex, and neocortex (26). Kainic acid and pilocarpine models are therefore useful for study of potential agents for SE termination, neuroprotection, and antiepileptogenesis.

The alumina gel–4-deoxypyridoxine hydrochloride model provokes frequent discrete seizures (27). Because the inter-seizure interval is relatively fixed, the relative efficacy of antiepileptic drugs (AEDs) can be measured. Injection of bicuculline into the deep prepiriform cortex produces prolonged convulsions that remain responsive to diazepam even late in SE. This model can therefore be used to study pharmacoresponsive SE. The cobalt–homocysteine model induces secondarily generalized SE using intraperitoneal homocysteine thiolactone after application of powdered cobalt to the dura (28,29). This model closely approximates the EEG characteristics (30) and pharmacological responsiveness (31) of human SE.

Other models of SE are induced by electrical stimulation of the brain. These models have the advantage that the inducing stimulus immediately ceases when the electrical stimulation is stopped, so there are no potential drug interactions. Electrogenic models of self-sustaining SE are an outgrowth of kindling studies (32), in which a subthreshold electrical stimulus is applied repetitively to the experimental animal. Self-sustained SE can be induced by electrical stimulation of the perforant path or of the amygdala. In general, electrical stimulation induces limbic seizures, but secondarily generalized seizures may also occur (33). Neuronal damage and rates of development of epilepsy are much lower in rats with only partial rather than generalized SE (33). Neuropathological changes and development of epilepsy in this model are similar to development of mesial temporal sclerosis in humans.

Febrile seizures are modeled by hyperthermia-induced seizures in immature rats. Prolonged hyperthermic seizures lead to acute hippocampal injury (34) and enhanced long-term hippocampal excitability (35). Hypothermic seizures also cause chronic neurotransmitter alterations, such as enhancement of presynaptic GABA release and inhibition of hippocampal pyramidal cell excitability (36).

Hypoxia-induced neonatal seizures have also been modeled in immature rats (37). Rats exposed to global hypoxia between postnatal days 7 and 15 develop acute seizures and later spontaneous seizures.

Hypoxia-induced seizures cause chronic increased excitability in hippocampal pyramidal neurons (38), analogous to the development of epilepsy after neonatal seizures in humans.

Although each animal model has its strengths and limitations, there are several findings common to most models. First, prolonged SE leads to neuronal damage in the neocortex, cerebellum, and hippocampus (39). Similar changes can be seen in paralyzed and mechanically ventilated baboons with SE, indicating that the neuronal damage is primarily due to ongoing seizure activity rather than to the systemic effects of prolonged motor movements such as lactic acidosis and hyperthermia (40). The location and severity of neuronal damage depends on the duration of seizure activity, with damage to the substantia nigra and globus pallidus after 30 minutes and in the neocortex, amygdala, thalamus, and hippocampus after longer seizures. There appears to be a time threshold below which histologic damage is not seen. The hippocampus, CA1, CA3, and the hilus are most severely affected across multiple animal models (41–44). Much of the seizure-induced neuronal damage and death is mediated by excessive release of glutamate (29). AEDs that shorten the seizure discharge consistently protect against ischemic cell change (44,45).

Second, survivors of SE show multiple physiological and structural alterations, including mossy fiber sprouting, loss of inhibitory interneurons, and enhanced excitation (43). These chronic changes are likely responsible for late cognitive dysfunction and the development of epilepsy after SE. Epilepsy, for example, can develop even if there are nearly no visible structural changes (46).

Third, prolonged hippocampal seizures result in changes in receptors, particularly a reduction in GABA receptor inhibition. As SE progresses, GABA becomes less effective in activating chloride channels.

Drugs acting via a benzodiazepine GABAergic mechanism (e.g., diazepam) lose potency when given late in SE, whereas efficacy of GABA and barbiturates is preserved (47–49). The mechanism of modulation of the $GABA_A$ receptor during seizures is not known but may involve changes in receptor subunit composition (50). At the same time, N-methyl-D-aspartate (NMDA) antagonists may become more effective (49,51). These and other physiological changes during SE may be responsible for refractory SE (RSE) or for prolonged SE that does not respond to therapy.

Fourth, the age at which SE occurs affects long-term outcome. Although immature (neonatal) animals are very susceptible to the development of SE (17), they appear to be more resistant to neuronal damage (52) and late epilepsy (53) after SE than adult animals. Decreased neurotoxicity may be mediated by relative resistance of the immature brain to the toxic effects of glutamate (54) because of lower density of synapses and decreased calcium entry into cells. Functional alterations after neonatal SE, such as altered synaptogenesis and decreased neurogenesis, can result in impaired learning and increased seizure susceptibility (55).

Finally, several drugs have been demonstrated to have neuroprotective or antiepileptogenic effects in some animal models of SE. These effects appear to be highly model dependent (56). For example, valproate appeared to be both neuroprotective and antiepileptogenic in the kainate model (57) but not in the pilocarpine model (58). Neuroprotective efficacy does not necessarily prevent the development of epilepsy. Although the NMDA antagonist MK-801 prevents neuronal damage in the kainate SE model, it does not prevent epileptogenesis (46). On the other hand, treatment with MK-801 did reduce the percentage of animals developing epilepsy if the drug was administered within 2 hours of electrically induced SE (51). Diazepam also reduced the percentage of animals developing epilepsy and ameliorated seizure severity (59). The implication of these findings for clinical trials of neuroprotective and antiepileptogenic drugs is that potential treatments must be tested in several animal models and that both acute and chronic outcomes should be included in the analysis.

Potential therapeutic agents should be tested in an appropriate animal model. The following guidelines indicate important areas for preclinical drug development (56,60,61):

- Test the new therapeutic agent in the animal model most closely resembling the human population in which the drug will be used. As discussed below, there are multiple different types of SE, and different animal models may be needed for each subpopulation. For example, a drug used to terminate ongoing pharmacoresponsive SE could be tested in either the alumina gel or cobalt–homocysteine model, whereas a drug designed to prevent epilepsy after SE might best be tested in a kainic acid, pilocarpine, or self-sustaining SE model.
- Determine the target pathophysiological mechanism and the time course of this mechanism in animal models. Confirm that the proposed agent has the expected effect on this mechanism.
- Study pharmacokinetics across multiple animal species, both genders, and a variety of age groups. Generate dose–response curves for efficacy and toxicity in several animal models. If possible, study several outcome measures in several species. Dose–response and toxicity curves allow selection of a target concentration for phase I and II human clinical trials. Measure brain concentrations of the drug to ensure adequate central nervous system (CNS)

penetration and plasma concentrations for translation to human studies.

- Establish animal intensive care units to parallel the current treatment of human SE, particularly RSE.
- Determine optimal timing and duration of treatment. Administer the study drug at a time point that could realistically be achieved in a human study. Many neuroprotection studies use "pretreatment," administration of the study agent before the induction of SE. This is clearly not possible in human studies. Because most patients with SE present to medical attention quickly, time windows for drug initiation can be relatively short. The duration of acute therapy to stop SE and prevent recurrence is usually on the order of minutes to hours. The duration of therapy that would be necessary for neuroprotection and antiepileptogenesis, however, is not known.
- Include at least two outcome measures in animal trials, such as clinical seizure cessation, EEG, neuropathology, neuroimaging measures, development of late seizures, or behavioral outcome. Longterm outcome is particularly important for trials of neuroprotection and antiepileptogenesis. At least one measure should be a functional outcome measure with some relevance to outcomes in humans.
- Perform animal studies as randomized blinded trials.
- Test new agents in combination with or in direct comparison with currently available therapies. Combination therapies may be synergistic or may have unexpected drug interactions. Experimental data suggest that the most effective treatment of SE may be polytherapy, either combining drugs with different mechanisms of action or using drugs that have multiple mechanisms of action (31).
- Develop clinically relevant biomarkers for translation to human studies.

EPIDEMIOLOGY AND NATURAL HISTORY

Some types of SE are much better studied than others. GCSE is most common and has high morbidity and mortality, and most prospective studies have focused on this subtype. The study of SE epidemiology is problematic, because it occurs not only in people with epilepsy, but also in the context of prior brain injuries or acute systemic or neurological insults (10). Estimates of the annual incidence of SE in the United States range from 65,000 (10) to more than 150,000 (62). Retrospective data from Rochester, Minnesota from 1965 to 1984 found an age-adjusted incidence of SE of 18.3 per 100,000 (63). A prospective population-based study in Richmond, Virginia (62) found a higher incidence of 41 per 100,000 patients per year. This population was more ethnically mixed, possibly accounting for the higher incidence of SE. EPISTAR, a population-based survey of SE in the French-speaking part of Switzerland, showed an incidence of about 10 cases per 100,000 people annually (64). The population in this study, however, was principally white, and cases of anoxic encephalopathies were excluded. The age-adjusted incidence in Germany was similar to that in Rochester at 17.1 per 100,000 (65). The incidence of SE is bimodally distributed, occurring most frequently during the first year of life and after the age of 60 years (62,63,65). SE is more common in men, with a ratio of 2:1 (65).

In adults, the major causes of SE are low levels of AEDs (34%) and cerebrovascular disease (22%), including acute or remote stroke and hemorrhage (62,63,66). Other common causes are alcohol withdrawal, metabolic abnormalities, anoxia or hypoxia, tumors, and CNS infections, whereas drug overdose and traumatic brain injury are less common etiologies (62,63,66). In children, fever secondary to non-CNS infections, low AED levels, and remote neurological injury are the most common causes (62,63,66).

The main determinants of mortality and morbidity after SE are age, etiology, and duration of SE (10,67). Death occurs in 5% to 50% of patients (67), with differences in mortality rates accounted for by variability in etiologies and age across studies. Most mortality is secondary to the underlying etiology of SE rather than seizures themselves. Age is an important predictor of mortality after SE. Children have low mortality, whereas persons over age 65 may have mortality rates greater than 50%. In Richmond, Virginia, overall 30-day mortality was 22%, 3% in children and 26% in adults (62). Other studies show that mortality in children ranges from 3% to 15% (68–71). Good outcomes in children are related to the resilience of the pediatric CNS as well as the preponderance of etiologies with favorable prognosis (infection with fever, AED withdrawal). Mortality in the elderly is higher, ranging from 38% to 50% (62). Long-term mortality is also increased, three times higher than expected in the general population (72). The main risk factors for long-term mortality are myoclonic SE, prolonged (>24 hour) duration of SE, and acute symptomatic etiology.

The etiology of SE has a significant impact on morbidity and mortality. Four main etiologic groups have been described: (a) acute symptomatic SE, in which SE occurs within 1 week of an acute medical or neurological insult (e.g. CNS infection, stroke, acute diffuse encephalopathy, and toxic/metabolic insults); (b) progressive symptomatic SE in the presence of progressive CNS conditions such as tumors and degenerative neurological diseases; (c) remote symptomatic SE in patients with a history of a prior CNS insult or epilepsy; and (d) idiopathic/cryptogenic SE, in the absence of any clear precipitating factors or prior insults (63). Short-term mortality is highest when SE occurs due to an acute insult (62,63,73). In particular,

anoxic encephalopathy and stroke have high mortality rates (74). In one study of NCSE (75), acute medical and neurological etiologies had a mortality of 27%, cryptogenic etiologies 18%, and previous epilepsy 3%.

The duration of SE also greatly influences morbidity and mortality. Most SE lasts less than 24 hours (75%), with 38% lasting 30 minutes to 2 hours and 38% 2 to 24 hours, and only 25% continuing for more than 24 hours (63). SE duration greater than 2 hours is associated with a significant increase in mortality (67,76). Patients with continuous SE have higher mortality than those with intermittent seizures, possibly because of more prolonged duration of actual seizures (77). Seizures lasting 10 to 29 minutes have lower mortality (4.4%) than SE lasting more than 30 minutes (22%) (15).

Seizure type and severity may influence outcome. In the VA Cooperative Study (9), 27% of patients with overt GCSE and 65% of those with subtle GCSE died within 30 days. In a study of nonconvulsive SE, patients with severe mental status impairment had a mortality of 39%, compared with only 7% in patients with mild obtundation (75).

Morbidity is common after SE, but it is often difficult to determine which complications are directly attributable to SE and which are caused by the underlying etiology. Prolonged SE is more likely to result in long-term deficits (see Defining Relevant Treatment Populations, below). Chronic encephalopathy and brain atrophy may occur in 6–15% of adult patients with GCSE (78–80), presumably because of diffuse cortical injury and neuronal death. Developmental deterioration has been reported in 34% of children with SE lasting from 30 to 720 minutes (81). New focal neurological signs occur in 2.2–15% of children, mostly in those with acute or progressive neurological diseases (68,71,80,82). Neurological deficits therefore cannot be attributed to SE alone. Epilepsy is the most common

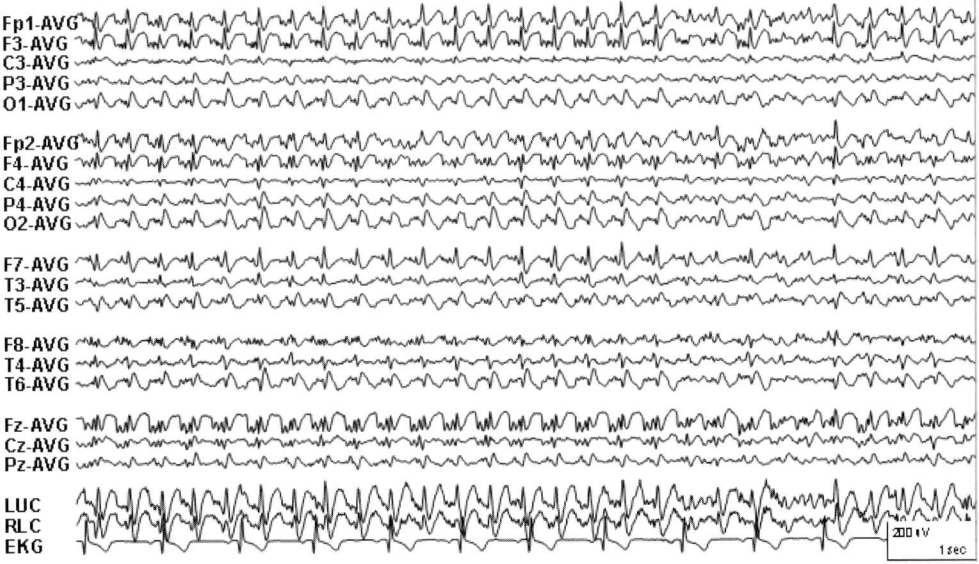

FIGURE 5.1 Subtle generalized convulsive status epilepticus. A 45-year-old woman had three generalized tonic-clonic seizures and then failed to regain consciousness. Examination showed irregular twitching of her face and eyelids. EEG shows widespread 3-Hz spikes and polyspikes.

long-term complication of convulsive SE (83). New onset of epilepsy has been reported in 20–36% of survivors of GCSE (68,71,80,81,84). Again, it is difficult to determine whether epilepsy results from SE itself or from the effects of an acute brain injury. In a study of acute symptomatic seizures (85), the risk of developing epilepsy after acute symptomatic SE was 40%, versus 10% after a single brief acute symptomatic seizure, suggesting that SE itself is epileptogenic. In some cases, an episode of SE may simply be the first seizure in a patient with a genetic or remote symptomatic etiology for epilepsy.

Generalized Convulsive SE

GCSE is characterized by paroxysmal or continuous tonic and/or clonic movements with coma or profound impairment of consciousness. Typically, the seizures begin as individual discrete seizures, which gradually merge to produce a continuous ictal state (30). Most GCSE is secondarily generalized, with seizures beginning in one part of the brain and spreading to generalized convulsions. As SE progresses, motor manifestations become more subtle or may disappear entirely (86). EEG is usually necessary to make the diagnosis of subtle GCSE (Fig. 5.1).

GCSE is the most common form of SE. Prospective population-based studies indicate that approximately 40–50% of SE is GCSE, either primary generalized or secondarily generalized (62,64,65,72). One epidemiological study (62) found that 69% of SE in adults and 64% in children was partial onset, followed by secondarily generalized SE in 43% of adults and 36% of children.

GCSE has high morbidity and mortality. Mortality ranges between 10% and 30% (67,87,88), with higher rates in the elderly, those with acute symptomatic SE, and those with SE lasting more than 2 hours. Morbidity includes cognitive impairment, focal neurological deficits, and epilepsy.

Nonconvulsive SE

NCSE can be divided into two subcategories: absence (generalized) NCSE and

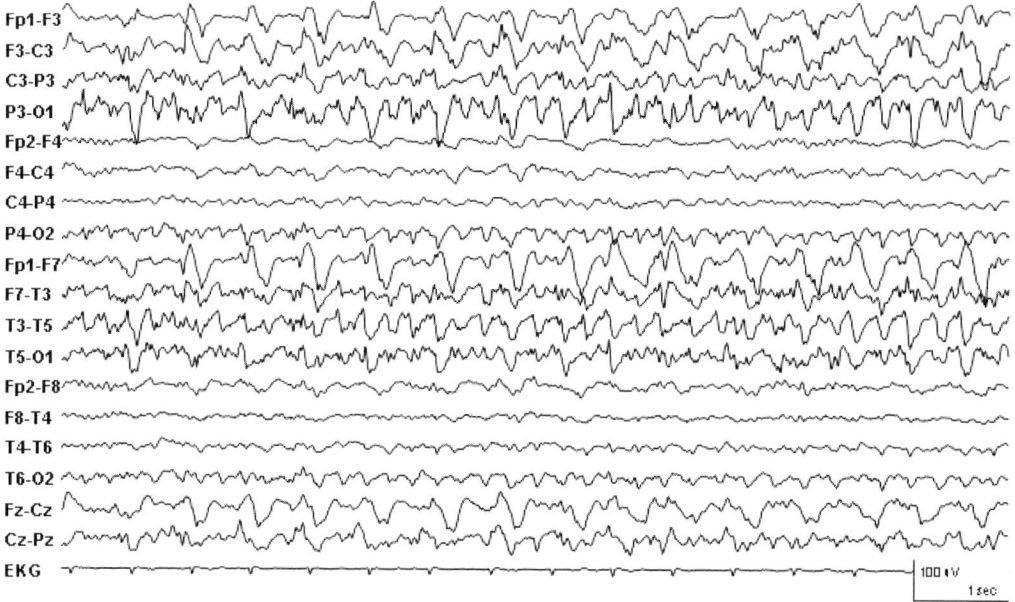

FIGURE 5.2   Electrographic partial status epilepticus. A 72-year-old woman had a left hemisphere stroke and became progressively obtunded over 48 hours. EEG showed rhythmic delta activity and intermixed beta activity over the left hemisphere, which resolved after treatment with intravenous lorazepam.

partial NCSE. Absence SE, sometimes called an epileptic "twilight state" (19), is characterized by mild impairment of consciousness and a characteristic EEG pattern of generalized spikes at 3 Hz. Most absence SE occurs in patients with a history of absence seizures or primary generalized epilepsy. Atypical absence SE, characterized by more variable impairment of consciousness and spikes at repetition rates less than 3 Hz on EEG, occurs in patients with the Lennox-Gastaut syndrome.

Partial NCSE is divided into simple partial, complex partial, and electrographic partial SE. Simple partial SE is characterized by continuous or repetitive focal motor, sensory, special sensory, autonomic, or psychic symptoms without impairment of consciousness (12). Complex partial SE is characterized by alteration of consciousness, often with oral or limb automatisms or focal motor movements (89). Impairment of consciousness ranges from mild confusion to coma. Electrographic partial SE occurs in comatose patients without overt clinical seizure activity in whom

EEG reveals focal ictal discharges lasting 30 minutes or more (90) (Fig. 5.2).

Slightly more than half of all SE is NCSE; 30% is complex partial SE, 20% simple partial SE, and 5% absence SE (62,64,65,72). Variable definitions complicate the estimate of the incidence of NCSE. Some authors include only absence SE as NCSE, calling other types complex partial SE, whereas others exclude electrographic partial SE (43,75,91,92). NCSE is likely underdiagnosed, as an EEG is necessary for ascertainment (90,93).

Outcomes of patients with NCSE are highly variable. Patients with absence SE tend to have good prognoses regardless of the duration of seizure activity, whereas those with partial NCSE have widely varying outcomes (93–97). A retrospective review of 100 patients with NCSE (75) showed a mortality of 18% in the group with acute medical problems but only 3% in those with NCSE secondary to epilepsy. The level of consciousness during NCSE may also predict outcome (93). Severe mental status impairment is associated with higher morbidity (39%) than milder cognitive impairment (7%) (75).

Refractory SE

RSE is defined as SE that fails to respond to treatment, but a precise definition is not universally accepted. SE becomes more difficult to treat as its duration increases (98). It is not clear whether some SE is intrinsically refractory or whether elapsed time before treatment decreases the probability that treatment will be effective.

The reported prevalence of RSE varies greatly (9–31%) based on the definition of "refractory" (99–101). Some consider failure of two AEDs to constitute RSE (99,102–104), whereas others require three drugs (2,101,105). The minimal duration of seizure activity for RSE also varies from none (2,101,102,105,106) to 1 hour (107) or 2 hours (103,104). In the VA Cooperative Study (9), RSE (continued seizures after two AEDs) occurred in 38% of patients with overt SE and 82% of patients with subtle SE. In a retrospective study (99), RSE occurred in 31% of 83 episodes of SE in 74 patients and was more common in those with NCSE or focal motor seizures at onset. The mean duration of RSE was 20 hours.

Outcome of RSE is extremely poor, with mortality at almost 50% and only a small fraction of patients returning to their premorbid functional baseline (100,103,108). RSE is associated with increased mortality, increased functional deterioration, and increased hospital length of stay (99). Similarly, severe RSE in children is associated with high mortality (32%) and functional deterioration in all previously normal survivors (109). In a series of children with seizures lasting more than 30 minutes (81), only 23% of the survivors were normal at follow-up, 34% showed developmental deterioration, and 36% developed new onset epilepsy.

Acute Repetitive Seizures

The epidemiology of ARS is largely unknown. Anecdotally, many patients with epilepsy experience ARS, characterized by an increase in the frequency, severity, or duration of typical seizures or development of a new more severe seizure type (3,110). Commonly, these are precipitated by physiological stresses, missed AED doses, sleep deprivation, or illnesses. Seizure clusters are more common in children (110). Only one study (4) examined the prevalence of ARS and their association with SE. In a retrospective interview, 36 of 76 patients (47%) with medically intractable complex partial seizures reported a history of seizure clusters, defined as three or more complex partial seizures within a 24-hour period, and 21 (28%) had a history of SE. SE occurred in 16 of 36 patients (44%) with clustered seizures, and in 5 of 40 patients (12.5%) with nonclustered seizures. The prevalence of ARS in patients with less severe epilepsy or other seizure types is not known. ARS are often associated with more prolonged postictal states or transient neurological deficits (Todd's phenomenon).

Febrile SE

Febrile seizures are defined as seizures occurring in children between the ages of 6 months and 5 years with no precipitating factors other than fever (temperature >100.4°F) (111,112). Between 2% and 5% of children under age 5 in the United States experience at least one febrile seizure (113,114). Most febrile seizures are benign, with little morbidity or mortality, but prolonged febrile seizures are associated with the later development of epilepsy. Approximately 4% to 5% of all febrile seizures last more than 30 minutes, thereby meeting criteria for febrile SE (10,114,115). An estimated 25,000 to 60,000 children are affected by complex febrile seizures and 4000 to 10,000 children by febrile SE per year in the United States (116).

Simple febrile seizures are generalized tonic-clonic seizures that occur in neurologically normal children and last less than 15 minutes (5). They do not recur during a 24-hour period. Complex febrile seizures, on the other hand, have focal onset, occur in children with previous neurological deficits, are prolonged (>15 minutes), or recur within a 24-hour period.

Risk factors for recurrence of febrile seizures include age <12 months at onset, a history of epilepsy or febrile seizures in first-degree relatives, many episodes of fever, a low rectal temperature during the first seizure, or, less consistently, an initial complex febrile seizure (117–119). Patients with three or more risk factors have an extremely high recurrence risk (50–100%) (117).

The morbidity and mortality of febrile seizures, even febrile SE, is low. Simple febrile seizures are not associated with developmental delay or cognitive dysfunction (120). Acute morbidity and mortality of febrile SE is also low. Twenty-seven children with a febrile convulsion lasting 30 minutes or longer showed no difference in mean IQ compared with control siblings who had not had convulsions (115). There were no deaths and no new cases of neurological deficits among 180 children with febrile SE lasting 30 to 59 minutes (58%), 60 to 119 minutes (24%), or greater than 120 minutes (18%) (116).

Febrile seizures are associated with an increased risk of development of epilepsy, but a causal relationship has not been proven. In the general population, the risk of developing epilepsy by age 7 years is approximately 1% (114). Children with a single simple febrile seizure have a minimally increased incidence of epilepsy. The risk of developing epilepsy is approximately 2.4% among those younger than 12 months old at the first simple febrile seizure and those with multiple simple febrile seizures (6). In contrast, the risk of development of epilepsy increases to 30% to 50% in patients with one or more

complex features, particularly seizures with focal features in a child with abnormal neurological development (6,121,122). Prolonged febrile seizures may be associated with the development of mesial temporal sclerosis (or hippocampal sclerosis) and intractable mesial temporal lobe epilepsy in later life (123), but this does not prove a causal association. The average latency between febrile seizures and later onset of epilepsy is 8 to 11 years, and long follow-up is essential to determine the cumulative risk (124). Use of AEDs as seizure prophylaxis in children with febrile seizures does not prevent the future development of epilepsy (125,126).

Several hypotheses exist for the relationship between febrile SE and mesial temporal sclerosis (124). Febrile SE in a previously normal brain may result in an acute hippocampal injury and later in mesial temporal sclerosis and intractable epilepsy. Alternatively, mesial temporal sclerosis (caused by some other brain injury) may be the cause of complex febrile seizures (6). Finally, hippocampal dysgenesis or migrational abnormalities may both lower the seizure threshold and predispose to hippocampal injury by prolonged seizures (124).

Neonatal Seizures

Neonatal seizures are a distinct seizure subtype, with different etiologies, treatment, and outcome than other childhood seizures. The neonatal period is defined as the first 28 days of life of a full-term infant or 44 weeks conceptional age (gestational age plus chronological age) of a preterm infant. The unique clinical presentation of neonatal seizures is an expression of the immaturity of the neonatal CNS.

Seizures occur in 0.1% to 0.2% of neonates (127,128) and may be underdiagnosed if EEG is not used (129). Estimates of seizure occurrence in infants in an intensive care setting range from 2.3% (130) to 20% (131), with higher rates with decreasing conceptional age and birth weight. The seizures may be tonic (extension of the arms and legs), clonic (jerking of a single limb or random migration of jerking from limb to limb), myoclonic (single flexion movements of the extremities), or subtle (abnormal eye movements, sucking, rowing, pedaling, or swimming movements) (132,133). EEG is essential both to detect electrographic seizures and to confirm that abnormal movements are seizures (133). Most neonatal seizures are due to an acute CNS insult such as hypoxic-ischemic encephalopathy, hemorrhage (intraventricular, intracerebral, or subarachnoid), or cerebral infarction. Other common causes are hypoglycemia, hypocalcemia, metabolic abnormalities, CNS infections, drug withdrawal, pyridoxine dependency, cerebral malformations, and benign familial epilepsy.

The prognosis of neonatal seizures is often poor (134). Reported sequelae include death, focal neurological deficits, mental retardation, and epilepsy. Early studies reported mortality in 20% to 42% and morbidity in 3% to 27% (131,135). Later studies reported similar mortality (30%) but developmental delay in 67% and cerebral palsy in 63% (136). Epilepsy may develop in more than 50% of children with a history of neonatal seizures (136,137). Etiology is a strong predictor of prognosis, with more abnormal outcomes in association with hypoxic-ischemic encephalopathy, CNS infection, and cerebral hemorrhage (7,134,136,138). There is still no clear consensus as to whether there are long-term neurological or cognitive sequelae of neurological seizures per se (139), because the effects of neonatal seizures on brain development are difficult to differentiate from those of the brain lesions causing them. Experimental evidence suggests that seizures themselves may be damaging to the developing brain (55).

MECHANISMS OF DISEASE AND PATHOLOGY

The central pathophysiology of SE is failure of the normal mechanisms that terminate an isolated seizure (11,43,56). Excessive excitation, ineffective inhibition, or a combination of the two may play a role in the origin of prolonged seizures, but the exact mechanisms are poorly understood (11,43,56). Excessive excitation can be modeled by domoic acid, a glutamate analogue (11). Patients who accidentally ingest mussels contaminated with domoic acid develop prolonged SE (140). Several inhibitory mechanisms may play a role in SE, including calcium-dependent potassium currents, magnesium blockade of NMDA channels, and inhibitory neurotransmitters such as adenosine and GABA (43). As SE becomes prolonged, functional changes in the GABA receptor lead to failure of GABA-mediated inhibition (47,141). In most cases, SE results from a combination of these excitatory and inhibitory abnormalities.

SE-induced neuronal damage is mediated by excitotoxicity or excessive release of excitatory amino acids such as glutamate (142,143). Glutamate release causes an influx of sodium, depolarization of the cell membrane, and cell swelling. Excessive calcium enters neurons via glutamate-activated NMDA, kainate, and α-amino-3-hydroxy-5-methylisoxazole-4-proprionic acid AMPA receptors. Intracellular calcium induces a cascade of events leading to cell damage or neuronal death.

GCSE induces a variety of physiological changes that increases the risk of cell damage, such as fever, acidosis, hypertension, hyperglycemia, cardiac arrhythmias, and respiratory distress (43) (Table 5.3) Continuous convulsive seizure activity carries a high energy demand. At the onset of SE, increased demand is met by increased cerebral blood flow and oxygenation (56). As SE persists, however, these compensatory mechanisms fail and

TABLE 5.3   Complications of SE

Metabolic
  Lactic acidosis
  Hypercapnia
  Hypoglycemia
  Hyperkalemia
  Hyponatremia
  Cerebrospinal fluid/serum leukocytosis
Autonomic
  Hyperpyrexia
  Failure of cerebral autoregulation
  Vomiting
  Incontinence
Renal
  Acute renal failure from rhabdomyolysis
  Myoglobinuria
Cardiac/respiratory
  Hypoxia
  Hypotension
  Cardiac arrhythmia
  Pulmonary edema
  Pneumonia

neurons become ischemic (144). Compensatory mechanisms begin to fail at approximately 30 minutes into SE. Hyperpyrexia caused by extreme muscle activity exacerbates neuronal damage in animal models of convulsive SE (21,145). Several animal models show that lactic acidosis and hyperglycemia also exacerbate cell damage and cell death (21,22,40). Finally, respiratory depression and hypotension result in hypoxia and ischemia (11). Similar mechanisms may also cause neuronal injury in NCSE, but these are less well studied.

Neuropathological changes induced by SE, such as cell loss and reactive gliosis, are most pronounced in the neocortex, hippocampus, thalamus, and cerebellum (146). These changes are found in both animal models of SE and humans dying from SE. One study found decreased hippocampal density in patients with SE compared with patients with epilepsy but no history of SE or compared with normal control subjects (147). SE causes cell death in hippocampal fields CA1, CA3, dentate granule cell layer, and the dentate hilus (41), a distribution similar to that seen in human mesial temporal sclerosis. More prolonged seizures cause more severe neuronal damage (98), with a threshold for damage at approximately 30 minutes. Seizures lasting more than 30 to 40 minutes result in neuronal death and epileptogenesis (146,148). Neuropathological changes are exacerbated by systemic abnormalities such as hypertension, hypoglycemia, hypoxia, and hyperpyrexia (40,149). Prolonged seizure activity itself, however, can cause neuronal damage in the absence of systemic abnormalities (17,150,151).

The development of epileptic seizures is a common consequence of convulsive SE (83). Typically, seizures begin after a latent period that may last days to weeks in animal models and from weeks to months in humans. A variety of structural and functional changes leads to hyperexcitability and spontaneous seizures, including network and synaptic reorganization, mossy fiber sprouting, and neurogenesis (83).

DIAGNOSIS AND SUBPOPULATIONS

Severity of Injury Measures

Seizure Type

Seizure type has a significant impact on diagnosis and outcome (86). Seizure type is classified by a combination of clinical examination and EEG. GCSE can usually be diagnosed on clinical grounds and is therefore rapidly detected and treated. As discussed previously, GCSE also has high morbidity and mortality, warranting aggressive therapy. NCSE, on the other hand, usually requires an EEG for diagnosis. NCSE may persist for hours or days before a diagnosis is made, increasing the chance of neurological injury (92). Other seizure types are rarer, and studies of outcome are often contradictory (92,152,153). In a multivariate regression analysis of risk factors for mortality, seizure type was not significantly associated with mortality (67).

Among patients with GCSE, the presence or absence of convulsive movements is a predictor of both treatment response and outcome. In the VA Cooperative Study (9), 56% of patients with overt GCSE responded to first-line therapy, as compared with 15% with subtle GCSE. Thirty-day mortality was 26.8% for overt GCSE and 64.9% for subtle GCSE. It is not clear whether the poor treatment response and increased mortality are related to the longer duration of SE before treatment is initiated or to a more severe underlying neurological injury.

Level of Consciousness

Within each seizure type, patients with greater impairment of consciousness may have poorer outcomes (93). In NCSE, severe mental status impairment is associated with higher morbidity (39%) than milder cognitive impairment (7%) (75). The Glasgow Coma Scale may be used to quantify level of consciousness.

Duration

Most studies of SE have demonstrated increased morbidity and mortality with increasing duration of seizures (15,66, 67,73,76). The 19% 30-day mortality rate for SE lasting 30 or more minutes is much higher than the 2.6% rate in seizures lasting 10 to 29 minutes (15). SE usually lasts less than 24 hours (38%, 30 minutes to 2 hours, and 38%, 2 to 4 hours), but 25% lasts at least 24 hours (63). Multivariate regression analysis demonstrates that seizure duration, etiology, and age are statistically significant predictors of mortality when controlling for all other variables (67). The odds ratio for mortality of seizures lasting more than 1 hour was 9.7. Continuous SE has higher mortality than intermittent SE in adults, even controlling for age, etiology, seizure type, and SE duration. Therefore, the actual duration of seizure time may be the most important variable (77).

The duration of SE before treatment may influence the response to treatment (98). If treatment is initiated within 30 minutes, 82% of patients respond to first-line therapy, compared with less than 40% of patients if treatment is delayed for more than 2 hours (76). The response rates to a second conventional AED is considerably better if it is administered earlier in the course of SE (99).

For the purpose of clinical trial design, SE can be classified into four groups: (a) impending SE in patients with ARS or prolonged seizures lasting less than 10 minutes; (b) early SE, lasting 10 to 29 minutes; (c) established SE, lasting 30 to 60 minutes; and (d) prolonged SE, lasting more than 60 minutes. Nearly 50% of patients in group b stop seizing spontaneously (15).

Response to Treatment

RSE is defined as SE that does not respond to two or three AEDs. In general, patients with more difficult-to-treat SE have poorer outcomes, including increased mortality, poor functional outcome, and more prolonged hospitalization (99,100,104).

Etiology

SE can be divided into four main groups according to etiology (68,154). Idiopathic and cryptogenic SE occur in the absence of an acute precipitating CNS insult or systemic metabolic dysfunction. Remote symptomatic SE occurs without acute provoking factors in a patient with a history of a prior CNS insult that is known to increase the risk of subsequent epilepsy, such as stroke, traumatic brain injury, or CNS infection. Acute symptomatic SE occurs during an acute illness with a known CNS insult or systemic metabolic dysfunction. Progressive SE occurs in association with a progressive neurological disorder, such as degenerative diseases, malignancies, and some neurocutaneous syndromes.

Etiology has a significant effect on morbidity and mortality. Short-term mortality is highest (34%) when SE is due to an acute insult (62,63,73). Ischemia may have a synergistic effect with SE; highest mortality rates are seen in patients with hypoxic-ischemic insults or cerebrovascular disease (74). Mortality is not elevated for those with idiopathic or cryptogenic GCSE. Mortality rates for NCSE were higher in patients with an acute etiology (27%) versus epilepsy (3%) and cryptogenic (18%) groups (75).

Concomitant Medical Illnesses

The number and severity of comorbid illnesses also influence prognosis. SE in hospitalized patients with multiple medical problems has a mortality of 61% (155). APACHE II (Acute Physiology and Chronic Health Evaluation II) scores can be used to predict mortality and control for differences in severity of comorbid conditions between treatment groups.

Defining Relevant Treatment Populations

Table 5.4 outlines several possible SE populations for treatment trials. The first, seizure type, defines the major subtypes of SE. Overt GCSE is common and easily recognized. Fifty to 65% of patients are expected to respond to first-line therapy. Treatment algorithms for initial therapy are well established. Trials in this population would need to include the current standard of care for treatment of SE (see later text on current disease management and controversies). This patient population might be most appropriate for trials of new drugs that have shown promising efficacy in other types of SE (e.g., RSE) or for older AEDs with new formulations. Overt GCSE patients might also be candidates for trials of potential neuroprotective agents. Subtle GCSE is less common and requires an EEG for diagnosis. Patients with subtle GCSE do

TABLE 5.4 Potential populations for trials of acute seizures and SE (11,105,259)

Seizure types
GCSE
NCSE
Complex partial SE
Absence SE
Presentation
Subtle GCSE
RSE (failed two or more AEDs)
Duration
ARS
Seizures <10 minutes
Seizures 10–29 minutes
Specific etiologies
Acute symptomatic SE
Remote symptomatic SE
Cryptogenic SE
Febrile SE
Special populations
Children
Neonates
Women (eclampsia)
Elderly patients
Hospitalized patients

not respond well (approximately 15%) to traditional AEDs. Such patients might be good candidates for trials of new drugs that have shown efficacy in animal models of refractory or prolonged SE.

NCSE requires an EEG for diagnosis. This significantly increases the time before therapy can be initiated and makes recruitment more difficult. The individual subtypes of NCSE are relatively rare and may require a large number of sites to allow for recruitment of sufficient numbers of patients to demonstrate treatment efficacy. Randomized trials of conventional AEDs would aid in management of NCSE patients.

There is no consensus on treatment for patients with RSE. Randomized trials could study commonly used therapies, such as intravenous midazolam, propofol, and pentobarbital, as well as potential neuroprotective agents. Major barriers to trials for RSE are standardization of initial therapy, low numbers of potential subjects at each site, and need for an EEG to determine treatment efficacy.

Trials for rapid cessation of acute seizures should treat patients with prolonged or repetitive seizures lasting between 5 and 30 minutes, because these patients have the best treatment response. Morbidity and mortality in acute seizures are low, so the most appropriate outcome measure would be time to termination of seizures. Trials of ARS are critically important to reduce progression to SE, with its higher morbidity and mortality.

SE caused by specific etiologies has different responses to therapy and different outcomes. Acute symptomatic SE has a poor prognosis and might be a specifically targeted subgroup for new therapies. Febrile SE has a nearly uniformly good prognosis acutely, but some subgroups have a high risk for the development of epilepsy. Such patients might be candidates for neuroprotection or antiepileptogenesis trials. These trials require longer follow-up periods than previous SE trials. Finally, subgroups according to age, particularly neonates, pediatric patients, and the elderly, might be appropriate for clinical trials. SE in women with eclampsia is another potential subgroup that is not discussed further here.

Defining Good and Poor Prognoses

Good outcome has most commonly been defined by cessation of seizures within a certain time period after drug administration, absence of seizure recurrence within the next 24 hours, and return to prior neurological function. Other outcome measures include 30-day mortality, functional deterioration as defined by a decline in Glasgow Outcome Score, focal neurological deficits, cognitive disability, and development of epilepsy. Prospective treatment trials have used only seizure cessation and seizure-free interval as primary outcome measures.

Patients with good prognosis include those with short duration of seizures, overt GCSE, children, and those with an idiopathic or remote symptomatic etiology. Children have a much lower mortality than adults. Mortality rates for children range between 3.6% and 7% (68, 82,156–158). Almost all cases of fatal outcome are associated with either acute CNS insults or progressive neurological disorders rather than SE itself. Poor prognosis is associated with prolonged SE, extremely young (neonatal) or old age, RSE, and an acute symptomatic etiology, particularly hypoxic-ischemic encephalopathy.

Relevant Prognostic Factors for Defining Inclusion and Exclusion Criteria

Based on the above discussion, inclusion criteria should focus on seizure type by clinical criteria, seizure type by EEG criteria, duration of seizures before enrollment, age, and response to previous therapy. Because etiology is often not known at the time of initial presentation, specific etiologies are difficult to use as inclusion criteria. Exclusion criteria should be designed to exclude patients with excessively good or poor prognoses or those at potentially high risk from the proposed study agent. For example, patients with GCSE after a hypoxic-ischemic insult have an extremely poor prognosis, with nearly 100% mortality. Such patients should not be included or should be randomized separately from other patients. Patients with hypotension, respiratory compromise, or hypoxia may be at higher risk from adverse effects of certain treatments and may need to be excluded. Because SE is a relatively rare condition requiring recruitment of patients in emergency situations, excessively restrictive inclusion criteria or broad exclusion criteria further limit enrollment of subjects.

Subsetting Patients for Planned and Ad Hoc Comparisons

Properly designed inclusion and exclusion criteria allow recruitment of a fairly homogeneous population, improving the likelihood of showing a treatment effect with a reasonable sample size. Subsets should be designed to further minimize and account for variability in outcome. This might include, for example, analyzing patients separately according to whether they present with overt versus subtle GCSE. Fifty to 65% of patients with overt GCSE respond to first-line treatment and mortality is 26.8%, versus a 15% response rate and more than 60% mortality with subtle GCSE. Randomizing patients separately according to type of GCSE and analysis of subgroups would significantly improve the power of a study of GCSE. Similarly, stratifying patients by age, etiology, seizure duration before treatment, concomitant medical conditions, or APACHE II scores would reduce variability in the various treatment groups (159). No more than two or three subsets should be planned, or the resultant groups may be too small to show significant differences in outcome.

CURRENT DISEASE MANAGEMENT AND CONTROVERSIES

Immediate diagnosis and effective treatment is necessary to reduce the morbidity and mortality of SE. A standardized protocol should be available in the emergency department to facilitate early treatment (Table 5.5) Successful management of SE requires drugs that can be administered extremely rapidly (i.e., intravenously), are effective and well tolerated, and remain in the brain to prevent recurrence of SE (160).

For all types of SE and acute seizures, the first steps in management are assessment of the patient's airway, oxygenation, and blood pressure (2). Oxygen should be administered by nasal cannula or a nonrebreathing mask, and the patient should be ventilated with a bag valve mask or intubated if there is evidence of respiratory compromise. Short-acting neuromuscular blocking agents should be used to facilitate intubation, so that paralysis does not impair assessment of ongoing seizures. A screening neurological examination should be performed to check for signs of focal intracranial lesions. Intravenous access should be established with an isotonic saline infusion. Serum electrolytes, blood urea nitrogen, glucose, complete blood count, AED levels, blood gas, and toxicology screen should be obtained to screen for possible SE etiologies. If hypoglycemia is present or blood glucose cannot be determined, 50 ml of 50% glucose should be administered. In adults, 100 mg thiamine should be coadministered with glucose to prevent Wernicke's encephalopathy. Hyperthermia can exacerbate neuronal damage and should be treated aggressively with active cooling. Specific AED treatment should be started as soon as possible (details below). Computed tomography of the head and lumbar puncture should be performed once the patient is stabilized. Electrical seizure activity can continue even when clinical signs of seizure activity have ended. EEG monitoring can be performed for any patient who remains unconscious after initial AEDs, receives a long-acting paralytic agent, or requires prolonged therapy for RSE.

Current Management Guidelines

Generalized Convulsive SE

Five prospective randomized studies compared treatments for SE (8,9,161–163). An early randomized double-blind trial comparing intravenous lorazepam (4 mg) and diazepam (10 mg) in 79 patients with all types of SE (163) found no difference in

TABLE 5.5   Treatment algorithm for SE

0–5 min	Diagnose SE
	Airway management:
	    O_2 via nasal cannula or mask
	    Assess oxygenation (O_2 saturation or blood gas)
	    Consider intubation at any point in protocol
	Vital signs, ECG monitoring
	Physical and neurological examinations
	Intravenous access, normal saline infusion
	Blood for glucose, electrolytes, alcohol, AED levels, blood gas, toxicology
	    screen, complete blood count

6–9 min 10–20 min	If hypoglycemic or no blood glucose available, administer glucose Adults: 100 mg thiamine, 50 ml of 50% glucose; children: 2 ml/kg of 25% glucose Administer benzodiazepine (if seizure continues) Lorazepam 0.1 mg/kg (rate 2 mg/min IV) OR Diazepam 0.2 mg/kg (rate 5 mg/min IV) OR Diastat PR if no IV access 1–5 yrs: 0.5 mg/kg/dose 6–11 yrs: 0.3 mg/kg/dose >12 yrs: 0.2 mg/kg/dose Repeat once if convulsion does not stop after 5–10 minutes
20–40 min	Fosphenytoin 20 mg PE/min (rate <150 mg PE/min) OR Phenytoin 20 mg/kg (adults: rate <50 mg/min; children: <1 mg/kg/min) Monitor ECG and blood pressure during infusion If seizures continue, give additional fosphenytoin 5–10 mg PE/kg OR phenytoin 5–10 mg/kg
40–60 min	Transfer to intensive care unit, intubate Midazolam 0.1–0.3 mg/kg IV bolus (0.05–1 mg/kg/hr maintenance) OR Propofol 1–2 mg/kg IV bolus (1–15 mg/kg/h maintenance) NOT IN CHILDREN OR Phenobarbital 20 mg/kg IV (rate 50 mg/min or 3 mg/kg/min in children) Monitor blood pressure, consider EEG
60+ min	Intubate patient or call anesthesiology to intubate Pentobarbital 5–10 mg/kg load (0.5–10 mg/kg/h maintenance) EEG monitoring Neuro exam, CT/LP, repeat labs, make decision on chronic treatment

CT, computed tomography; ECG, electrocardiogram; LP, lumbar puncture; PE, phenytoin equivalent.

the frequency of success between the two treatments, possibly because of the small study sample size. Lorazepam terminated SE in 78% of patients versus diazepam in 58%. The relative duration of action of the two benzodiazepines could not be compared, because all patients received a loading dose of phenytoin shortly after benzodiazepine administration. A comparison of phenobarbital with diazepam followed by phenytoin in 36 patients with GCSE showed no statistically significant difference in the frequency of success between the two treatments, although phenobarbital showed a trend toward more rapid control of seizures (162). Treiman and colleagues (161) compared lorazepam and phenytoin in the initial treatment of 87 cases of overt or subtle GCSE. Lorazepam successfully terminated SE 77.2% of the time, whereas phenytoin was successful only 49.8% of the time.

Two large randomized controlled trials (8,9) confirmed that benzodiazepines should be the initial drug therapy in patients with GCSE. A VA Cooperative Study (9) compared four intravenous drug regimens in the initial management of overt (384 patients) or subtle (134 patients) GCSE: phenytoin alone (18 mg/kg), diazepam (0.15 mg/kg) followed by phenytoin (18 mg/kg), phenobarbital alone (15 mg/kg), and lorazepam alone (0.1 mg/kg). Overall, the success rate in patients with overt GCSE was 56% but was only 15% in patients with subtle GCSE. Lorazepam showed the highest success rate, terminating seizures in 64.9% of patients with overt GCSE, then phenobarbital (58.2%), diazepam followed by phenytoin (55.8%), and phenytoin alone (43.6%), but these differences were not statistically significant. In subtle GCSE, phenobarbital had the highest response rate (24.2%), followed by lorazepam (17.9%), diazepam plus phenytoin (8.3%), and phenytoin monotherapy (7.7%). There were no differences in SE recurrence rates over 12 hours, 30-day outcome measures, or complication rates among the four treatment groups. The authors concluded that lorazepam was the easiest medication to use among these four first-line agents. The Prehospital Treatment of Status Epilepticus Study (8) randomized 205 patients with an out-of-hospital diagnosis of SE to treatment with intravenous diazepam, lorazepam, or placebo, administered by paramedics. SE was terminated before emergency department arrival in 59% of patients treated with lorazepam and 43% treated with diazepam, compared with 21% with placebo. Complication rates were low (11% lorazepam, 10% diazepam, and 23% placebo). Patients whose seizures were not terminated before emergency department arrival were more likely to require intensive care

unit admission (73% vs. 32%). Finally, a retrospective trial of lorazepam (4 mg) versus diazepam (10 mg) (164) showed that the two drugs were equal in aborting SE but that lorazepam patients had lower seizure recurrence rates over 12 hours.

There is no randomized controlled trial examining second-line therapy for GCSE. The Epilepsy Foundation's Working Group on Status Epilepticus (2) and others (9,105,165) recommend phenytoin 18–20 mg/kg. A survey of neurologists on treatment of GCSE indicated that 95% of respondents would use phenytoin or fosphenytoin (a phenytoin prodrug) as second-line therapy (166). No randomized trials comparing success rates of phenytoin versus fosphenytoin have been performed. Another possible option is phenobarbital, which has also been shown to be efficacious in the treatment of GCSE (2,9,162). Phenobarbital loading, however, usually necessitates intubation, particularly if the patient received a benzodiazepine initially, and often produces prolonged depression of consciousness. Third-line choices in the SE treatment survey included phenobarbital (43%), continuous intravenous AEDs (cIV-AEDs, 19%), and valproate (16%) (166). Patients are less likely to respond to a second agent or third agent if a first-line therapy fails (167), and many neurologists now proceed to cIV-AEDs if initial treatment does not abort seizures. Surveys of neurologists, however, showed that most still prefer barbiturates over cIV-AEDs (166,168). In a randomized study comparing four treatments for GCSE (9), 38% of patients with overt GCSE and 82% with subtle GCSE had seizures uncontrolled by a second AED, and only 2% and 5%, respectively, stopped seizing after a third agent. Second-line agents may be more effective if they are administered earlier in the course of SE (99).

Several new AED formulations are available, but their place in the management of SE has not been established. Intravenous valproate may be useful in some patients with SE (169–174), but its place in the overall treatment algorithm and optimal dosing regimen is not yet clear.

Nonconvulsive SE

There are no randomized controlled trials evaluating the treatment of NCSE, and treatment decisions are extrapolated from trials of GCSE or based on the severity of the clinical presentation (175). The likelihood of successful treatment and the prognosis is strongly related to the subtype of NCSE (typical absence SE, complex partial SE, or nonconvulsive SE in patients with acute neurological conditions and stupor or coma). It is still debated whether absence SE and complex partial SE cause neuronal damage in humans (93), and therefore less aggressive treatment is usually advocated for NCSE. Most morbidity from NCSE appears to be due to the underlying illness rather than to the NCSE itself (94). NCSE should still be terminated quickly, however, to return patients to their baseline neurological status, prevent physical injury, and prevent progression to GCSE. Treatment options include oral benzodiazepines, intravenous benzodiazepines, additional doses of the patient's typical AED, and intravenous phenytoin or valproate (170). Medications that can cause respiratory compromise or mental status depression should usually be avoided unless NCSE is refractory to initial treatments (175). NCSE in patients with acute medical or neurological insults such as traumatic brain injury or stroke may exacerbate neuronal damage (176), but it has not been determined whether treatment of such seizures improves neurological outcome (152). One retrospective study of NCSE in critically ill elderly patients (177) showed increased mortality in patients who received benzodiazepines, highlighting the potential adverse effects of treatment, particularly respiratory compromise requiring intubation and mechanical ventilation.

Refractory SE

There is currently no consensus regarding treatment of RSE (99). Guidelines give a variety of options (2,102,105,167), including cIV-AEDs (pentobarbital, midazolam, propofol, thiopental), valproate, and inhalational anesthetics. Pentobarbital is the best-established cIV-AED for treatment of RSE (106,178–181) but often causes hemodynamic instability by depressing myocardial function. Midazolam aborted 80% of episodes of RSE in a retrospective review of 33 patients, but breakthrough seizures and posttreatment seizures occurred in more than 50% of patients (108). Most breakthrough seizures were subclinical, seen only on EEG. A small prospective study of 16 patients with RSE (103) showed that propofol aborted SE in 63% of patients versus 82% treated with pentobarbital. Propofol more rapidly controlled seizures (mean 2.6 minutes vs. 123 minutes for pentobarbital). There were no significant differences in the overall outcome or side effects, but the number of patients treated was very small. Several case reports, however, have reported hyperlipidemia, systemic acidosis, multiorgan failure, and death related to prolonged propofol infusions (182).

Most neurologists (72%) responding to an SE treatment survey (72%) chose a cIV-AED for RSE, usually pentobarbital (36%), propofol (17%), or midazolam (16%). Thiopental was preferred by anesthesiologists (82%) in an earlier survey (183). A systemic review of 28 studies including 193 patients compared the efficacy of midazolam ($n = 54$), propofol ($n = 33$), and pentobarbital ($n = 106$) (100) for treatment of RSE. The authors examined both immediate treatment failure (seizures 1–6 hours after starting cIV-AEDs) and mortality for each agent and for the titration goal (seizure suppression vs. EEG background suppression). Mortality was 48% and was not significantly associated with the choice of agent or titration goal. Pentobarbital showed a lower frequency of short-term treatment failure, breakthrough seizures, and changes to another therapy but a much higher frequency of hypotension requiring therapy. The authors concluded that pentobarbital appeared to be more effective in short-term seizure control but that this was associated with a higher likelihood of adverse effects and did not improve mortality. In addition, because intermittent rather than continuous EEG monitoring was typically used to assess pentobarbital treatment efficacy, some breakthrough seizures could have been missed.

The optimal goal of treatment for RSE is debated. Patients are usually intubated, and cIV-AEDs cause coma or obtundation, so clinical signs of seizure activity may no longer be present. For pentobarbital, a burst-suppression pattern on EEG is usually the goal, although seizures may still occur from a burst-suppression background (167). For midazolam and propofol, suppression of seizures as verified by continuous EEG monitoring is more often the goal. Most breakthrough seizures (89%) are subclinical, so EEG monitoring is essential (108). One retrospective study suggested that outcome of treatment with pentobarbital is improved if the EEG is completely suppressed (184), and the above meta-analysis suggested improved short-term seizure control with suppression of the EEG background (100). The high doses of cIV-AEDs required to produce a burst suppression or completely suppressed background, however, often result in hypotension requiring use of vasopressors. The optimal duration of seizure suppression before cIV-AEDs can be withdrawn is also unclear.

Seizures refractory to cIV-AEDs are usually treated with inhalational anesthetic agents such as isoflurane (106,185) and desflurane (186). Paraldehyde (187) and lidocaine (188) have also been reported to stop seizures in SE refractory to other agents.

Newer potential treatment options include NMDA antagonists and several of the newer AEDs. Ketamine is an NMDA antagonist that has shown some efficacy in the treatment of RSE (49,189). NMDA antagonists may have neuroprotective properties in animal models of SE, even when seizures are not completely controlled (190–192). Topiramate and levetiracetam, two newer AEDs, may also have some neuroprotective efficacy (58,193), but only topiramate has been used to treat RSE in small numbers of patients (194–196). Other potential neuroprotective treatments include calcium channel blockers, antioxidants, erythropoietin, and hypothermia.

Pediatric SE

There are no randomized controlled trials for the treatment of SE in children, and management is extrapolated from treatment of SE in adults. Treatment of SE in children may vary greatly from that in adults, however, including the necessity for weight-based dosing, different intravenous infusion rates, and difficulties obtaining intravenous access. A single open quasi-randomized study (197) compared lorazepam and diazepam (intravenous or rectal administration) in 102 patients with GCSE aged between 1 month and 16 years, but this study has several methodological flaws. Intravenous lorazepam was effective in 70% of 27 children and intravenous diazepam in 65% of 34 children. Patients received a second dose of the study medication if the seizure lasted more than 8 minutes after the first dose was given. The mean time to seizure cessation was similar for the two drugs. Lorazepam showed a slight advantage in secondary measures; recurrence rates over 24 hours were slightly lower in the lorazepam group (22% vs. 35%), fewer patients treated with lorazepam required additional AEDs to terminate GCSE, and respiratory depression was less common (4% vs. 21%). Rectal lorazepam was effective in 100% of 6 children and rectal diazepam in 32%

of 19 children. This study was small, unblinded, inadequately randomized, had unequal numbers of patients in the two treatment groups, used variable methods of drug administration, and did not use an intention-to-treat analysis.

A proposed four-step guideline for treatment of GCSE in children (197) begins with (a) intravenous lorazepam 0.1 mg/kg, (b) a second dose of lorazepam 0.1 mg/kg if the patient continues to seize after 10 minutes, followed by (c) phenytoin 18 mg/kg (or phenobarbital 20 mg/kg for patients already taking phenytoin) if the patient continues to seize after 10 minutes, and finally (d) general anesthesia with intravenous thiopental 4 mg/kg if seizures continue for 20 more minutes. If the patient has no intravenous access, treatment begins with rectal diazepam 0.5 mg/kg, followed by paraldehyde 0.4 ml/kg rectally. Paraldehyde is not available in United States.

A meta-analysis of treatment of RSE in 101 children reviewed the efficacy of diazepam, midazolam, thiopental, pentobarbital, and isoflurane (106). Diazepam was less efficacious than other treatments. Mortality appeared to be lower in midazolam-treated patients. Overall mortality was higher for those with symptomatic RSE (20%) than idiopathic RSE (4%).

Other options for RSE in children include intravenous valproate (173,174), intravenous propofol (198), inhalational anesthetics, and topiramate (196). Propofol is not currently recommended for use in children because of an excess in mortality in a controlled study comparing 1% and 2% propofol preparations with other forms of sedation in the pediatric intensive care unit (199).

For pediatric SE, the potential adverse effects of AEDs on the developing brain must be considered. Some AEDs may cause changes in cell growth and energy substrate utilization (134). Phenobarbital may cause changes in dendritic arborization (139).

Acute Repetitive Seizures

"Rescue" therapy can prevent prolonged or repetitive seizures from progressing to SE. Because treatment needs to be administered at home by caretakers, intravenous therapy is not an option, and alternative routes of administration are necessary. Most studies have focused on the benzodiazepines. Rectally administered diazepam is the most well-studied treatment of ARS in children and adults (200,201) and is currently the only option approved by the U.S. Food and Drug Administration. Primary endpoints in these two studies were termination of the seizure cluster and absence of recurrent seizures over 12 to 24 hours. Seventy-one percent of adults treated with rectal diazepam gel remained seizure free, compared with 28% of placebo patients (3). The only adverse effects were somnolence and dizziness; respiratory depression did not occur. Other modes of administration include buccal (202,203), intranasal (204,205), and intramuscular (206,207). All these choices demonstrate good efficacy, rapid seizure control, and low rates of adverse events. Seizure control is often faster than with intravenous benzodiazepines, because intravenous access does not need to be established before treatment is started.

A recent large-scale clinical trial of out of hospital treatment of adult patients with SE found that lorazepam and diazepam were far superior to placebo in the management of patients in this setting. There were strong trends favoring lorazepam over diazepam for several endpoints in the study (8).

Febrile SE

Most febrile seizures are brief and do not recur. Antipyretics at the first sign of fever do not reliably prevent febrile seizure recurrence (208). Oral diazepam prevents seizure recurrence in some studies but not in others (113,117,209). The benefits of preventing seizure recurrence must be weighed against the sedating effects of benzodiazepines.

Prolonged febrile seizures and febrile SE should be treated aggressively. The typical treatment is intravenous or rectal diazepam (210). Rectal diazepam can be used to abort prolonged seizures at home (211). Other intravenous drugs may be effective, but there are no controlled studies. Intranasal midazolam was as efficacious and safe as intravenous diazepam in one small randomized study (212). If seizures persist after adequate doses of a benzodiazepine, a full SE protocol should be initiated as for pediatric SE above.

Chronic use of several AEDs has been shown to be effective in preventing recurrence of febrile seizures, but the adverse effects often outweigh the short-term seizure prevention (213–215). Phenobarbital can cause hyperactivity, rash and hypersensitivity reactions, and possibly developmental delay. Valproate may cause hepatotoxicity, particularly in young children, thrombocytopenia, and pancreatitis. Phenytoin and carbamazepine are ineffective in preventing recurrent febrile seizures (113). Prophylactic treatment with benzodiazepines, phenobarbital, or valproate does not prevent the eventual development of epilepsy (216). It is not known whether newer AEDs would be more effective in preventing epileptogenesis. In general, prophylactic medications are not recommended for simple febrile seizures (215). Some patients with frequent complex febrile seizures or recurrent febrile SE may benefit from prophylactic treatment at the onset of fever or as soon as seizures begin.

Neonatal SE

The first step in management of neonatal seizures is correct diagnosis; some apparent seizures are nonepileptic, and EEG may be needed to confirm an electrographic correlate. Appropriate treatment of the underlying etiology should be initiated as soon as possible. Screening

studies should include sodium, blood urea nitrogen, glucose, calcium, magnesium, bilirubin, ammonia, blood gases, blood amino acids, urine organic acids, and a lumbar puncture as well as neuroimaging with ultrasound or computed tomography. The seizures are typically treated with IV-AEDs, including phenobarbital, phenytoin, diazepam, lorazepam, or midazolam (217). The goals of AED therapy are to stop neonatal seizures, prevent recurrence, and minimize late sequelae (neurological deficits, cognitive dysfunction, and epilepsy). An EEG is needed to confirm seizure cessation, because electrographic seizures may persist after clinical seizures are controlled. How aggressively electrographic seizures should be treated is debated, because electrographic seizures may be extremely resistant to therapy with multiple AEDs and the influence of such subclinical seizures on ultimate outcome is unclear. Higher doses and number of AEDs are associated with higher rates of adverse effects, such as CNS depression, hypotension, and respiratory compromise.

Phenobarbital is the most commonly used AED and may also be effective for prophylaxis of neonatal seizures in infants at high risk (218). Typical loading doses are 20 mg/kg, with goal serum levels 25–40 µg/ml or higher. Phenobarbital may impair brain growth, but separating the effects of AEDs from that of the underlying etiology and neonatal seizures themselves is difficult (139). Choice of second- and third-line agents is variable (138). One prospective study of phenobarbital and phenytoin showed no significant difference between treatments; both drugs were relatively ineffective, controlling seizures only in 43% and 45% of neonates, respectively (219). When used together, phenytoin and phenobarbital were slightly more effective, successfully controlling seizures in approximately 60% of patients (219). This study used EEG to confirm successful control of seizures. Phenytoin loading dose is 20 mg/kg at a rate of 1 mg/kg/min. Free levels of phenytoin should be followed, because phenytoin binding is variable in neonates. Refractory neonatal seizures have been treated with diazepam 0.3 to 1 mg/kg (220), lorazepam 0.05 to 0.15 mg/kg (217), continuous midazolam 0.1 to 0.4 mg/kg/h (221), and valproate 20 to 25 mg/kg (222). Less commonly used treatments include lidocaine, carbamazepine, primidone, paraldehyde, clonazepam, and vigabatrin.

Specific Drugs

Table 5.6 summarizes the mechanisms of action, routes of administration, dosages, and adverse effects of the most commonly used drugs for treatment of SE.

Benzodiazepines

Benzodiazepines are effective and rapidly acting drugs for the treatment of acute seizures and SE. The most commonly used benzodiazepines include diazepam (Valium), lorazepam (Ativan), and midazolam (Versed). Benzodiazepines enhance inhibition by binding to the benzodiazepine binding site of the GABA receptor.

Diazepam is given at a dose of 1.5 to 2.5 mg/kg at a rate of 5 mg/min and can be repeated every 5 to 10 minutes up to three doses. Diazepam can be administered intravenously, intramuscularly, or rectally. A prepackaged rectal diazepam gel (Diastat) is available for rapid out-of-hospital administration. Diazepam enters the brain rapidly but redistributes to fat stores in other areas of the body after 15 to 20 minutes. It may therefore have a less prolonged duration of action and is often followed by another AED to prevent seizure recurrence (8). Lorazepam is given at a dose of 0.1 mg/kg at a rate of 2 mg/min and can be repeated at 5- to 10-minute intervals up to 0.3 mg/kg. Lorazepam binds more tightly to GABA receptors and is less lipid soluble than diazepam. It therefore has a longer duration of action

and lower risk of recurrent seizures (163). Routes of administration include intravenous, rectal, sublingual, and intramuscular. Typical dosing for midazolam is 0.2 mg/kg by intravenous, nasal, buccal, or intramuscular routes (223). Midazolam is more commonly used for initial treatment of SE in Europe than in the United States. Side effects of sedation and respiratory depression are rare. When used for RSE, loading doses of 0.05 to 0.4 mg/kg may be needed to stop seizures, with maintenance doses of 0.1–2 mg/kg/h titrated to seizure suppression (108,224). Prolonged use of midazolam is limited by tachyphylaxis, and doses may need to be increased to maintain efficacy.

The major side effects of all benzodiazepines include depression of mental status, respiratory depression, and hypotension. Local tissue irritation may occur with rapid administration.

Phenytoin and Fosphenytoin

Phenytoin and its prodrug, fosphenytoin, are commonly used second-line treatments for SE (166,168). Both drugs work by blocking repetitive firing of sodium channels. Phenytoin is poorly water soluble and is formulated with a propylene glycol vehicle with a high pH. Intravenous phenytoin can cause local tissue irritation, phlebitis, and rarely the purple glove syndrome, a type of severe tissue necrosis. Arrhythmias and hypotension often occur, particularly in the elderly and at high rates of intravenous administration. Electrocardiogram and blood pressure should be monitored during phenytoin infusions. To decrease adverse effect rates, phenytoin is typically given at a dose of 20 mg/kg at a rate of no more than 50 mg/min or 1 mg/kg/min in children.

Fosphenytoin is a water-soluble phosphorylated phenytoin prodrug. It is converted to phenytoin by serum phosphatases with a half-life of approximately 10 minutes (225,226). Fosphenytoin is given at a dose of 20 mg phenytoin

TABLE 5.6 Common drugs for treatment of SE

	Mechanism of action	Routes	Loading dose	Maintenance dose	Half-life	Levels	Adverse effects	Advantages
Lorazepam	GABA$_A$ agonist	Intravenous Rectal Sublingual Intramuscular	0.1 mg/kg at 2 mg/min	Repeat every 10 min to maximal dose 0.3 mg/kg	8–25 h	0.2–0.5 μg/ml	Respiratory depression Hypotension Decreased level of consciousness	Longer acting than diazepam
Diazepam	GABA$_A$ agonist	Intravenous Rectal (gel) Intramuscular	0.15 mg/kg at 5 mg/min	Repeat every 10 min to maximal dose 0.3 mg/kg	28–54 h	0.2–0.8 μg/ml	Respiratory depression Hypotension Decreased level of consciousness	Short duration of action because of redistribution; usually need second AED
Midazolam	GABA$_A$ agonist	Intravenous Buccal Nasal Intramuscular	0.2 mg/kg (IV)	0.1–2 mg/kg/h	3 h		Respiratory depression Hypotension Decreased level of consciousness	Short duration of action, tachyphylaxis, high cost
Phenytoin	Na$^+$ channel blocker	Intravenous	20 mg/kg at 50 mg/min or 1 mg/kg/min in children	1.5 mg/kg every 8 h	24 h	20–25 μg/ml	Hypotension Cardiac arrhythmias Infusion site reactions/ purple glove syndrome	Multiple drug interactions (highly protein bound, induces P450 hepatic enzymes)
Fosphenytoin	Na$^+$ channel blocker	Intravenous Intramuscular	20 mg PE/kg at 150 mg/min	1.5 mg PE/kg every 8 h	24 h	20–25 μg/ml	Hypotension Cardiac arrhythmias	High cost, lower risk infusion site reactions than phenytoin, paresthesias
Phenobarbital	GABA$_A$ agonist	Intravenous	20 mg/kg at 50–75 mg/min	2–4 mg/kg daily	90 h	30–50 μg/ml	Respiratory depression Hypotension Decreased level of consciousness	Usually requires intubation, multiple drug interactions
Propofol	GABA$_A$ agonist	Intravenous	3–5 mg/kg	2–15 mg/kg/h	2 h	2.4–4 μg/ml	Respiratory depression Hypotension Decreased level of consciousness	High cost, lipid load, multiorgan failure, especially in children

Drug	Mechanism	Route	Loading dose	Maintenance dose	Half-life	Therapeutic level	Side effects	Comments
Valproate	Na$^+$ channel blocker, enhances GABA, Ca^{2+} channel blocker	Intravenous	20–40 mg/kg	4-8 mg/kg TID	15 h	70–150 µg/ml	Hypotension	Unclear efficacy compared with standard agents, pancreatitis, multiple drug interactions (P450 hepatic enzyme inhibitor)
Pentobarbital	GABA$_A$ agonist	Intravenous	10–15 mg/kg	0.5–10 mg/kg/h	10–20 h	10–20 µg/ml	Respiratory depression Hypotension Decreased level of consciousness	Multiple drug interactions
Thiopental	GABA$_A$ agonist	Intravenous	100–200 mg/kg	3–5 mg/kg/h	12–36 h	15–50 µg/ml	Respiratory depression Hypotension Decreased level of consciousness	
Lidocaine	Na$^+$ channel blocker	Intravenous	1.5–2 mg/kg at 50 mg/min	3–4 mg/kg/h			Drowsiness, confusion, cardiac arrhythmia	Low risk respiratory failure of decreased level of consciousness
Ketamine	Noncompetitive NMDA glutamate receptor antagonist	Intravenous	2 mg/kg	2–7.5 mg/kg/h	2–3h		Respiratory depression Hypotension Decreased level of consciousness	Potentially neuroprotective, possible neurotoxicity (cingulate and cerebellar)
Topiramate	Na$^+$ channel blocker, enhances GABA, Ca^{2+} channel blocker, glutamate antagonist	Oral	100–800 mg	200–800 mg/day	21 h		Sedation, cognitive dysfunction, metabolic acidosis	

Ca, calcium; Na, sodium; PE, phenytoin equivalents.

equivalents/kg at a rate of 150 mg phenytoin equivalents/kg. Fosphenytoin does not cause local skin necrosis and may be given intramuscularly, reaching therapeutic levels in about 30 minutes (227). It rapidly controlled GCSE in 93.8% of patients in one small study (228). Fosphenytoin is much more expensive than phenytoin but has a lower risk of local tissue complications (229). Adverse effects include pruritus and perineal paresthesias at high rates of administration. Cardiac arrhythmias and hypotension can occur.

Phenobarbital

Phenobarbital is typically a third-line agent for treatment of SE. Phenobarbital has its primary effect on the GABA$_A$ receptor complex. Prolonged sedation and respiratory depression are common adverse effects and usually require intubation, especially if benzodiazepines have been used earlier in treatment. Hypotension may also occur, particularly with rapid infusion rates. Its propylene glycol diluent can cause hypotension by myocardial depression. The usual loading dose is 15 to 20 mg/kg, infused at a rate of no more than 100 mg/min.

Valproate

The parenteral formulation of valproate, Depacon, is approved by the U.S. Food and Drug Administration for rapid intravenous loading or replacement of oral therapy and not specifically for treatment of SE. A broad spectrum of efficacy and low rates of adverse effects, however, make valproate an attractive option for treatment of SE (169). Valproate may be effective for GCSE, myoclonic SE, and NCSE (169,171,172). Adverse effects include gastrointestinal distress, lethargy, and local skin irritation. Valproate appears to be well tolerated, even in patients with cardiovascular instability. Typical loading doses are 25 mg/kg and up to 30–40 mg/kg in children (173).

Pentobarbital and Thiopental

Pentobarbital is an anesthetic barbiturate that enhances inhibition via GABA receptors. Typical loading doses are 5–12 mg/kg to induce a burst-suppression EEG pattern and then maintenance doses of 0.5–10 mg/kg/h. Adverse effects include sedation, respiratory depression, and hypotension as well as susceptibility to infection and inadequate temperature regulation. Thiopental is more commonly used in Europe (230). It may accumulate in tissues and therefore have an extremely prolonged duration of sedation and respiratory depression. Again, the goal is a burst-suppression EEG, usually requiring doses of 5–8 mg/kg/h.

Propofol

Propofol is a cIV-AED that has benzodiazepine-like effects. Several retrospective studies document that propofol is effective in the treatment of RSE. A typical loading dose is 3–5 mg/kg, followed by a maintenance infusion at 1–15 mg/kg/h to suppress seizures. Abrupt discontinuation may precipitate withdrawal seizures. Adverse effects include CNS depression, respiratory suppression, metabolic acidosis, hyperlipidemia, hypertension, and occasional bradycardia. Multiple case reports of significant morbidity and mortality with prolonged propofol use have decreased its use for RSE (182).

Inhalational Anesthetics

Isoflurane and desflurane produce a burst-suppression pattern on EEG and can be titrated over time to control seizures (186). Both drugs often induce hypotension.

Topiramate

Topiramate is a new AED with multiple mechanisms of action, including sodium and calcium channel blockade, enhancement of GABA activity, and inhibition of kainate-evoked currents. Topiramate may have neuroprotective efficacy in experimental SE (193), limiting hippocampal neuronal death. Several case reports have explored the efficacy of topiramate for the treatment of RSE (65,194,195). Doses range from 200 to 1600 mg per day in adults and 5–6 mg/kg per day in children. Topiramate is only available in an oral formulation and therefore cannot be used for initial treatment of GCSE.

Ketamine

Ketamine is an NMDA receptor antagonist that is neuroprotective in experimental SE and has shown efficacy in human RSE in several case reports (49,189). Ketamine may produce vivid dreams or frightening hallucinations. Ketamine has been administered at a loading dose of 2 mg/kg and maintenance infusions of 2–7.5 mg/kg/h. One case report highlighted the possible neurotoxic effects of ketamine, including cingulate and cerebellar damage (231). At this point, efficacy and long-term effects of ketamine remain uncertain.

ISSUES IN CLINICAL PROTOCOL DEVELOPMENT

Time to Treatment

The duration of SE before treatment is initiated is a prime determinant of the response to therapy. Similarly, morbidity and mortality increase significantly as the duration of SE increases. Rapid institution of treatment is therefore essential in SE trials. Multiple factors may influence the time before treatment can be initiated, including transportation time, availability of intravenous access, need for stabilization of airway and blood pressure, and availability of EEG for diagnosis. Most SE protocols emphasize the initiation of treatment as soon as possible, usually within 5 minutes after seizure onset. One study of treatment access time (232) for outpatient SE revealed that initiation of treatment occurred at least 40 minutes after seizure

onset and occasionally exceeded 4 hours. Time from onset of SE to the arrival of the emergency medical team averaged 30 minutes (range, 15 to 140 minutes), time from arrival of emergency medical team to arrival in the emergency department averaged 30 minutes (range, 10 to 40 minutes), and time from arrival in the emergency department to initiation of an SE protocol averaged 35 minutes (range, 15 to 80 minutes). When the patient required an EEG for diagnosis of SE, treatment could be delayed up to 6.5 hours. Another survey study on emergency EEG estimated a mean response time of 3 ± 4 hours from the time of request to time of expert interpretation of the EEG. EEG may be necessary to confirm the diagnosis of SE, classify the type of SE, and exclude nonepileptic psychogenic seizures (233).

The above statistics indicate the need for a well-trained team in the emergency room, the neurology department, and the clinical neurophysiology laboratory for patients to be enrolled rapidly in SE trials. In a clinical trial, time to obtain informed consent must also be included in time to treatment estimates. Because SE is a life-threatening condition with high mortality, some protocols may be able to use a waiver of informed consent procedure initially, with later written informed consent from the patient or surrogate for continuation of participation in the study.

Operational Definitions of SE

As described above, SE has a variety of clinical and EEG manifestations and can affect widely disparate populations. The duration of seizures before the diagnosis of SE also varies widely from study to study. A clear operational definition for diagnosis of SE is therefore essential to ensure that an appropriate and homogeneous population is enrolled into the trial. For example, the VA Cooperative Study (9) defined overt GCSE as two or more convulsions without complete recovery of consciousness between the convulsions or continuous convulsive activity for more than 10 minutes. EEG may be necessary for the diagnosis of subtle GCSE, in which case a clear description of the EEG patterns meeting diagnostic criteria must be established.

Site Selection

SE is a relatively rare condition, and a large number of centers may be needed for adequate enrollment. The VA Cooperative Study (9) included 16 VA medical centers and six affiliated university hospitals but still failed to fully meet enrollment goals. Once the inclusion criteria for the trial are determined, the frequency of cases of SE at each site should be determined prospectively. A feasibility trial may be useful to determine whether patients can be identified and treatment initiated within the desired treatment window. Many more patients will be screened than are eligible for participation in the study. Reasons for lack of enrollment in the VA Cooperative Trial (9) included failure to notify the study team, patient not in generalized SE, patient previously enrolled, contraindication to administration of one of the study drugs, age less than 18 years, pregnancy, or presence of a neurological emergency. Each site will require an SE team consisting of a site investigator, emergency room co-investigator, study coordinator(s), and EEG technician, who will need to be available 24 hours a day, 7 days a week to enroll patients. For out-of-hospital treatment of acute seizures and SE, families, caretakers, or paramedics must be intensively trained to recognize SE, administer study medications, and monitor patients for treatment success and adverse effects.

Sample Size

The most important issue in sample size determination is the decision of what difference in efficacy is clinically relevant.

Because GCSE is a life-threatening disorder, a 10% improvement in the success rate of SE treatment could be clinically meaningful. To find such a small difference, a large sample size is required. Recruitment for SE trials is difficult, and very large sample sizes may be nearly impossible to accrue. For example, recruitment in the VA Cooperative Study was only 91% of expected for overt GCSE and 50% for subtle GCSE, even though two recruitment years were added and poorly recording centers were removed from the study (234).

Treatment Selection

For most SE treatment trials, new therapies need to be given as adjunctive therapy to currently available treatments. Because SE has high morbidity and mortality if treatment is delayed, a placebo control group is not ethically acceptable.

Dosing Regimen Decisions

SE is often undertreated because of low AED doses (105). In a survey of neurologists regarding treatment of SE, undertreatment was common; 39% recommended giving a lorazepam dose less than 0.1 mg/kg and 21% a phenytoin/fosphenytoin dose less than 18 mg/kg. In the VA Cooperative Study (9), drug doses were slightly lower than those typically used for treatment of SE (phenobarbital 15 mg/kg, phenytoin 18 mg/kg, lorazepam 0.1 mg/kg, and diazepam 0.15 mg/kg), and serum concentrations were in the middle of the usual therapeutic range. In addition, the infusion times for the four drugs were different (mean 4.7 ± 7.2 min for lorazepam, 16.6 ± 11.5 min for phenobarbital, 33 ± 20.1 min for phenytoin, and 42 ± 38.1 min for diazepam plus phenytoin). Variability in infusion times was necessitated by the potential for adverse effects at higher infusion rates, particularly for phenytoin and phenobarbital. Drugs were prepackaged in a Tubex and five treatment vials per treatment box to allow blinded administration at a rate of 1 ml/min. A nomogram was used to determine the appropriate dose for each subject. If the patient failed the first treatment, subsequent treatments were given to control SE as rapidly as possible.

For new SE trials, doses should be calculated to give a high therapeutic serum level after intravenous loading. Escape criteria should be clearly defined for patients who fail initial therapy, and a plan for additional treatments should be in place. Because seizure recurrence after initial successful treatment can adversely affect outcome, maintenance doses should be planned to maintain high therapeutic levels. Each treatment should be administered as rapidly as possible, because early termination of SE improves outcome.

Drug Interactions

Many patients with SE have a history of epilepsy and are on other AEDs. Many AEDs are highly protein bound (phenytoin, valproate) and are strong inducers (phenobarbital, phenytoin, carbamazepine) or inhibitors (valproate) of the hepatic P450 enzyme system. Potential drug interactions should be explored for new SE agents before use in a randomized clinical trial.

Optimal Duration of Treatment

There is little information on the optimal duration of treatment for SE. Treatments to stop SE are usually given as intravenous loading doses, with maintenance doses given over 12 to 24 hours as dictated by the pharmacokinetics of the drug being used. The best regimen to prevent seizure recurrence has not been established (166) and likely differs depending on the clinical scenario. For some patients, such as those with epilepsy, the drug used to stop SE can often be discontinued, whereas others with acute medical disorders may require continued AEDs for days to weeks. Many

of the AEDs cause sedation and must be weaned for the patient to return to a normal mental status. Maintenance AEDs, if needed, are usually given at much lower doses than those required to stop SE. Most trials for SE have used regimens designed to stop SE and to prevent seizure recurrence during the next 12 to 24 hours.

There is also little information about the duration of treatment for potentially neuroprotective therapies. Some drugs may be effective in preventing cell death with one or several doses at the time of SE, as evidenced by animal trials. Many require pretreatment with the putative neuroprotective agent. For prevention of late sequelae such as cognitive dysfunction and epilepsy, there may be benefit from continuing treatment for weeks to months. Thus far, no antiepileptogenesis trials in humans have showed a decrease in the development of epilepsy after brain insults, but none has specifically targeted SE (126). The treatment duration should be explored in animal models to guide therapy in human trials.

OUTCOME MEASURES

Most SE studies have used early outcome measures such as efficacy in seizure cessation and early seizure recurrence rather than late outcomes such as cognitive outcome and epilepsy. This is based both on the feasibility and financial aspects of performing long-term randomized outcome assessments and on wide variability in outcomes after SE and lack of well-validated outcome measures. Late outcome measurements are confounded by multiple factors, including the duration of SE before treatment, type of SE, underlying etiology, and insensitivity of clinical measurement tools. Late outcome measures are essential, however, for assessment of potential neuroprotective or antiepileptogenic therapies. Both early outcome measures and suggestions for late outcome measures are discussed later.

The most appropriate outcome measure depends on several factors, most importantly the seizure duration before enrollment in the study. Possible outcomes for patients with seizures lasting less than 10 minutes include percentage of seizures stopping within a certain time interval after treatment, progression to SE, and death. For established SE (>30 minutes) and RSE, the percentage of seizures stopping within a certain time interval after treatment, death, and long-term sequelae (epilepsy, neurological deficits, and cognitive dysfunction) are appropriate primary and secondary endpoints.

Seizure Cessation

The first and most critical outcome measure is the cessation of ongoing seizure activity. For treatment trials, a time point for seizure control is usually set at which point the patient is considered to have failed treatment and moves on to the next therapy. Trials have used efficacy time points of 20 minutes (9) after the start of therapy or time of arrival to the emergency room after out-of-hospital treatment (235). Another possible endpoint could be the time interval from drug administration to seizure termination. The time point must be carefully chosen to take into account the time required for the study drug to be administered and to reach effective concentrations in the brain. This may have been one reason for the apparently poorer efficacy of phenytoin in the VA Cooperative Trial (9), because slow infusion requirements prevented full loading of phenytoin in some patients by the 20-minute outcome assessment time. The high morbidity and mortality of prolonged SE, however, justifies a short window for efficacy before the patient is considered a treatment failure and moves on to the next therapy.

Seizure cessation is ascertained clinically by the end of visible seizure manifestations and return to normal consciousness.

For GCSE, cessation of convulsive movements is usually a good indicator of the end of SE, but only 17% of patients with overt GCSE and no patients with subtle GSCE returned to normal consciousness within 12 hours in the VA Cooperative Study (9). Twenty to 48% of patients may continue to have subclinical electrographic seizures after clinical seizures are controlled and 14% have persistent NCSE (9,236). These patients are usually comatose and show no overt signs of seizure activity. Persistent electrographic seizures after cessation of clinical seizures predict poorer outcome, with 52% mortality in patients with persistent NCSE after GCSE versus 13% without ictal discharges after SE (236). Therefore, patients who do not rapidly return to normal consciousness after termination of seizures should have an EEG to verify seizure control, and EEG monitoring should be continued for at least 1 hour. For patients with RSE, cessation of all electrographic seizure activity or attainment of a burst-suppression EEG pattern (Fig. 5.3) may be an appropriate outcome. Need for EEG confirmation of seizure cessation adds to the complexity and cost of SE studies, and further research is needed to optimize methods of continuous EEG monitoring.

Absence of Seizure Recurrence

Most treatment studies include an analysis of seizure recurrence within the first hour after treatment and then a more prolonged assessment of efficacy such as seizure recurrence rates within 12 to 24 hours after successful treatment of SE. Again, seizure recurrence may need to be ascertained using EEG. In one retrospective study of midazolam for RSE, breakthrough seizures occurred in 56% of patients and were clinically subtle or electrographic only in 89% (108). It has not been established whether isolated recurrent seizures after control of SE impact long-term outcome.

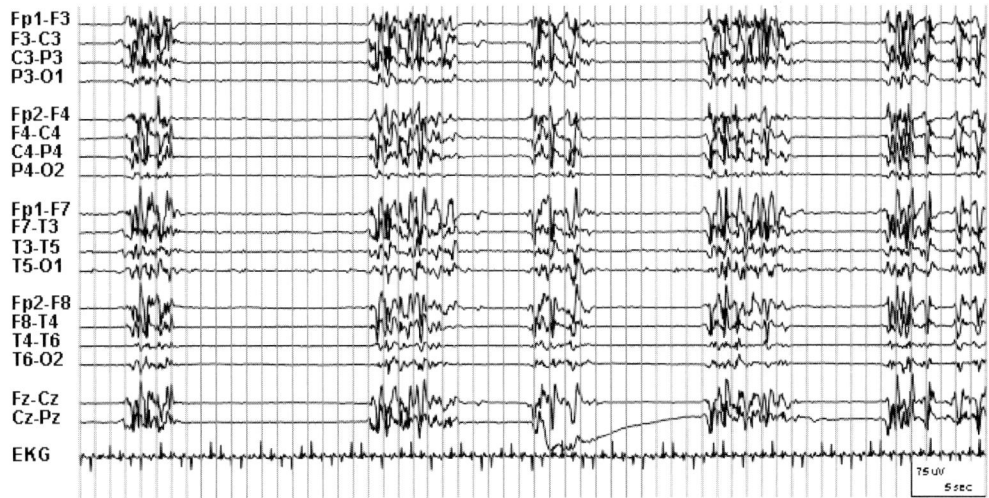

| Fp1-F3 |
| F3-C3 |
| C3-P3 |
| P3-O1 |
| Fp2-F4 |
| F4-C4 |
| C4-P4 |
| P4-O2 |
| Fp1-F7 |
| F7-T3 |
| T3-T5 |
| T5-O1 |
| Fp2-F8 |
| F8-T4 |
| T4-T6 |
| T6-O2 |
| Fz-Cz |
| Cz-Pz |
| EKG |

75 uV
5 sec

FIGURE 5.3   Burst-suppression EEG pattern in pentobarbital coma.

Mortality

Mortality of SE is 8–30% in adults and 3% in children. Most of the mortality is due to the underlying etiology, with only 2% attributable to SE itself (67). Therefore, 30-day mortality has been used as an outcome measure for SE. The primary determinants of mortality are duration of seizures (76), age (67), and etiology (73,76,159,237). Because most deaths are not directly a consequence of SE, effective treatments for SE may not directly impact mortality. On the other hand, neuroprotective treatments may lessen the overall neurological injury and improve outcomes, including mortality.

Hospital and Intensive Care Unit Length of Stay

Length of intensive care unit and hospital stay are often determined by the underlying etiology of SE and complications of therapy rather than by initial treatment efficacy. Patients with symptomatic SE usually have longer hospital courses than those with idiopathic SE. Use of AEDs that cause respiratory suppression or mental status depression may require prolonged mechanical ventilation for airway protection, whether or not seizures are fully controlled. One retrospective study showed no difference in mortality but increased duration of hospitalization in patients with RSE (99).

Functional Status at Discharge or 30 Days After SE

No prospective trials have used functional status as a primary or secondary outcome measure, but several retrospective and observational studies have tried to examine functional status. Approximately 10% of survivors of SE are left with disabling neurological deficits (76,79,238), using a variety of definitions. In a large hospital-based series of SE, poor outcome, defined as "severe neurological deficits requiring long-term hospitalization or full supportive care," occurred in 11% of 132 patients. Less severe functional deterioration, as measured by a decline in Glasgow Outcome Score, may occur in nearly 25% of SE survivors, especially those with acute symptomatic etiologies (159).

Because the etiologies of SE are so diverse and outcomes widely disparate, a broad measure of outcome is most appropriate for initial long-term outcome studies. Two such measures are the

Glasgow Outcome Scale (GOS) and Glasgow Expanded Outcome Scale (GEOS), which are the standard outcome measures in traumatic brain injury trials. These scales are well validated and can be administered quickly and reliably with a minimum of training. Validated telephone versions are available. The GOS is a five-point scale assessing the level of disability based on required help in daily living, whereas the GEOS adds two additional levels of disability. One advantage is that GOS can be easily compared with level of functioning before SE. GOS and GEOS are not good for detecting small differences in patients with relatively good outcome. Outcome can be measured at the time of discharge or at uniform time points such as 30 days or, preferably, 3, 6, or 12 months. Most patients reach their pre-SE state before discharge and follow-up after discharge is difficult, so measurement at time of discharge may be the easiest measure. Patients who do not recover by the time of discharge, however, may need to be followed for longer periods of time to determine whether persistent deficits exist. Late outcome assessments add significantly to the complexity and cost of randomized trials but are essential for assessment of putative neuroprotective agents.

Cognitive Outcome

Cognitive decline after SE can be evaluated by neuropsychometric testing (67,239), but there are several limitations to this approach. Many patients with SE have preexisting neurological disorders or epilepsy, with baseline cognitive deficits. In addition, SE is often caused by an acute brain insult such as trauma, stroke, or encephalitis, which may itself cause cognitive impairment. The effects of the acute brain injury are difficult to separate from those caused by SE. Most patients have not had prior neuropsychological testing. Subtle deficits are therefore very difficult to detect. In children, most SE occurs under the age of 3 years, and estimating premorbid intellectual function is difficult. At this point, detailed cognitive assessments are not well established as a late outcome measure for SE.

Older studies of SE revealed that SE was commonly associated with cognitive deficits, whereas newer studies show better cognitive outcomes. A retrospective study of cognitive outcome in 239 children with SE showed that 23% had mental retardation, which was attributed to SE itself (70). Another retrospective study revealed that 37% of children who were neurologically normal before SE developed deficits in mental ability after SE (240). Deficits are often associated with the underlying etiology; IQ scores were 25–30% lower in children with SE and brain lesions than for idiopathic SE (241). Idiopathic SE can cause personality and intellectual changes in 8–25% of patients (79,238), especially with longer durations of SE. A low morbidity in more recent studies was associated with idiopathic SE (68).

Only a few observational studies have used neuropsychological tests to examine cognitive outcome after SE. Dodrill and Wilensky (78) performed neuropsychological testing at 5-year interval in 143 patients with intractable epilepsy. Nine patients with intervening SE had lower intellectual ability and some specific neuropsychological deficits than matched control subjects with only brief seizures, but most of these were present on the baseline testing before SE. The authors concluded that SE has only a slight adverse effect on cognitive abilities in patients with epilepsy. The National Collaborative Perinatal Project examined 27 children with prolonged febrile convulsions and found no differences in cognitive function at age 7 years compared with their siblings (120).

Late Seizures and Epilepsy

Epilepsy is a clinically important outcome of SE, but measurement of epilepsy outcome is fraught with difficulties. First, many patients with SE have preexisting epilepsy. Forty-two percent of patients in the Richmond prospective study had seizures before their episode of SE (62). Only patients without a preceding history of epilepsy can be used to determine the risk of developing epilepsy. Second, the etiology of SE is often a brain insult, which in and of itself increases the rate of late epilepsy, such as stroke or traumatic brain injury. Assessment of the impact of SE itself, independent of etiology, may be difficult. Third, epilepsy is usually a late outcome. About 40% of patients with acute symptomatic SE develop a subsequent unprovoked seizure in the 10 years after SE (85). The risk of developing epilepsy by age 20 was 17% in children with complicated febrile seizures (242). Following patients for such prolonged periods in a prospective clinical trial is not feasible.

Finally, determining whether a single episode is a seizure or not may be very difficult. Trials must be blinded to avoid bias in deciding whether an unusual movement, fall, or episode of loss of consciousness was, in fact, a seizure. No prospective trials have used late seizures or epilepsy as an outcome measure after SE.

Other Measures and Analyses

Other potential outcomes measures, such as quality of life assessments, pharmacoeconomic analyses, and neuroimaging surrogate markers, have not been used in prospective SE treatment trials.

SURROGATE ENDPOINTS

There are no validated surrogate endpoints for the outcome of SE. Several possible biomarkers of injury and outcome may be useful in developing new treatments for SE, particularly in the development of neuroprotective strategies. Use

of these markers requires further study to determine optimal patient populations and assessment times and to validate association with clinical outcomes. Validation of biomarker use as surrogate markers requires a true neuroprotective strategy, so that a treatment effect on both the biomarker and the outcome can be assessed.

Neuroimaging Markers

Several neuroimaging changes have been reported after GCSE. Acute abnormalities in T2-weighted signal intensity, proton magnetic spectroscopy metabolite concentrations and ratios, and diffusion-weighted images are commonly seen after SE. In experimental animals, T2 signal abnormalities are first seen 12 to 24 hours after SE (243). Increases in lactate on magnetic resonance spectroscopy correlate to the degree of histological damage (243,244). Diffusion imaging early after SE shows a decrease in apparent diffusion coefficient, possibly reflecting underlying neuronal swelling (243,245). In children with prolonged febrile convulsions, hippocampal volumes are acutely increased and T2 relaxation times prolonged (246). Progressive hippocampal atrophy and hippocampal sclerosis can occur after SE, sometimes progressing to radiological mesial temporal sclerosis (146,247–249). Similar changes occur 4 to 8 months after febrile SE (250). Whether this is associated with the later development of epilepsy or neurological outcome has not been established.

Neuroimaging markers may aid in the prediction of late outcomes. The progression of damage in animal models can easily be followed using serial neuroimaging. Progressive structural alterations (prolonged T2, prolonged T1 rho, and elevated diffusivity) were seen in the amygdala and hippocampus after SE induced by electrical stimulation of the amygdala (251). Similarly, acute changes might predict the development of epilepsy after SE. Increases in T2 relaxation time in piriform and entorhinal cortex predicted the development of chronic epilepsy and 21-day-old rats (252). Similar progression has not yet been demonstrated after SE in humans.

Potential problems with use of neuro-imaging markers include the possibility that preexisting structural abnormalities are predisposed to SE rather than being a consequence of SE itself (124,253). In addition, the optimal time points for neuroimaging assessments have not been defined. Structural changes appear to be less common in adults than children (254).

Cerebrospinal Fluid and Serum Markers

Neuron-specific enolase (NSE), an enzyme involved in glycolytic energy metabolism in the brain, is released from neurons during injury and is often elevated after GCSE (255–258). NSE has therefore been proposed as a potential biomarker of neuronal injury in SE. These elevations are not specific to patients with SE, however. One study of patients with cryptogenic or remote symptomatic SE showed a marked increase in cerebrospinal fluid NSE in 9 of 11 patients compared with control subjects. Cerebrospinal fluid NSE was a better marker than serum NSE (258). NSE levels were highest in patients with subclinical SE, possibly related to the underlying etiology. Elevations of NSE in patients without obvious acute neurological deficits suggest that SE may be causing permanent changes in the human brain that are not detectable by our current outcome measures.

Using current methods, assessment of a single cerebrospinal fluid sample for NSE is of limited utility. Advances in biochemistry may lead to neuroimaging methods that can be used to noninvasively to monitor patients after SE.

SAFETY EVALUATION

Medical Complications of SE

Common physiological changes during SE (see Mechanisms of Disease and Pathology and Table 5.3) include tachycardia, bradycardia, other cardiac arrhythmias, hypotension, hyperthermia, respiratory failure, hypoxia, hypercapnia, hypoglycemia, metabolic acidosis, and other electrolyte disturbances. These changes become much more frequent and severe in RSE. Because SE may require prolonged intensive care, patients are at increased risk for pulmonary infection, urinary tract infection, cellulitis, renal failure, hepatic failure, gastritis, pulmonary edema, and neuropathy.

Many similar complications can be caused by drugs currently used for treatment of SE. The most common complications are mental status depression, respiratory suppression, hypotension, and cardiac arrhythmias. Respiratory suppression requiring mechanical ventilation occurred in 18.9% and pressor support was needed in 32.6% of patients treated for SE in one study (9).

Although SE commonly has multiple potential medical complications, similar changes can be caused by drug administration. It is therefore difficult to determine whether a specific adverse event is caused by SE itself or by treatment, and all adverse events likely need to be treated as possible drug adverse events.

CONCLUSION

SE is an acute medical and neurological emergency with high morbidity and mortality. Various subtypes of SE differ in presentation, populations affected, and treatment response. Rapid response to therapy is associated with better outcome, but 60% of patients fail to respond to current standard therapies. More effective acute treatments are needed and must be tested head-to-head against currently available therapies. For patients with prolonged SE, neuroprotective and anti-epileptogenic drugs must be developed to prevent late cognitive effects, neurological deficits, and epilepsy.

References

1. Theodore WH, et al. The secondarily generalized tonic-clonic seizure: a videotape analysis. Neurology 1994;44:1403–1407.

2. Treatment of convulsive status epilepticus. Recommendations of the Epilepsy Foundation of America's Working Group on Status Epilepticus. JAMA 1993;270:854–859.

3. Cereghino JJ, Cloyd JC, Kuzniecky RI. Rectal diazepam gel for treatment of acute repetitive seizures in adults. Arch Neurol 2002;59:1915–1920.

4. Haut SR, et al. The association between seizure clustering and convulsive status epilepticus in patients with intractable complex partial seizures. Epilepsia 1999;40:1832–1834.

5. Knudsen FU. Febrile seizures: treatment and prognosis. Epilepsia 2000;41:2–9.

6. Annegers JF, et al. Factors prognostic of unprovoked seizures after febrile convulsions. N Engl J Med 1987;316:493–498.

7. Painter MJ, Alvin J. Neonatal seizures. Curr Treat Options Neurol 2001;3:237–248.

8. Alldredge BK, et al. A comparison of lorazepam, diazepam, and placebo for the treatment of out-of-hospital status epilepticus. N Engl J Med 2001;345:631–637.

9. Treiman DM, et al. A comparison of four treatments for generalized convulsive status epilepticus. Veterans Affairs Status Epilepticus Cooperative Study Group. N Engl J Med 1998;339:792–798.

10. Hauser WA. Status epilepticus: epidemiologic considerations. Neurology 1990;40(5 suppl 2): 9–13.

11. Lothman E. The biochemical basis and pathophysiology of status epilepticus. Neurology 1990;40 (5 suppl 2):13–23.

12. Proposal for revised clinical and electroencephalographic classification of epileptic seizures. From the Commission on Classification and Terminology of the International League Against Epilepsy. Epilepsia 1981;22:489–501.

13. Alldredge BK, Lowenstein DH. Status epilepticus: new concepts. Curr Opin Neurol 1999;12:183–190.

14. Lowenstein DH, Bleck T, Macdonald RL. It's time to revise the definition of status epilepticus. Epilepsia 1999;40:120–122.

15. DeLorenzo RJ, et al. Comparison of status epilepticus with prolonged seizure episodes lasting from 10 to 29 minutes. Epilepsia 1999;40:164–169.

16. Shinnar S, et al. How long do new-onset seizures in children last? Ann Neurol 2001; 49:659–664.

17. Holmes GL. Seizure-induced neuronal injury: animal data. Neurology 2002;59(suppl 5): S3–S6.

18. Treiman DM. Generalized convulsive status epilepticus in the adult. Epilepsia 1993;34 (suppl 1):S2–S11.

19. Kaplan PW. Behavioral manifestations of non-convulsive status epilepticus. Epilepsy Behav 2002;3:122–139.

20. Sarkisian MR. Overview of the current animal models for human seizure and epileptic disorders. Epilepsy Behav 2001;2:201–216.

21. Meldrum BS, Horton RW. Physiology of status epilepticus in primates. Arch Neurol 1973;28:1–9.

22. Tomlinson FH, Anderson RE, Meyer FB. Effect of arterial blood pressure and serum glucose on brain intracellular pH, cerebral and cortical blood flow during status epilepticus in the white New Zealand rabbit. Epilepsy Res 1993;14: 123–137.

23. Lothman EW, Collins RC. Kainic acid induced limbic seizures: metabolic, behavioral, electroencephalographic and neuropathological correlates. Brain Res 1981;218:299–318.

24. Covolan L, Mello LE. Temporal profile of neuronal injury following pilocarpine or kainic acid-induced status epilepticus. Epilepsy Res 2000;39:133–152.

25. Leite JP, Garcia-Cairasco N, Cavalheiro EA. New insights from the use of pilocarpine and kainate models. Epilepsy Res 2002;50:93–103.

26. Ben-Ari Y, et al. A new model of focal status epilepticus: intra-amygdaloid application of kainic acid elicits repetitive secondarily generalized convulsive seizures. Brain Res 1979; 163:176–179.

27. Lockard JS, et al. A monkey model for status epilepticus: carbamazepine and valproate compared to three standard anticonvulsants. Adv Neurol 1983;34:411–419.

28. Walton NY, Treiman DM. Experimental secondarily generalized convulsive status epilepticus induced by D,L-homocysteine thiolactone. Epilepsy Res 1988;2:79–86.

29. Treiman DM, Walton NY. Experimental models of status epilepticus. In: Ehgel, J, Pedley TA, eds. Epilepsy: A Comprehensive Textbook. Philadelphia: Lippencott-Raven, 1998:443–455.

30. Treiman DM, Walton NY, Kendrick C. A progressive sequence of electroencephalographic changes during generalized convulsive status epilepticus. Epilepsy Res 1990;5:49–60.

31. Walton NY, Treiman DM. Rational polytherapy in the treatment of status epilepticus. Epilepsy Res 1996;11(suppl):123–139.

32. Taber KH, McNamera JJ, Zornetzer SF. Status epilepticus: a new rodent model. Electroencephalogr Clin Neurophysiol 1977;43:707–724.

33. Brandt C, et al. Epileptogenesis and neuropathology after different types of status epilepticus induced by prolonged electrical stimulation of the basolateral amygdala in rats. Epilepsy Res 2003;55:83–103.

34. Toth Z, et al. Seizure-induced neuronal injury: vulnerability to febrile seizures in an immature rat model. J Neurosci 1998;18:4285–4294.

35. Dube C, et al. Prolonged febrile seizures in the immature rat model enhance hippocampal excitability long term. Ann Neurol 2000; 47:336–344.

36. Chen K, Baram TZ, Soltesz I. Febrile seizures in the developing brain result in persistent modification of neuronal excitability in limbic circuits. Nat Med 1999;5:888–894.

37. Jensen FE. An animal model of hypoxia-induced perinatal seizures. Ital J Neurol Sci 1995;16:59–68.

38. Jensen FE, et al. Acute and chronic increases in excitability in rat hippocampal slices after perinatal hypoxia in vivo. J Neurophysiol 1998; 79:73–81.

39. Meldrum, BS, Brierley JB. Prolonged epileptic seizures in primates. Ischemic cell change and its relation to ictal physiological events. Arch Neurol 1973;28:10–17.

40. Meldrum BS, Vigouroux RA, Brierley JB. Systemic factors and epileptic brain damage. Arch Neurol 1973;29:82–87.

41. Ben-Ari Y. Limbic seizure and brain damage produced by kainic acid: mechanisms and relevance to human temporal lobe epilepsy. Neuroscience 1985;14:375–403.

42. Cavalheiro EA. The pilocarpine model of epilepsy. Ital J Neurol Sci 1995;16:33–37.

43. Fountain NB, Lothman EW. Pathophysiology of status epilepticus. J Clin Neurophysiol 1995; 12:326–342.

44. Fujikawa DG. The temporal evolution of neuronal damage from pilocarpine-induced status epilepticus. Brain Res 1996;725:11–22.

45. Lemos T, Cavalheiro EA. Suppression of pilocarpine-induced status epilepticus and the late development of epilepsy in rats. Exp Brain Res 1995;102:423–428.

46. Ebert U, Brandt C, Loscher W. Delayed sclerosis, neuroprotection, and limbic epileptogenesis after status epilepticus in the rat. Epilepsia 2002;43(suppl 5):86–95.

47. Kapur J, Macdonald RL. Rapid seizure-induced reduction of benzodiazepine and Zn2+ sensitivity of hippocampal dentate granule cell GABAA receptors. J Neurosci 1997;17: 7532–7540.

48. Walton NY, Treiman DM. Motor and electroencephalographic response of refractory experimental status epilepticus in rats to treatment with MK-801, diazepam, or MK-801 plus diazepam. Brain Res 1991;553:97–104.

49. Borris DJ, Bertram EH, Kapur J. Ketamine controls prolonged status epilepticus. Epilepsy Res 2000;42:117–122.

50. Brooks-Kayal AR, et al. Selective changes in single cell GABA(A) receptor subunit expression and function in temporal lobe epilepsy. Nat Med 1998;4:1166–1172.

51. Prasad A, Williamson JM, Bertram EH. Phenobarbital and MK-801, but not phenytoin, improve the long-term outcome of status epilepticus. Ann Neurol 2002;51:175–181.

52. Berger ML, et al. Maturation of kainic acid seizure-brain damage syndrome in the rat. III. Postnatal development of kainic acid binding sites in the limbic system. Neuroscience 1984; 13:1095–1104.

53. Sankar R et al. Epileptogenesis after status epilepticus reflects age- and model-dependent plasticity. Ann Neurol 2000;48:580–589.

54. Liu Z, et al. Age-dependent effects of glutamate toxicity in the hippocampus. Brain Res Dev Brain Res 1996;97:178–184.

55. Holmes GL, et al. Consequences of neonatal seizures in the rat: morphological and behavioral effects. Ann Neurol 1998;44:845–857.

56. Loscher W. Animal models of epilepsy for the development of antiepileptogenic and disease-modifying drugs. A comparison of the pharmacology of kindling and post-status epilepticus models of temporal lobe epilepsy. Epilepsy Res 2002;50:105–123.

57. Bolanos AR, et al. Comparison of valproate and phenobarbital treatment after status epilepticus in rats. Neurology 1998;51:41–48.

58. Klitgaard H, et al. Electrophysiological, neurochemical and regional effects of levetiracetam in the rat pilocarpine model of temporal lobe epilepsy. Seizure 2003;12:92–100.

59. Nissinen JPT, et al. Diazepam treatment has a disease-modifying effect on the developing epilepsy after status epilepticus in rat. Epilepsia 2001;42(suppl):131.

60. STAIR. Recommendations for standards regarding preclincal neuroprotective and restorative drug development. Stroke 1999;30:2752–2758.

61. Narayan RK, et al. Clinical trials in head injury. J Neurotrauma 2002;19:503–557.

62. DeLorenzo R, et al. A prospective, population-based epidemiologic study of status epilepticus in Richmond, Virginia. Neurology 1996; 46:1029–1035.

63. Hesdorffer D, et al. Incidence of status epilepticus in Rochester, Minnesota 1965–1984. Neurology 1998;50:735–741.

64. Coeytaux A, et al. Incidence of status epilepticus in French-speaking Switzerland: (EPISTAR). Neurology 2000;55:693–697.

65. Knake S, et al. Incidence of status epilepticus in adults in Germany: a prospective, population-based study. Epilepsia 2001;42:714–718.

66. Cascino GD, et al. Morbidity of nonfebrile status epilepticus in Rochester, Minnesota 1965–1984. Epilepsia 1998;39:829–832.

67. Towne AR, et al. Determinants of mortality in status epilepticus. Epilepsia 1994;35:27–34.

68. Maytal J, et al. Low morbidity and mortality of status epilepticus in children. Pediatrics 1989;83:323–331.

69. Lacroix J, et al. Admissions to a pediatric intensive care unit for status epilepticus: a 10-year experience. Crit Care Med 1994;22:827–832.

70. Aicardi J, Chevrie JJ. Convulsive status epilepticus in infants and children. A study of 239 cases. Epilepsia 1970;11:187–197.

71. Metsaranta P, et al. Outcome after prolonged convulsive seizures in 186 children: low morbidity, no mortality. Dev Med Child Neurol 2004;46:4–8.

72. Logroscino G, et al. Long-term mortality after a first episode of status epilepticus. Neurology 2002;58:537–541.

73. Logroscino G, et al. Short-term mortality after a first episode of status epilepticus. Epilepsia 1997;38:1344–1349.

74. Waterhouse EJ, et al. Synergistic effect of status epilepticus and ischemic brain injury on mortality. Epilepsy Res 1998;29:175–183.

75. Shneker BF, Fountain NB. Assessment of acute morbidity and mortality in nonconvulsive status epilepticus. Neurology 2003;61:1066–1073.

76. Lowenstein D, Alldredge B. Status epilepticus at an urban public hospital in the 1980s. Neurology 1993;43:483–488.

77. Waterhouse EJ, et al. Prospective population-based study of intermittent and continuous convulsive status epilepticus in Richmond, Virginia. Epilepsia 1999;40:752–758.

78. Dodrill CB, Wilensky AJ. Intellectual impairment as an outcome of status epilepticus. Neurology 1990;40(5 suppl 2):23–27.

79. Aminoff MJ, Simon RP. Status epilepticus. Causes, clinical features and consequences in 98 patients. Am J Med 1980;69:657–666.

80. Eriksson KJ, Koivikko MJ. Status epilepticus in children: aetiology, treatment, and outcome. Dev Med Child Neurol 1997;39:652–658.

81. Barnard C, Wirrell E. Does status epilepticus in children cause developmental deterioration and exacerbation of epilepsy? J Child Neurol 1999;14:787–794.

82. Dunn DW. Status epilepticus in children: etiology, clinical features, and outcome. J Child Neurol 1988;3:167–173.

83. Lothman EW, Bertram EH 3rd. Epileptogenic effects of status epilepticus. Epilepsia 1993;34(suppl 1):S59–S70.

84. Awaya Y, Iwamoto H, Fukuyama Y. A long-term follow-up study of first episodes of idiopathic status convulsivus in childhood: in relation to subsequent epilepsy (second report). Jpn J Psychiatry Neurol 1992;46:303–306.

85. Hesdorffer DC, et al. Risk of unprovoked seizure after acute symptomatic seizure: effect of status epilepticus. Ann Neurol 1998;44:908–912.

86. Treiman DM. Electroclinical features of status epilepticus. J Clin Neurophysiol 1995;12:343–362.

87. Wu YW, et al. Incidence and mortality of generalized convulsive status epilepticus in California. Neurology 2002;58:1070–1076.

88. Sagduyu A, Tarlaci S, Sirin H. Generalized tonic-clonic status epilepticus: causes, treatment, complications and predictors of case fatality. J Neurol 1998;245:640–646.

89. Kaplan PW. Nonconvulsive status epilepticus. Semin Neurol 1996;16:33–40.

90. Towne AR, et al. Prevalence of nonconvulsive status epilepticus in comatose patients. Neurology 2000;54:340.

91. Kaplan PW. Assessing the outcomes in patients with nonconvulsive status epilepticus: nonconvulsive status epilepticus is underdiagnosed, potentially overtreated, and confounded by comorbidity. J Clin Neurophysiol 1999;16:341–352; discussion 353.

92. Krumholz A. Epidemiology and evidence for morbidity of nonconvulsive status epilepticus. J Clin Neurophysiol 1999;16:314–322; discussion 353.

93. Kaplan PW. No, some types of nonconvulsive status epilepticus cause little permanent neurologic sequelae (or: "the cure may be worse than the disease"). Neurophysiol Clin 2000;30:377–382.

94. Drislane FW. Evidence against permanent neurologic damage from nonconvulsive status epilepticus. J Clin Neurophysiol 1999;16:323–331; discussion 353.

95. Scholtes FB, Renier WO, Meinardi H. Nonconvulsive status epilepticus: causes, treatment, and outcome in 65 patients. J Neurol Neurosurg Psychiatry 1996;61:93–95.

96. Shirasaka Y. Lack of neuronal damage in atypical absence status epilepticus. Epilepsia 2002;43:1498–1501.

97. Wong M, Wozniak DF, Yamada KA. An animal model of generalized nonconvulsive status epilepticus: immediate characteristics and long-term effects. Exp Neurol 2003;183:87–99.

98. Walton NY, Treiman DM. Response of status epilepticus induced by lithium and pilocarpine to treatment with diazepam. Exp Neurol 1988;101:267–275.

99. Mayer SA, et al. Refractory status epilepticus: frequency, risk factors, and impact on outcome. Arch Neurol 2002;59:205–210.

100. Claassen J, et al. Treatment of refractory status epilepticus with pentobarbital, propofol, or midazolam: a systematic review. Epilepsia 2002;43:146–153.

101. Bleck TP. Advances in the management of refractory status epilepticus. Crit Care Med 1993;21:955–957.

102. Jagoda A, Riggio S. Refractory status epilepticus in adults. Ann Emerg Med 1993;22:1337–1348.

103. Stecker MM, et al. Treatment of refractory status epilepticus with propofol: clinical and pharmacokinetic findings. Epilepsia 1998;39:18–26.

104. Prasad A, et al. Propofol and midazolam in the treatment of refractory status epilepticus. Epilepsia 2001;42:380–386.

105. Lowenstein DH, Alldredge BK. Status epilepticus. N Engl J Med 1998;338:970–976.

106. Gilbert DL, Gartside PS, Glauser TA. Efficacy and mortality in treatment of refractory generalized convulsive status epilepticus in children: a meta-analysis. J Child Neurol 1999;14:602–609.

107. Hanley DF, Kross JF. Use of midazolam in the treatment of refractory status epilepticus. Clin Ther 1998;20:1093–1105.

108. Claassen J, et al. Continuous EEG monitoring and midazolam infusion for refractory non-convulsive status epilepticus. Neurology 2001;57:1036–1042.

109. Sahin M, et al. Outcome of severe refractory status epilepticus in children. Epilepsia 2001;42:1461–1467.

110. Mitchell WG. Status epilepticus and acute repetitive seizures in children, adolescents, and young adults: etiology, outcome, and treatment. Epilepsia 1996;37(suppl 1):S74–S80.

111. Practice parameter: the neurodiagnostic evaluation of the child with a first simple febrile seizure. American Academy of Pediatrics. Provisional Committee on Quality Improvement, Subcommittee on Febrile Seizures. Pediatrics 1996;97:769–772; discussion 773–775.

112. Baumann RJ, Duffner PK. Treatment of children with simple febrile seizures: the AAP practice parameter. American Academy of Pediatrics. Pediatr Neurol 2000;23:11–17.

113. Shinnar S, Glauser TA. Febrile seizures. J Child Neurol 2002;17(suppl 1):S44–S52.

114. Verity CM, Golding J. Risk of epilepsy after febrile convulsions: a national cohort study. BMJ 1991;303:1373–1376.

115. Nelson KB, Ellenberg JH. Prognosis in children with febrile seizures. Pediatrics 1978;61:720–727.

116. Shinnar S, et al. Short-term outcomes of children with febrile status epilepticus. Epilepsia 2001;42:47–53.

117. Knudsen FU. Recurrence risk after first febrile seizure and effect of short term diazepam prophylaxis. Arch Dis Child 1985;60:1045–1049.

118. Offringa M, et al. Risk factors for seizure recurrence in children with febrile seizures: a pooled analysis of individual patient data from five studies. J Pediatr 1994;124:574–584.

119. Berg AT, et al. Predictors of recurrent febrile seizures: a metaanalytic review. J Pediatr 1990;116:329–337.

120. Ellenberg JH, Nelson KB. Febrile seizures and later intellectual performance. Arch Neurol 1978;35:17–21.

121. Nelson KB, Ellenberg JH. Predictors of epilepsy in children who have experienced febrile seizures. N Engl J Med 1976;295:1029–1033.

122. Verity CM, Ross EM, Golding J. Outcome of childhood status epilepticus and lengthy febrile convulsions: findings of national cohort study. BMJ 1993;307:225–228.

123. Kuks JB, et al. Hippocampal sclerosis in epilepsy and childhood febrile seizures. Lancet 1993;342:1391–1394.

124. Lewis DV, et al. Do prolonged febrile seizures produce medial temporal sclerosis? Hypotheses, MRI evidence and unanswered questions. Prog Brain Res 2002;135:263–278.

125. Wallace SJ, Smith JA. Prophylaxis against febrile convulsions. Br Med J 1980;280:863–864.

126. Temkin NR. Antiepileptogenesis and seizure prevention trials with antiepileptic drugs: meta-analysis of controlled trials. Epilepsia 2001;42:515–524.

127. Ronen GM, Penney S, Andrews W. The epidemiology of clinical neonatal seizures in Newfoundland: a population-based study. J Pediatr 1999;134:71–75.

128. Lanska MJ, et al. A population-based study of neonatal seizures in Fayette County, Kentucky. Neurology 1995;45:724–732.

129. Clancy RR. The contribution of EEG to the understanding of neonatal seizures. Epilepsia 1996;37(suppl 1):S52–S59.

130. Scher MS, et al. Electrographic seizures in preterm and full-term neonates: clinical correlates, associated brain lesions, and risk for neurologic sequelae. Pediatrics 1993;91:128–134.

131. Bergman I, et al. Outcome in neonates with convulsions treated in an intensive care unit. Ann Neurol 1983;14:642–647.

132. Scher MS. Controversies regarding neonatal seizure recognition. Epileptic Disord 2002;4:139–158.

133. Mizrahi EM, Kellaway P. Characterization and classification of neonatal seizures. Neurology 1987;37:1837–1844.

134. Mizrahi EM, Plouin P, Kellaway P. Neonatal seizures. In: Engel J Jr, Pedley TA, eds. Epilepsy: A Comprehensive Textbook. Philadelphia: Lippencott-Raven, 1998:647–663.

135. Lombroso CT. Prognosis in neonatal seizures. Adv Neurol 1983;34:101–113.

136. Legido A, Clancy RR, Berman PH. Neurologic outcome after electroencephalographically proven neonatal seizures. Pediatrics 1991; 88:583–596.

137. Clancy RR, Legido A. Postnatal epilepsy after EEG-confirmed neonatal seizures. Epilepsia 1991;32:69–76.

138. Legido A, Clancy RR, Berman PH. Recent advances in the diagnosis, treatment, and prognosis of neonatal seizures. Pediatr Neurol 1988;4:79–86.

139. Holmes GL, Khazipov R, Ben-Ari Y. New concepts in neonatal seizures. Neuroreport 2002;13:A3–A8.

140. Teitelbaum JS, et al. Neurologic sequelae of domoic acid intoxication due to the ingestion of contaminated mussels. N Engl J Med 1990;322:1781–1787.

141. Macdonald RL, Kapur J. Acute cellular alterations in the hippocampus after status epilepticus. Epilepsia 1999;40(suppl 1):S9–S20; discussion S21–S2.

142. Lipton SA, Rosenberg PA. Excitatory amino acids as a final common pathway for neurologic disorders. N Engl J Med 1994;330:613–622.

143. Meldrum B. Excitotoxicity and epileptic brain damage. Epilepsy Res 1991;10:55–61.

144. Shorvon S. Does convulsive status epilepticus (SE) result in cerebral damage or affect the course of epilepsy—the epidemiological and clinical evidence? Prog Brain Res 2002;135:85–93.

145. Blennow G, et al. Epileptic brain damage: the role of systemic factors that modify cerebral energy metabolism. Brain 1978;101:687–700.

146. Corsellis JA, Bruton CJ. Neuropathology of status epilepticus in humans. Adv Neurol 1983; 34:129–139.

147. DeGiorgio CM, et al. Hippocampal pyramidal cell loss in human status epilepticus. Epilepsia 1992;33:23–27.

148. Sloviter RS. Status epilepticus-induced neuronal injury and network reorganization. Epilepsia 1999;40(suppl 1):S34–S39; discussion S40–S41.

149. Soderfeldt B, et al. Influence of systemic factors on experimental epileptic brain injury. Structural changes accompanying bicuculline-induced seizures in rats following manipulations of tissue oxygenation or alpha-tocopherol levels. Acta Neuropathol (Berl) 1983;60:81–91.

150. Nevander G, et al. Status epilepticus in well-oxygenated rats causes neuronal necrosis. Ann Neurol 1985;18:281–290.

151. Wasterlain CG, et al. Pathophysiological mechanisms of brain damage from status epilepticus. Epilepsia 1993;34(suppl 1):S37–S53.

152. Ruegg SJ, Dichter MA. Diagnosis and treatment of nonconvulsive status epilepticus in an intensive care unit setting. Curr Treat Options Neurol 2003;5:93–110.

153. Kaplan PW. Prognosis in nonconvulsive status epilepticus. Epileptic Disord 2000;2:185–193.

154. Hauser WA, Kurland LT. The epidemiology of epilepsy in Rochester, Minnesota 1935 through 1967. Epilepsia 1975;16:1–66.

155. Delanty N, et al. Status epilepticus arising de novo in hospitalized patients: an analysis of 41 patients. Seizure 2001;10:116–119.

156. Dulac O, et al. Status epilepticus in the infant. Semeiologic, etiologic and prognostic aspects. Rev Electroencephalogr Neurophysiol Clin 1985; 14:255–262.

157. Yager JY, Cheang M, Seshia SS. Status epilepticus in children. Can J Neurol Sci 1988;15:402–405.

158. Phillips SA, Shanahan RJ. Etiology and mortality of status epilepticus in children. A recent update. Arch Neurol 1989;46:74–76.

159. Claassen J, et al. Predictors of functional disability and mortality after status epilepticus. Neurology 2002;58:139–142.

160. Hirsch LJ, Claassen J. The current state of treatment of status epilepticus. Curr Neurol Neurosci Rep 2002;2:345–356.

161. Treiman DM, et al. Lorazepam versus phenytoin in the treatment of major motor status epilepticus: a preliminary report. Epilepsia 1983; 24:520.

162. Shaner DM, et al. Treatment of status epilepticus: a prospective comparison of diazepam and phenytoin versus phenobarbital and optional phenytoin. Neurology 1988;38:202–207.

163. Leppik IE, et al. Double-blind study of lorazepam and diazepam in status epilepticus. JAMA 1983;249:1452–1454.

164. Cock HR, Schapira AH. A comparison of lorazepam and diazepam as initial therapy in convulsive status epilepticus. Q J Med 2002;95:225–231.

165. Bleck TP. Management approaches to prolonged seizures and status epilepticus. Epilepsia 1999;40(suppl 1):S59–S63; discussion S64–S66.

166. Claassen J, Hirsch LJ, Mayer SA. Treatment of status epilepticus: a survey of neurologists. J Neurol Sci 2003;211:37–41.

167. Bleck TP. Refractory status epilepticus in 2001. Arch Neurol 2002;59:188–189.

168. Karceski S, Morrell MJ. The expert consensus guideline series: Treatment of epilepsy. Epilepsy Behav 2001;2:A1–A50.

169. Sinha S, Naritoku DK. Intravenous valproate is well tolerated in unstable patients with status epilepticus. Neurology 2000;55:722–724.

170. Kaplan PW. Intravenous valproate treatment of generalized nonconvulsive status epilepticus. Clin Electroencephalogr 1999;30:1–4.

171. Hodges BM, Mazur JE. Intravenous valproate in status epilepticus. Ann Pharmacother 2001; 35:1465–1470.

172. Hovinga CA, et al. Use of intravenous valproate in three pediatric patients with nonconvulsive or convulsive status epilepticus. Ann Pharmacother 1999;33:579–584.

173. Uberall MA, et al. Intravenous valproate in pediatric epilepsy patients with refractory status epilepticus. Neurology 2000;54:2188–2189.

174. Yu KT, et al. Safety and efficacy of intravenous valproate in pediatric status epilepticus and acute repetitive seizures. Epilepsia 2003; 44:724–726.

175. Walker MC. Diagnosis and treatment of nonconvulsive status epilepticus. CNS Drugs 2001;15:931–939.

176. Vespa PM, et al. Increased incidence and impact of nonconvulsive and convulsive seizures after traumatic brain injury as detected by continuous electroencephalographic monitoring. J Neurosurg 1999;91:750–760.

177. Litt B, et al. Nonconvulsive status epilepticus in the critically ill elderly. Epilepsia 1998;39: 1194–1202.

178. Krishnamurthy KB, Drislane FW. Relapse and survival after barbiturate anesthetic treatment of refractory status epilepticus. Epilepsia 1996;37:863–867.

179. Lowenstein DH, Aminoff MJ, Simon RP. Barbiturate anesthesia in the treatment of status epilepticus: clinical experience with 14 patients. Neurology 1988;38:395–400.

180. Rashkin MC, Youngs C, Penovich P. Pentobarbital treatment of refractory status epilepticus. Neurology 1987;37:500–503.

181. Yaffe K, Lowenstein D. Prognostic factors of pentobarbital therapy for refractory generalized status epilepticus. Neurology 1993;43:895–900.

182. Niermeijer JM, Uiterwaal CS, Van Donselaar CA. Propofol in status epilepticus: little evidence, many dangers? J Neurol 2003;250:1237–1240.

183. Walker MC, Smith SJ, Shorvon SD. The intensive care treatment of convulsive status epilepticus in the UK. Results of a national survey and recommendations. Anaesthesia 1995;50:130–135.

184. Krishnamurthy KB, Drislane FW. Depth of EEG suppression and outcome in barbiturate anesthetic treatment for refractory status epilepticus. Epilepsia 1999;40:759–762.

185. Kofke WA, et al. Electrographic tachyphylaxis to etomidate and ketamine used for refractory status epilepticus controlled with isoflurane. J Neurosurg Anesthesiol 1997;9:269–272.

186. Sharpe MD, et al. Prolonged desflurane administration for refractory status epilepticus. Anesthesiology 2002;97:261–264.

187. Appleton R, Martland T, Phillips B. Drug management for acute tonic-clonic convulsions including convulsive status epilepticus in children. Cochrane Database Syst Rev, 2002:CD001905.

188. De Giorgio CM, et al. Lidocaine in refractory status epilepticus: confirmation of efficacy with continuous EEG monitoring. Epilepsia 1992;33: 913–916.

189. Sheth RD, Gidal BE. Refractory status epilepticus: response to ketamine. Neurology 1998;51: 1765–1766.

190. Bertram EH, Lothman EW. NMDA receptor antagonists and limbic status epilepticus: a comparison with standard anticonvulsants. Epilepsy Res 1990;5:177–184.

191. Fujikawa DG. Neuroprotective effect of ketamine administered after status epilepticus onset. Epilepsia 1995;36:186–195.

192. Stewart LS, Persinger MA. Ketamine prevents learning impairment when administered immediately after status epilepticus onset. Epilepsy Behav 2001;2:585–591.

193. Niebauer M, Gruenthal M. Topiramate reduces neuronal injury after experimental status epilepticus. Brain Res 1999;837:263–269.

194. Towne AR, et al. The use of topiramate in refractory status epilepticus. Neurology 2003; 60:332–334.

195. Reuber M, Evans J, Bamford JM. Topiramate in drug-resistant complex partial status epilepticus. Eur J Neurol 2002;9:111–112.

196. Kahriman M, et al. Efficacy of topiramate in children with refractory status epilepticus. Epilepsia 2003;44:1353–1356.

197. Appleton R, et al. The treatment of convulsive status epilepticus in children. The Status Epilepticus Working Party, Members of the Status Epilepticus Working Party. Arch Dis Child 2000;83:415–419.

198. Brown LA, Levin GM. Role of propofol in refractory status epilepticus. Ann Pharmacother 1998;32:1053–1059.

199. Wooltorton E. Propofol: contraindicated for sedation of pediatric intensive care patients. CMAJ 2002;167:507.

200. Cereghino JJ, et al. Treating repetitive seizures with a rectal diazepam formulation: a randomized study. The North American Diastat Study Group. Neurology 1998;51:1274–1282.

201. Dreifuss FE, et al. A comparison of rectal diazepam gel and placebo for acute repetitive seizures. N Engl J Med 1998;338:1869–1875.

202. Camfield PR. Buccal midazolam and rectal diazepam for treatment of prolonged seizures in childhood and adolescence: a randomised trial. J Pediatr 1999;135:398–399.

203. Scott RC, Besag FM, Neville BG. Buccal midazolam and rectal diazepam for treatment of prolonged seizures in childhood and adolescence: a randomised trial. Lancet 1999; 353:623–626.

204. Scheepers M, et al. Is intranasal midazolam an effective rescue medication in adolescents and adults with severe epilepsy? Seizure 2000; 9:417–422.

205. Jeannet PY, et al. Home and hospital treatment of acute seizures in children with nasal midazolam. Eur J Paediatr Neurol 1999;3:73–77.

206. Chamberlain JM, et al. A prospective, randomized study comparing intramuscular midazolam with intravenous diazepam for the treatment of seizures in children. Pediatr Emerg Care 1997;13:92–94.

207. Towne AR, DeLorenzo RJ. Use of intramuscular midazolam for status epilepticus. J Emerg Med 1999;17:323–328.

208. Camfield PR, et al. The first febrile seizure—antipyretic instruction plus either phenobarbital or placebo to prevent recurrence. J Pediatr 1980;97:16–21.

209. Rosman NP, et al. A controlled trial of diazepam administered during febrile illnesses to prevent recurrence of febrile seizures. N Engl J Med 1993;329:79–84.

210. Knudsen FU, Vestermark S. Prophylactic diazepam or phenobarbitone in febrile convulsions: a prospective, controlled study. Arch Dis Child 1978;53:660–663.

211. Camfield CS, et al. Home use of rectal diazepam to prevent status epilepticus in children with convulsive disorders. J Child Neurol 1989; 4:125–126.

212. Lahat E, et al. Comparison of intranasal midazolam with intravenous diazepam for treating febrile seizures in children: prospective randomised study. BMJ 2000;321:83–86.

213. Rantala H, Tarkka R, Uhari M. Preventive treatment for recurrent febrile seizures. Ann Med 2000;32:177–180.

214. Rantala H, Tarkka R, Uhari M. A meta-analytic review of the preventive treatment of recurrences of febrile seizures. J Pediatr 1997; 131:922–925.

215. Practice parameter: long-term treatment of the child with simple febrile seizures. American Academy of Pediatrics. Committee on Quality Improvement, Subcommittee on Febrile Seizures. Pediatrics 1999;103(6 Pt 1):1307–1309.

216. Knudsen FU, et al. Long term outcome of prophylaxis for febrile convulsions. Arch Dis Child 1996;74:13–18.

217. Deshmukh A, et al. Lorazepam in the treatment of refractory neonatal seizures. A pilot study. Am J Dis Child 1986;140:1042–1044.

218. Hall RT, Hall FK, Daily DK. High-dose phenobarbital therapy in term newborn infants with severe perinatal asphyxia: a randomized, prospective study with three-year follow-up. J Pediatr 1998;132:345–348.

219. Painter MJ, et al. Phenobarbital compared with phenytoin for the treatment of neonatal seizures. N Engl J Med 1999;341:485–489.

220. Gamstorp I, Sedin G. Neonatal convulsions treated with continuous, intravenous infusion of diazepam. Ups J Med Sci 1982;87:143–149.

221. Sheth RD, et al. Midazolam in the treatment of refractory neonatal seizures. Clin Neuropharmacol 1996;19:165–170.

222. Gal P, et al. Valproic acid efficacy, toxicity, and pharmacokinetics in neonates with intractable seizures. Neurology 1988;38:467–471.

223. Yoshikawa H, et al. Midazolam as a first-line agent for status epilepticus in children. Brain Dev 2000;22:239–242.

224. Igartua J, et al. Midazolam coma for refractory status epilepticus in children. Crit Care Med 1999;27:1982–1985.

225. Brown J. Fosphenytoin and status epilepticus. Hosp Med 1999;60:70–71.

226. Ramsay RE, DeToledo J. Intravenous administration of fosphenytoin: options for the management of seizures. Neurology 1996;46(6 suppl 1):S17–S19.

227. Browne TR, et al. Bioavailability of ACC-9653 (phenytoin prodrug). Epilepsia 1989;30(suppl 2): S27–S32.

228. Knapp LE, Kugler AR. Clinical experience with fosphenytoin in adults: pharmacokinetics, safety, and efficacy. J Child Neurol 1998;13(suppl 1): S15–S18; discussion S30–S32.

229. O'Brien TJ, et al. Incidence and clinical consequence of the purple glove syndrome in patients receiving intravenous phenytoin. Neurology 1998;51:1034–1039.

230. Parviainen I, et al. High-dose thiopental in the treatment of refractory status epilepticus in intensive care unit. Neurology 2002;59: 1249–1251.

231. Ubogu EE, et al. Ketamine for refractory status epilepticus: a case of possible ketamine-induced neurotoxicity. Epilepsy Behav 2003;4:70–75.

232. Jordan KG. Status epilepticus. A perspective from the neuroscience intensive care unit. Neurosurg Clin North Am 1994;5:671–686.

233. Wilner AN, Bream PR. Status epilepticus and pseudostatus epilepticus. Seizure 1993; 2:257–260.

234. Collins JF. Data and safety monitoring board issues raised in the VA Status Epilepticus Study. Control Clin Trials 2003;24:71–77.

235. Lowenstein DH, et al. The prehospital treatment of status epilepticus (PHTSE) study: design and methodology. Control Clin Trials 2001;22: 290–309.

236. DeLorenzo RJ, et al. Persistent nonconvulsive status epilepticus after the control of convulsive status epilepticus. Epilepsia 1998;39:833–840.

237. Logroscino G, et al. Time trends in incidence, mortality, and case-fatality after first episode of status epilepticus. Epilepsia 2001;42:1031–1035.

238. Oxbury JM, Whitty CW. Causes and consequences of status epilepticus in adults. A study of 86 cases. Brain 1971;94:733–744.

239. Krumholz A, et al. Complex partial status epilepticus accompanied by serious morbidity and mortality. Neurology 1995;45:1499–1504.

240. Fujiwara T, et al. Status epilepticus in childhood: a retrospective study of initial convulsive status and subsequent epilepsies. Folia Psychiatr Neurol Jpn 1979;33:337–344.

241. Lindsay J, Ounsted C, Richards P. Long-term outcome in children with temporal lobe seizures. III. Psychiatric aspects in childhood and adult life. Dev Med Child Neurol 1979;21:630–636.

242. Annegers JF, et al. The risk of epilepsy following febrile convulsions. Neurology 1979;29:297–303.

243. Ebisu T, et al. MR spectroscopic imaging and diffusion-weighted MRI for early detection of kainate-induced status epilepticus in the rat. Magn Reson Med 1996;36:821–828.

244. Najm IM, et al. MRS metabolic markers of seizures and seizure-induced neuronal damage. Epilepsia 1998;39:244–250.

245. Zhong J, et al. Barbiturate-reversible reduction of water diffusion coefficient in flurothyl-induced status epilepticus in rats. Magn Reson Med 1995;33:253–256.

246. Scott RC, et al. Magnetic resonance imaging findings within 5 days of status epilepticus in childhood. Brain 2002;125(Pt 9):1951–1959.

247. Tien RD, Felsberg GJ. The hippocampus in status epilepticus: demonstration of signal intensity and morphologic changes with sequential fast spin-echo MR imaging. Radiology 1995; 194:249–256.

248. Wieshmann UC, et al. Development of hippocampal atrophy: a serial magnetic resonance imaging study in a patient who developed epilepsy after generalized status epilepticus. Epilepsia 1997;38:1238–1241.

249. Nohria V, et al. Magnetic resonance imaging evidence of hippocampal sclerosis in progression: a case report. Epilepsia 1994;35:1332–1336.

250. Scott RC, et al. Hippocampal abnormalities after prolonged febrile convulsion: a longitudinal MRI study. Brain 2003;126:2551–2557.

251. Pitkanen A, et al. Progression of neuronal damage after status epilepticus and during spontaneous seizures in a rat model of temporal lobe epilepsy. Prog Brain Res 2002;135:67–83.

252. Roch C, et al. Predictive value of cortical injury for the development of temporal lobe epilepsy in 21-day-old rats: an MRI approach using the lithium-pilocarpine model. Epilepsia 2002;43: 1129–1136.

253. VanLandingham KE, et al. Magnetic resonance imaging evidence of hippocampal injury after prolonged focal febrile convulsions. Ann Neurol 1998;43:413–426.

254. Salmenpera T, et al. MRI volumetry of the hippocampus, amygdala, entorhinal cortex, and perirhinal cortex after status epilepticus. Epilepsy Res 2000;40:155–170.

255. DeGiorgio CM, et al. Neuron-specific enolase, a marker of acute neuronal injury, is increased in complex partial status epilepticus. Epilepsia 1996;37:606–609.

256. DeGiorgio CM, et al. Serum neuron-specific enolase in human status epilepticus. Neurology 1995;45:1134–1137.

257. Correale J, et al. Status epilepticus increases CSF levels of neuron-specific enolase and alters the blood-brain barrier. Neurology 1998; 50:1388–1391.

258. DeGiorgio CM, et al. Serum neuron-specific enolase in the major subtypes of status epilepticus. Neurology 1999;52:746–749.

259. Shorvon SD. Status Epilepticus: Its Clinical Features and Treatment in Children and Adults. Cambridge: Cambridge University Press, 1994.

Clinical Trials in Neuro-Ophthalmology

Madhura Tamhankar and Laura Balcer

OPTIC NEURITIS TREATMENT TRIAL

Optic neuritis is an acute inflammatory disease of the optic nerve. The form of optic neuritis most familiar to neurologists, acute demyelinating optic neuritis, most frequently occurs in the setting of multiple sclerosis (MS) and may be its first manifestation (1). Although not a true neuro-ophthalmologic emergency, optic neuritis frequently presents in the emergency setting, and clinical trials have established not only the clinical profile of this disorder, but also guidelines for appropriate neuro-imaging and initiation of immunomodulatory therapies in those at greatest risk for MS.

Acute demyelinating optic neuritis, referred to herein as optic neuritis, is characterized by sudden vision loss, which can vary from mild deficit in the field of vision to complete loss of light perception. Most patients improve without treatment (2). However, although visual acuity may recover to 20/20, many patients continue to suffer from visual disability in terms of abnormal contrast sensitivity, color vision, stereopsis, light-brightness sense and visual field (3).

Until the late 1980s, oral corticosteroids were the most commonly prescribed treatment for optic neuritis. A few nonrandomized studies in the late 1980s and early 1990s suggested rapid recovery of vision with intravenous corticosteroids (4). To resolve the controversies concerning the benefits of corticosteroids in optic neuritis and to establish appropriate treatment guidelines, the Optic Neuritis Treatment Trial (ONTT) was initiated (5). The primary objectives of ONTT were to answer the following questions:

1. Does the treatment of optic neuritis with oral prednisone or intravenous methylprednisolone reduce the residual visual function that is present after resolution?
2. Does either treatment speed visual recovery compared with placebo (natural history)?
3. Are the complications of the treatment insignificant in relation to the magnitude of the treatment effect?

The ONTT was a randomized multicenter trial that studied 454 patients with optic neuritis over 10 years (6). In addition to determining the potential role for intravenous corticosteroids in the treatment of optic neuritis, the ONTT provided important information with regard to the following:

1. To establish the clinical profile of acute optic neuritis with regard to

presenting symptoms, signs, and deficits in visual function;
2. To determine the value of magnetic resonance imaging (MRI), chest roentgenography, and serological studies in the evaluation and treatment of patients;
3. To document the clinical characteristics of optic neuritis as an aid in designing future studies.

Epidemiology and Natural History

The ONTT provided the most extensive information to date about the clinical features and natural history of acute demyelinating optic neuritis. Patients with optic neuritis are typically young (aged 20–50 years) and female (7). The incidence of acute monosymptomatic demyelinating optic neuritis is approximately 3 per 100,000 per year (compared with 1/100,000 per year in lower risk populations), with a prevalence of optic neuritis at 115 per 100,000 (2,7).

Patients with acute demyelinating optic neuritis typically experience loss of central visual acuity and ocular/periorbital pain. The ocular pain may precede or occur concomitantly with visual loss, is typically exacerbated by eye movements, and usually lasts no longer than a few days.

These clinical features reported by >90% of ONTT participants are usually subacute in onset and progress over hours to days. Worsening or progression of visual loss beyond 2 weeks or failure of visual recovery to begin within a 2- to 4-week period are considered atypical for acute demyelinating optic neuritis and should raise suspicion for alternative etiologies. Although visual loss is usually monocular, involvement of both eyes occurs occasionally, particularly in children.

Examination of the patient with acute demyelinating optic neuritis reveals features consistent with optic nerve dysfunction. Baseline visual function tests, performed within 8 days after symptom onset in the ONTT, demonstrated abnormalities of visual acuity, visual fields, contrast sensitivity, and color vision in affected eyes. Severity of visual loss in affected eyes varied from mild visual field defects to marked loss of visual acuity (3% of ONTT participants had no light perception). An afferent pupillary defect was present in almost all cases of acute demyelinating optic neuritis. Approximately two-thirds of patients had a normal optic disc appearance (retrobulbar optic neuritis). Visual recovery was excellent in most patients with acute optic neuritis as well as recurrent optic neuritis, although persistent visual symptoms (abnormal contrast sensitivity, color vision, stereopsis, light-brightness sense, and visual fields) were reported months to years after an acute attack of optic neuritis (8).

During and even beyond the recovery of vision after acute optic neuritis, patents often experienced temporary worsening of symptoms with exposure to heat (hot shower or exercise). This is referred to as Uhthoff's symptom. Patients with optic neuritis may also experience positive visual phenomena consisting of flashing bright lights or photopsias in the affected eye. These symptoms may be precipitated by eye movement or by sound and were reported by 30% of ONTT participants.

Diagnosis

The diagnosis of acute demyelinating optic neuritis is based on appropriate history and clinical course as described previously.

Current Disease Management

The most commonly used treatment for acute demyelinating optic neuritis, intravenous methylprednisolone (1g/day for 3 days) followed by oral prednisone (1 mg/kg per day for 11 days followed by a 4-day taper), may hasten the speed of visual recovery by 2–3 weeks when started within 1–2 weeks of the onset of symptoms. Intravenous methylprednisolone was also found to lessen the risk of development of MS within the first 2 years of follow-up in monosymptomatic patients, especially in those determined to be at high risk by brain MRI (two or more white matter lesions) (9).

Design Overview

The organizational structure of ONTT (5) consisted of study headquarters, Data Coordinating Center, and the Visual Field Reading Center, along with 15 clinical centers throughout the United States. Between June 1988 and July 1991; 457 subjects were recruited into the ONTT at these centers.

Eligibility criteria for enrollment included the diagnosis of acute unilateral optic neuritis with visual symptoms of 8 days or less, patient age between 18 and 46 years, no previous history of optic neuritis or ophthalmoscopic signs of optic atrophy in the affected eye, no evidence of systemic disease (other than MS) that might be associated with the optic neuritis, and no previous treatment with corticosteroids for optic neuritis in the fellow eye. Patients in the ONTT were randomized to one of the following three treatment groups:

1. Oral prednisone (1 mg/kg per day) for 14 days with 4-day taper (20 mg on day 15, 10 mg on days 16 and 18);
2. Intravenous methylprednisolone sodium succinate (250 mg every 6 hours for 3 days) followed by oral prednisone (1 mg/kg per day) for 11 days followed by a 4-day taper;
3. Oral placebo for 14 days.

Baseline tests of visual function performed within 8 days after onset of symptoms included visual acuity, color vision, contrast sensitivity, and visual fields. Methodology used to assess vision was as follows:

1. Visual acuity: a retro illuminated Snellen chart at 4 meters;

2. Color vision: pseudoisochromatic Ishihara plates (11 plates used) and the Farnsworth-Munsell 100-hue test (FM-100);
3. Contrast sensitivity: Pelli-Robson chart at 1 meter;
4. Visual field: Humphrey Field Analyzer program 30-2. The mean deviation was used as the outcome measure. The severity and pattern of field loss were determined for each field by the Visual Field Reading Center.

Additional testing included a standardized neurological examination, MRI of the brain, blood tests (glucose, antinuclear antibodies, and fluorescent treponemal antibody absorption tests) and a chest roentgenogram. From the results of the neurological history and examination, each patient was classified according to the clinical criteria of Poser et al. (10) as having "no," "possible," "probable," or "definite" MS. All MRI scans were read at a central reading center by a standardized method. Each scan was assigned five grades based on the severity of changes typical for MS ranging from 0 (normal) to IV (most severe). At the 6-month visit all study patients were given a questionnaire designed to assess the impact of an episode of optic neuritis on their quality of life.

Follow-up visits were scheduled on or about days 4, 15, and 30; weeks 7, 13, and 19; months 6 and 12; and then yearly. The data collected at the 6-month visit was the primary outcome measure. At each visit an interval history, refraction, visual acuity, contrast sensitivity, and visual field were tested. At the 6-month visit and all subsequent visits, color vision testing and neurological examination was performed.

Issues in Clinical Protocol Development

The decision to evaluate two drugs in parallel was motivated by the lack of evidence of superiority of any single treatment. Oral prednisone was chosen because it was widely used in the treatment of optic neuritis before the ONTT. Intravenous methylprednisolone was included on the strength of several small series (3), which suggested rapid recovery in acute optic neuritis. Dosage of both steroid regimens was chosen based on dosage usually prescribed in clinical practice. Experimental data were not available to support a more systematic determination of the optimal dose. The investigators believed that if corticosteroids were beneficial, the effect would be more apparent and the benefit greatest if the treatment was restricted to the acute phase of the disease. Therefore, eligibility was restricted to an 8-day time window after the onset of symptoms.

Determination of Outcomes

The primary aim of ONTT was to examine the effect of intravenous and oral corticosteroid therapy on visual outcome in acute demyelinating optic neuritis. Analyses were also performed to determine the role of corticosteroid therapy in the development of clinically definite MS (CDMS). Each treatment was evaluated for its ability to decrease residual visual dysfunction after resolution of optic neuritis as well as its ability to speed visual recovery.

Visual field and contrast sensitivity were the primary measures of outcome. Visual acuity and color vision were secondary outcome measures. The development of CDMS was also a secondary endpoint in the ONTT.

Outcomes of the ONTT

The ONTT (2,6,8,9,11–19) has been the most comprehensive study to date regarding the treatment of acute demyelinating optic neuritis and has had a significant impact on the practice patterns of both neurologists and ophthalmologists. Follow-up data from the ONTT cohort has been extensive and has provided important information regarding baseline clinical features, long-term visual outcome, vision-specific health-related information quality of life, and the role of brain MRI in determining the prognosis for subsequent development of CDMS.

Visual acuity data collected at the 6-month follow-up visit served as the primary outcome for analysis (11). Patients ($n = 438$) who completed the 6-month visit were included in the analyses. Mean age of the patients was 32.0 ± 6.7 years; 77% were women. Through randomization, 150 patients were assigned to the placebo group, 151 to the intravenous group, and 156 to the prednisone group. The primary conclusions of the ONTT were as follows:

1. Intravenous methylprednisolone hastened the recovery of visual function in acute optic neuritis but did not affect visual outcome at 6 months or beyond (through 10 years of follow-up) compared with oral prednisone or placebo. This benefit for intravenous methylprednisolone was greatest within first 15 days of follow-up.

2. Patients treated with oral prednisone alone demonstrated an increased rate of recurrent attacks of optic neuritis in both the affected and fellow eyes compared with intravenous methylprednisolone and placebo group (30% in oral prednisone group at 2 years vs. 16% and 13% in placebo and intravenous groups, respectively). This finding, though curious, persisted throughout follow-up of the ONTT cohort (more than 10 years) (6).

3. Patients in the intravenous methylprednisolone group had a reduced rate of development of CDMS during the first 2 years follow-up, but this did not persist beyond 2 years and was seen only in patients who had abnormal brain MRI scans (defined as two or more white matter lesions) at the time of acute optic neuritis (12).

TABLE 6.1   Visual acuity at 6 months for patients with severe visual loss at baseline

Visual acuity at 6 months	Baseline vision				
	Total $n = 66$	NLP $n = 14$	LP $n = 14$	HM $n = 24$	CF $n = 14$
$\geq$20/20	40	3	7	19	11
20/25 to 20/40	14	5	3	4	2
20/50 to 20/190	5	3	0	1	1
$\leq$20/200	7	3	4	0	0

NLP, no light perception; LP, light perception; HM, hand motions; CF, counting fingers.
Adapted with permission from ref. 8.

4. Based on the data from ONTT and similar trials, there is no treatment for acute demyelinating optic neuritis that changes long-term visual outcome or visual prognosis compared with placebo (natural history).

Current Recommendations for Treatment of Acute Demyelinating Optic Neuritis

Intravenous methylprednisolone followed by oral corticosteroids should be considered in patients with acute demyelinating optic neuritis. Treatment with oral prednisone alone should be avoided.

Results of Clinical Severity Measures

Visual Acuity

Four hundred thirty-eight patients who completed the 6-month visit were included in the analyses. Visual acuity data were transformed to log minimal angle of resolution (MAR) scores for the purpose of analyses. In most patients, regardless of treatment, visual recovery was noted within first 2 weeks of study entry and began sooner in patients with intravenous corticosteroids than in the other two groups. In all three treatment groups, median visual acuity was better than 20/25 at day 15 and better than 20/20 (log MAR value $= 0$) at 6 months. The best predictor of visual outcome was the severity of visual loss and age. Visual recovery statistically was worse in the patients with

more severe visual loss at baseline (Table 6.1). As demonstrated in Table 6.1, 81.8% of patients whose initial visual acuity was counting fingers or worse had 20/40 or better visual acuity at 6 months.

Younger patients (aged 18–34 years) statistically had slightly better visual acuity at 6 months than older patients (aged 35–45 years). The actual difference in median visual acuities was only two or more letters correct on the 20/20 line in the younger compared with the older patients. Visual function was assessed for 319 of 387 patients (82%) who consented to continue in follow-up after 1997, more than 10 years after an initial episode of optic neuritis. In most patients, once visual acuity had stabilized after the initial episode of optic neuritis, visual acuity remained remarkably stable for more than 10 years. After 10 years, 69% of patients had visual acuity better than 20/20 and only 1% were worse than 20/200. As reported at 5 years of follow-up, visual function was worse in those patients with MS than in those without in the 10-year study. Lower scores for health-related quality of life (National Eye Institute Visual Function Questionnaire) were obtained in patients with abnormal visual acuity and in patients with MS as compared with the reference group.

Color Vision

Color vision was considered to be abnormal if one or more Ishihara plates (total of 11 plates) were identified incorrectly.

In patients with visual acuities worse than 20/20, abnormal color vision was seen in 92%, whereas in patients with 20/20 or better vision, abnormal results were obtained in 51.1% of study participants.

Contrast Sensitivity

Using the Pelli-Robson chart, 99% of patients with visual acuities worse than 20/20 had abnormal contrast sensitivity. Even in patients with 20/20 or better vision, contrast sensitivity remained abnormal in 87.2%. In affected eyes, contrast sensitivity was more often abnormal than visual acuity, visual field, or color vision.

Visual Field

Visual field defects in affected patients were classified in the ONTT. The most commonly observed field defects were diffuse (48%), altitudinal (15%), quadrant (96%), central, and cecocentral (15). As per the ONTT enrollment criteria for the affected eyes, all the visual fields were abnormal at baseline. Fifty-one percent of visual fields were normal at 6 months, and 55.9% were normal at 1 year. The mean values of Humphrey mean deviations for affected eyes were −22.88 db at baseline, −1.94 db at 6 months, and −1.62 db at 1 year. Moreover, 13.2% of patients demonstrated a chiasmal (5.1%) or retrochiasmal (8.9%) type of visual field defect at least once during the first year. Patients with a retrochiasmal defect had a higher incidence of abnormal brain 1MRI scans at baseline than the ONTT patients without these defects. Little change in visual function occurred between the 1-year follow-up and 5-year and 10-year follow-up examinations (6).

Other Measures and Analyses

Fellow Eye Abnormalities

Many patients in the ONTT were found to have mild asymptomatic visual function abnormalities in their fellow eyes upon entry into the ONTT (16). The prevalence of fellow eye abnormalities in 448 patients entered into the ONTT was also determined. Each patient was classified into one of two groups based on the presence (group I) or absence (group II) of a past history of optic neuritis in the fellow eye. Abnormalities in the fellow eye were found on measurement of visual acuity in 13.8%, contrast sensitivity in 15.4%, color vision abnormalities in 21.7%, and visual field in 48% of the study participants. The percentage of abnormal fellow eyes was higher on each measure ($p < 0.01$) in patients with a past history of optic neuritis in the fellow eye (group I) than in patients without such a past history.

Recurrences of Optic Neuritis

Further attacks of optic neuritis in either eye occurred in 35% of all patients and were twofold more common among patients diagnosed with MS at baseline or who developed MS during the follow-up period (Table 6.2). Two factors, including treatment group and MS status, predicted recurrences of optic neuritis over a 5-year

TABLE 6.2   Recurrences of optic neuritis among patients completing the follow-up examination according to MS status in ONTT

Involved eye	Total $n = 319$	Patients with MS $n = 148$	Patients without MS $n = 171$	p value
None	207 (65%)	77 (52%)	130 (76%)	
Affected, fellow, or both	112 (35%)	71 (48%)	41 (24%)	<0.001

p value from the Fisher's exact test.
Adapted with permission from ref. 6.

follow-up period (17). The cumulative probability of having a new episode of optic neuritis during the 5 years of follow-up was 19% for the affected eye, 17% for the fellow eye, and 30% for either eye. Consistent with the data reported after 2 years of follow-up, the probability of recurrent optic neuritis in either eye was almost twofold higher in the oral prednisone group than in either the placebo group ($p = 0.004$) or the intravenous group ($p = 0.003$). Two-thirds of the recurrences were observed in the first 2 years in each treatment group. The comparatively higher recurrence rate in the prednisone group was predominantly present among patients who did not have clinical or MRI evidence of MS.

Brain MRI

Brain MRI was performed in the ONTT (18) to ascertain whether the identification of brain lesions consistent with demyelination is predictive of development of CDMS. In the ONTT, MRI was performed using standardized protocol at all the centers within 24 hours of entry into the study and before the initiation of treatment. Five MRI classifications, or grades, were established based on the number of white matter lesions, with grade 0 being normal, grade I having changes not specific for demyelinating disease, and grades II through IV having lesions suggestive of demyelination. In the classification, all lesions were at least 3 mm in size. Scans were read at intervals during a 2-year period independently by two neuroradiologists. MRI was useful for establishing an alternative diagnosis to acute demyelinating optic neuritis in only two patients (one had a pituitary tumor and the other an ophthalmic artery aneurysm).

MRI scans from 418 patients were evaluated. Abnormal scans were present in 46.9% of the cohort. Patients with an abnormal scan tended to have more severe visual loss at the time of presentation ($p = 0.037$), a higher incidence of retrobulbar neuritis compared with papillitis ($p = 0.04$), and absence of history of preceding viral syndrome ($p = 0.018$). A grade IV scan was present for 21.3% of patients with a clinical diagnosis of no MS (59/277) and 76.9% (20/26) with definite MS.

Surrogate Endpoints: Development of CDMS

The presence of white matter lesions on brain MRI performed at the time of optic neuritis was the single most important predictor of the development of CDMS by 5 years and beyond. The cumulative probability of CDMS was 16% in patients with no MRI lesions, 37% with one to two lesions, and 51% with three or more lesions. Prior nonspecific neurological symptoms and prior optic neuritis in the fellow eye were the only other factors associated with an increased 5-year risk of MS in these patients with abnormal brain MRI scans.

Among the 189 patients who had no brain MRI lesions and no prior neurological symptoms or optic neuritis in the fellow eye (monofocal optic neuritis), CDMS developed within 5 years in only 25 patients (13%) and was more likely in women (rate ratio, 2.03; 95% confidence interval, 0.59 to 7.03) and in whites (rate ratio, 2.07; 95% confidence interval, 0.45 to 9.65). Three other clinical features were associated with a particularly low risk of CDMS in this subgroup: lack of pain, the presence of optic disk swelling, and less severe visual acuity loss. CDMS developed by 5 years in 0 of 19 patients (0%) whose visual loss was painless, in 8 of 90 patients (9%) in whom visual acuity at study entry was 20/40 or better, and in 6 of 78 patients (8%) who had a swollen optic disk. Among the patients with a swollen disk, CDMS did not develop in any patient who had severe disk edema ($n = 21$), disc or peripapillary hemorrhage ($n = 16$), or macular exudates ($n = 8$). Patients with severe disk edema followed a course consistent with optic neuritis. Further 10-year follow-up data revealed the same pattern as was observed at 5 years; in patients of both genders without brain lesions seen on MRI ($n = 191$), MS did not develop in any patients whose visual loss was painless ($n = 18$) or total (no light perception, $n = 6$) or in those who had ophthalmoscopic findings of severe disc swelling ($n = 22$), hemorrhage of the optic disc or surrounding retina ($n = 16$), or retinal exudates ($n = 8$).

The development of MS was strongly associated with the presence of one or more lesions on the baseline MRI of the brain at 10-year follow-up study (Table 6.3, Fig. 6.1). The 10-year risk of MS in the ONTT cohort was 38% (95% confidence interval, 33%–43%) (9). Patients ($n = 160$) who had one or more typical lesions on the baseline MRI scan of the brain had a 56% risk; those with no lesions ($n = 191$) had a 22% risk ($p < 0.001$, log rank test). There-fore, even the presence of one white matter lesion on the MRI scan (at least 3 mm in diameter) more than doubled the 10-year risk of MS (from 22% to 56%). However,

the presence of one or more lesions did not signify that the patient was destined to develop MS. Among patients with brain lesions seen on MRI, the 10-year probabil-ity of remaining free of MS was 44% (Table 6.3). Conversely, the absence of brain lesions seen on MRI did not eliminate the risk of developing MS. This natural history information is a critical input for estimating a patient's 10-year MS risk and for weighing the benefit of initiating prophylactic treatment at the time of optic neuritis or other initial demyelinating events in the central nervous system.

Among patients who had not developed MS at 5 years after study enrollment, the probability of being diagnosed as having MS between 5 and 10 years was 7% in the 142 patients with no brain lesions seen on MRI and 27% in the 89 patients with one or more lesions. In the presence of one or more brain lesions seen on MRI, a history of nonspecific neurological symptoms (usually transient numbness) or prior optic neuritis in the fellow eye further increased the 10-year risk of MS (70% vs. 50%, $p = 0.005$). However, among these 160 patients with one or more brain lesions

TABLE 6.3   Baseline factors predictive of MS according to the presence or absence of brain MRI lesions at study enrollment in ONTT

Variable	Patients who had no brain lesions on MRI			Patients who had ≥1 brain lesion on MRI		
	No. of patients	10-yr risk of MS* (%)	p	No. of patients	10-yr risk of MS* (%)	p
Overall	191	22	NA	160	56	NA
Gender						
Female	142	25		128	58	
Male	49	10	0.05	32	51	0.86
Optic disc						
Appearance						
Normal	110	28		105	60	
Edema	81	14	0.01	55	50	0.57
Pain						
Yes	173	24		148	57	
No	18	0		12	42	0.46

Values are Kaplan-Meier estimates of cumulative probability.
NA, not applicable.
Adapted in part with permission from ref. 19.

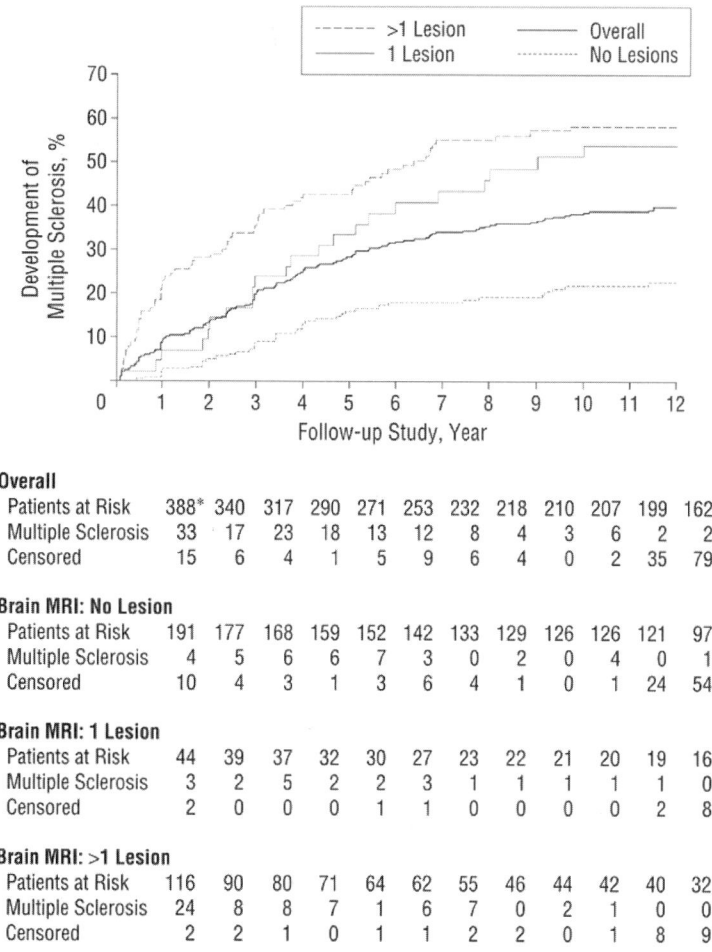

Overall											
Patients at Risk	388*	340	317	290	271	253	232	218	210	207	199	162
Multiple Sclerosis	33	17	23	18	13	12	8	4	3	6	2	2
Censored	15	6	4	1	5	9	6	4	0	2	35	79

Brain MRI: No Lesion

Patients at Risk	191	177	168	159	152	142	133	129	126	126	121	97
Multiple Sclerosis	4	5	6	6	7	3	0	2	0	4	0	1
Censored	10	4	3	1	3	6	4	1	0	1	24	54

Brain MRI: 1 Lesion

Patients at Risk	44	39	37	32	30	27	23	22	21	20	19	16
Multiple Sclerosis	3	2	5	2	2	3	1	1	1	1	1	0
Censored	2	0	0	0	1	1	0	0	0	0	2	8

Brain MRI: >1 Lesion

Patients at Risk	116	90	80	71	64	62	55	46	44	42	40	32
Multiple Sclerosis	24	8	8	7	1	6	7	0	2	1	0	0
Censored	2	2	1	0	1	1	2	2	0	1	8	9

FIGURE 6.1 The cumulative probability of MS was statistically significantly higher in patients with one or more lesions seen on the baseline MRI scan of the brain than in patients with no brain lesions ($p < 0.001$, log rank test) but was not significantly different comparing patients with a single brain lesion and patients with multiple lesions ($p = 0.22$, log rank test). The numbers of patients at risk are the numbers who had not developed MS at the beginning of each year. The "multiple sclerosis" rows indicate the number of patients classified as having MS during each yearly interval. The "censored" rows indicate the number of patients not developing MS whose last available follow-up data occurred during each yearly interval. Asterisk indicates 37 patients who had no baseline MRIs. (From ref. 19.)

seen on MRI, no demographic or any other clinical features of the optic neuritis were predictive of developing MS (Table 6.3).

Within the cohort of 388 patients followed up from the onset of an acute episode of optic neuritis, the 10-year risk of development of MS, based strictly on conventional clinical criteria, was 38%, compared with a 5-year risk of 30% (12). Thus, although the cohort continued to develop MS with each passing year, most did so within the first 5 years after the initial episode of acute optic neuritis. These results have applicability not only to optic neuritis but also to patients seen with an initial demyelinating event of the brainstem or spinal cord because the three presentations share a common pathogenesis and have been reported to have similar risks for MS (20).

Conclusions

Continued follow-up of the cohort of patients who were enrolled into the ONTT has provided a unique opportunity to assess the long-term course of vision after an episode of acute optic neuritis. It was found that in most patients, once visual acuity stabilized after the initial episode of optic neuritis, it remained remarkably stable for more than 10 years. The presence of even one brain MRI lesion was the strongest predictor of future development of CDMS in this cohort. The results of the ONTT can be used by clinicians to advise patients that long-term visual prognosis is good after an initial episode of optic neuritis. Follow-up of this cohort is expected to continue, and patients will be reexamined in 2006.

Implications for Present and Future Treatment of MS

Among patients at high risk for the development of CDMS as established by ONTT criteria, a randomized trial (the Controlled High Risk Avonex MS Prevention Study [CHAMPS]) demonstrated that treatment with interferon-β1a (21) after a first demyelinating event may significantly reduce the cumulative 3-year probability of CDMS versus placebo ($p = 0.002$). This trial enrolled 383 patients between the ages of 18 and 50 years who had an acute isolated first demyelinating event (50% had optic neuritis). Participants also had brain MRI findings consistent with CDMS as established by the ONTT (two or more white matter lesions, 3 mm or larger in diameter, at least one lesion periventricular

or ovoid). The CHAMPS study demonstrated efficacy for interferon-β1a not only in reducing the 3-year cumulative probability of CDMS (probability 0.50 for interferon-β1a vs. 0.35 for placebo) (Fig. 6.2), but also showed that treated patients had a reduced rate of accumulation of new but clinically silent lesions on brain MRI ($p = 0.001$ for new and enlarging T2 lesion at 18 months). Given the potential long-term benefits for early therapy in MS, results from the CHAMPS study represent perhaps the most significant development in the treatment of patients with acute monosymptomatic demyelinating optic neuritis (21).

ISCHEMIC OPTIC NEUROPATHY DECOMPRESSION TRIAL

Ischemic optic neuropathy (ION) is an acute presumably vascular optic neuropathy that presents with a sudden "stroke-like" vision loss in elderly patients. ION essentially occurs in two broad clinical settings: nonarteritic and arteritic. Arteritic ION, typically seen in giant cell arteritis (GCA), refers to optic nerve infarction that occurs secondary to vessels narrowed by vasculitis. Acute vision loss, the most feared complication of GCA, occurs in 7–60% of patients (22). Because many cases of blindness in GCA are preventable with immediate administration of corticosteroids, suspected patients require emergent intervention.

Nonarteritic ischemic optic neuropathy (NAION), on the other hand, is believed to result from acute ischemia to the optic nerve head. It is the most common cause of acute optic neuropathy in adults over age 50 (23–25) and frequently causes visual acuity and/or visual field loss. Many of the affected patients have associated vasculopathic risk factors such as diabetes or hypertension. Patients are particularly at risk if they have small crowded optic discs. Because NAION can affect both eyes in up

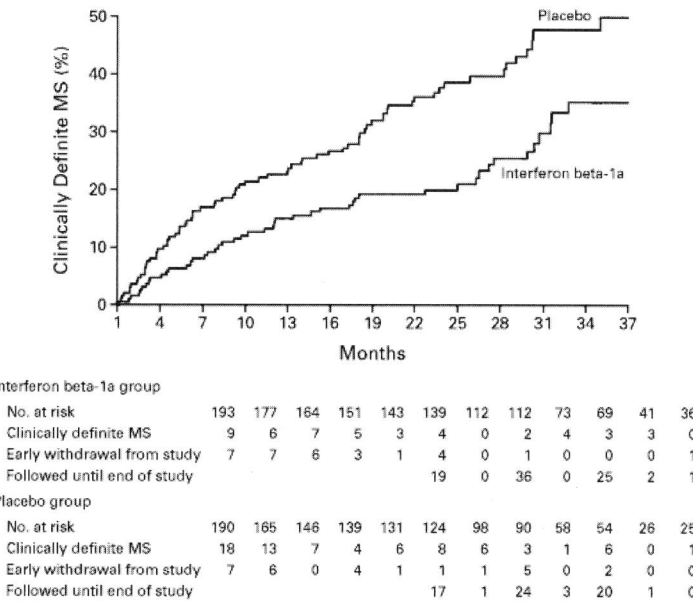

FIGURE 6.2 Kaplan-Meier estimates of the cumulative probability of the development of CDMS according to treatment group in CHAMPS. The cumulative probability of the development of CDMS during the 3-year follow-up period was significantly lower in the interferon-β1a group than in the placebo group ($p = 0.002$ by the Mantel log-rank test). The numbers of patients at risk are the numbers in whom CDMS had not developed at the beginning of each 3-month period. The endpoint was assessed beginning at 1 month, because according to the protocol that was the earliest possible time at which the endpoint could be reached. The "early-withdrawal" row indicates the number of patients in whom MS did not develop and whose follow-up ended before the study ended. Data were censored at the time of the patient's last completed neurological examination. (From ref. 21.)

to 40% of patients (26) and because final visual acuity declines to 20/200 in about 45% of affected eyes, this disorder can have devastating effects on the independence and quality of life for patients (27,28).

There is no known effective therapy for NAION. The effectiveness of optic nerve sheath decompression surgery (ONDS) as a possible form of treatment for progressive NAION was evaluated in a single, masked, multicenter, randomized clinical trial, the Ischemic Optic Neuropathy Decompression Trial (IONDT). The primary aims of this trial were (a) to assess the safety and efficacy of ONDS compared with careful follow-up alone in patients with NAION, (b) to describe the natural history of NAION in untreated patients,

and (c) to assess changes in quality of life of patients with NAION. Although ONDS was demonstrated to be an ineffective and potentially harmful therapy for NAION, the IONDT provided important natural history data that may provide a foundation for future epidemiological and treatment studies.

Epidemiology and Natural History

Most patients with NAION are 60–70 years of age with an annual incidence rate of 2.3 per 100,000 population/year and an increased rate in whites compared with African Americans or Hispanic individuals (25). There is an increased risk of NAION in patients with diabetes mellitus and systemic hypertension (29,30). In the IONDT, 47% of the patients had hypertension and 24% had diabetes mellitus. Nocturnal hypotension occurring as a consequence of treating arterial hypertension may be a separate and distinct risk factor, which may explain the high rate of vision loss that occurs upon awakening in some patients with NAION (31–33). The other important risk factor is the disc at risk or crowded optic nerve head. The fellow eye of patients with NAION is often found to have a small or absent cup. Patients with NAION do not have increased incidence of carotid disease on the affected side compared with age-matched control subjects (34).

Most patients describe a sudden onset monocular visual loss upon awakening (40% in IONDT). There are usually no prodromal ocular symptoms and no associated systemic symptoms. The presence of prodromal amaurosis fugax or diplopia, on the other hand, should increase the suspicion for GCA. Ocular pain is uncommon but does occur in about 10% of patients (35). Examination of patients with NAION is characterized by reduced visual acuity (at any level), dyschromatopsia, and an afferent pupillary defect. Visual field loss is most often inferior altitudinal in nature. The characteristic disc appearance is pale swelling with splinter hemorrhages and dilated capillaries on the disc surface. Up to one-fourth of patients have central scotoma. Vitreitis is usually absent, and its presence should increase clinical suspicion for alternative inflammatory and infectious causes.

Most patients have a fixed visual deficit, but a variable course with a significant incidence of progressive vision loss in the first month and a high rate of spontaneous visual recovery (43% in the IONDT cohort) can occur. A small percentage of patients have progressive or recurrent visual loss (36) beyond this period. As the disc edema resolves, optic atrophy develops and "luxury perfusion" may develop due to presumed up-regulation of flow in capillaries on the surface of the ischemic disc. There is a lifetime risk of 30–40% for second eye involvement (26).

Diagnosis

The diagnosis in most cases is clinical, based on the pattern of visual loss and disc appearance. In patients older than 55–60 years, an erythrocyte sedimentation rate and C-reactive protein are obtained to rule out visual loss secondary to GCA. Fluorescein angiography may be able to distinguish NAION swelling from other causes of disc swelling by demonstrating delayed optic nerve head filling. MRI does not generally have a role in the diagnosis of acute NAION but on occasion is necessary to exclude compressive and infiltrative conditions that may mimic NAION. Patients with NAION may have an increased number of white matter changes compared with age- and disease-matched control subjects (37). This may reflect the presence of vasculopathic risk factors.

Pathogenesis

The work of Hayreh (38) provided much information regarding the presumed pathogenesis of NAION. Anterior ION may result from insufficiency in the posterior ciliary artery circulation and in branches of the peripapillary choroidal system, causing infarction of the optic nerve head. How systemic microvascular disease, a crowded optic nerve head, and possibly nocturnal hypotension ultimately lead to optic nerve head infarction is not certain. Other features of NAION that support a vascular occlusive etiology include sudden onset, association with diabetes and hypertension, lack of evidence of inflammation in pathological specimens, and the fact that a similar syndrome can be created in animal models through vascular occlusion. NAION has also been reported in patients with hypercoagulable state secondary to antiphospholipid antibodies (39) and in homocystinuria. These possibilities should therefore be investigated in young patients who present with ION (39–45). ION, both posterior (without disc swelling) and anterior in localization, is also frequently reported postoperatively in patients undergoing surgical procedures.

Treatment

There is no known effective therapy for NAION. Corticosteroids, hyperbaric oxygen therapy (46), and aspirin (47) have not been demonstrated to affect the course of visual loss or recovery. Similarly, there is no known effective prophylaxis for second eye involvement. Patients are frequently placed on daily aspirin therapy (because they often have vascular risk factors), but there is no prospective proof of its efficacy.

Rationale for the IONDT

Until recently, the clinician's main task in managing patients with NAION was to exclude GCA and to control other factors such as elevated blood pressure that might affect risk for other vascular events. In 1989, it was first suggested, based on anecdotal experience, that ONDS might improve vision, particularly in patients with progressive form of NAION (48). It was postulated that progressive visual loss after the initial ischemic event occurs because of interference with rapid axoplasmic transport produced by cerebrospinal fluid pressure within the anatomically restricted confines of the perineural optic nerve space. During ONDS, one or more slits, or windows, are cut into the optic nerve sheath surrounding the optic nerve to decrease the perineural pressure exerted by the cerebrospinal fluid. Drainage of cerebrospinal fluid would then allow recovery of axoplasmic transport with subsequent improvement in vision. Other reports subsequently suggested a beneficial effect of ONDS on visual acuity and visual fields (49–51). None of the studies reporting improvement was

randomized controlled trials. Furthermore, the sample sizes were small, uniform visual testing procedures were not used, and progressive disease was not well defined. Because surgery was being performed more frequently, it became imperative to test the procedure in a randomized clinical trial. Accordingly, the IONDT was initiated.

Design Overview

The IONDT was a multicenter, randomized, controlled clinical trial sponsored by the National Eye Institute through cooperative agreements with the objective of assessing the safety and efficacy of ONDS for NAION (52). Patients diagnosed as having NAION were randomly assigned either to ONDS or to careful follow-up and were followed for at least 2 years.

The primary study outcome was a change of at least three lines of visual acuity, measured using the Early Treatment Diabetic Retinopathy Study [ETDRS] charts (Lighthouse Low Vision Products, Long Island, NY) at 6 months after randomization.

Additional outcome measures were as follows:

- Change of at least three lines of visual acuity at 1 year and at subsequent follow-up times after randomization;
- Change in visual field as measured by automated Humphrey perimetry;
- Occurrence of systemic and ophthalmic complications and intraoperative and postoperative complications (first postoperative week);
- Change in quality of life and other morbidity or mortality related to ONDS.

Inclusion Criteria

To be eligible for participation in the IONDT, patients were 50 years of age or older and were enrolled within 14 days of the onset of visual symptoms. Visual acuities of study participants were required to be 20/64 or worse in the affected eye but not worse than light perception. Visual field defects consistent with optic neuropathy, an afferent pupillary defect, and optic disc edema were also inclusion criteria.

Patients eligible for randomization within 14 days of onset of NAION were referred to as "regular-entry" patients. Eligible patients with visual acuity better than 20/64 (i.e., did not meet inclusion criteria above) were followed weekly for 30 days to determine whether visual acuity declined to 20/64 or worse. If such a decline in visual acuity occurred, then patients were considered eligible for randomization as "late-entry" participants. For the purpose of this study, late-entry patients were defined as having progressive NAION. When vision remained better than 20/64, patients were considered eligible for follow-up as part of a "natural-history" cohort.

A quality-of-life assessment was added to the protocol 12 months after recruitment began. Patients completed two instruments, the MOS 36-Item Short-Form Health Survey (SF-36) (53) and a form designed specifically for the IONDT administered at 6 months and 12 months.

Treatment and Follow-Up

Regular-entry patients assigned to ONDS were required to receive surgery within 14 days of the onset of symptoms and not more than 4 days from the time of randomization. Late-entry patients had to receive surgery not more than 4 days from the time of randomization. All patients enrolled in the trial were seen for follow-up visits at 1 week, 1 month, 3 months, 6 months, and 12 months after randomization and at 6-month intervals thereafter for at least 2 years. Natural-history patients were followed at 6 and 12 months and at 12-month intervals thereafter; procedures

at follow-up visits were the same as those for randomized patients.

Visual Function Assessments

Visual Acuity

Retroilluminated ETDRS charts were used to assess vision by masked visual acuity technicians. Each visual acuity score was converted to log MAR units. When visual acuity could be estimated using number of letters read on the ETDRS chart (patients were "on chart"), standard methods of calculating a log MAR score (54) were used. Patients with count fingers, hand motion, and light perception were considered "off chart" and were assigned log MAR scores of 2.0 for count fingers, 2.3 for hand motion, and 2.6 for light perception.

Visual Field

Visual field mean deviation at 3, 6, and 12 months was measured by certified visual field technicians masked to treatment assignment using automated static perimetry with program 24-2 and size III stimulus on the Humphrey field analyzer. As part of an ancillary study, 10 clinics also measured visual fields using a size V stimulus on the study eye in those participants who had substantially reduced performance with the size III stimulus. Substantially reduced performance was defined as fewer than 10 spots with a response greater than 5 dB or 5 dB or less response at each of the four cardinal points (12-degree diagonal from fixation point) on the standard 24-2 with a size III stimulus.

Fundus Photography

In 14 clinical centers, color 30-degree stereo fundus photographs of two standard fields (optic disc and macula) were taken at baseline and at the 6-month follow-up visits by certified fundus photographers masked to treatment assignment.

Outcomes of IONDT

Between 1992 and 1994, 418 patients with NAION were enrolled in the IONDT; 258 were randomized (127 to the surgery group and 131 to the careful follow-up group) and 160 were not randomized and constituted the natural-history cohort. Data collection for the IONDT and IONDT follow-up study ceased on January 21, 2001 when all enrolled patients had at least 5 years of follow-up (range, 0–7.4 years; median. 5.1 years) (55).

The baseline demographic and clinical characteristics of patients are shown in Table 6.4. Overall, 61% of patients were men, 95% were white, and age ranged from 50 to 89 years (median and mean age = 66 years). Randomized patients were older than nonrandomized patients ($p < 0.0001$) and a smaller proportion were men (55% compared with 71%; $p = 0.001$). Sixty-four percent of randomized patients reported having one or more baseline vascular risk factors (hypertension, diabetes mellitus, and myocardial infarction) more frequently than did nonrandomized patients (54%; $p = 0.04$).

Among all patients at baseline, 12% were current smokers, 37% previous smokers, and 38% had never smoked. No data with respect to smoking was available for 54 patients (13%). Randomized patients tended to report current smoking more often than nonrandomized patients (16% vs. 10%). At the baseline visit, 29% of all patients reported taking aspirin on a

TABLE 6.4 Demographic and clinical characteristics of eligible patients at baseline in IONDT

Patient characteristics	Eligible patients*		
	Randomized	Not randomized	Total
Total (%)	258 (100)	162 (100)	420 (100)
Men	143 (55.4)	116 (71.6)	259 (61.7)
Race			
White	244 (94.6)	154 (94.1)	398 (94.8)
Age at onset mean ± SD, yr			
All patients	68 ± 8.5	63 ± 8.2	66 ± 8.7
Medical history			
Hypertension	131 (50.8)	65 (40.6)	196 (46.9)
Diabetes mellitus	69 (26.7)	31 (19.4)	100 (23.9)
Myocardial infarction	28 (10.9)	18 (11.3)	46 (11.0)
Ocular history			
NAION (nonstudy eye)	39 (15.1)	22 (12.7)	61 (14.6)
Visual acuity in affected eyes			
>20/64	0	148 (91.4)	148 (35.2)
20/64 to 20/100	67 (26.0)	5 (3.1)	72 (17.1)
20/125 20/200	46 (17.8)	1 (1.9)	49 (11.7)
20/250 to 20/400	45 (17.4)	2 (1.2)	47 (11.2)
20/500 to 20/1000	48 (18.6)	3(1.9)	51 (12.1)
Cf, HM or LP	52 (20.2)	1 (0.6)	53 (12.0)
Humphrey visual field in affected eyes (mean deviation), mean ± SD, db	−21.36 ± 8.25	−14.42 ± 7.04	−18.63 ± 8.14

*Data are given as number (percent) of patients.

Adapted with permission from Ischemic Optic Neuropathy Decompression Trial Study Group. Characteristics of patients with non-arteritic anterior ischemic optic neuropathy eligible for the Ischemic Optic Neuropathy Decompression Trial. Arch Ophthalmol 1996;114:1366–1374.

regular basis for 1 month or longer and 4% reported taking anticoagulants.

Results of Clinical Severity Measures

Visual Acuity

At 6 months, 38 of 89 patients (42.7%) in the careful follow-up group had improved three or more lines of vision and an additional 40 of 89 patients (44.9%) reported no change in visual acuity. In the surgery group, the corresponding values were 30 of 92 patients (32.6%) and 40 of 92 patients (43.5%), respectively. There were no significant differences between surgery and careful follow-up at 3, 6, and 12 months with respect to visual improvement, although data were in favor of careful follow-up (Fig. 6.3).

Visual Field

There was no beneficial effect of surgery compared with careful follow-up in terms of change in mean deviation for visual field at 3, 6, or 12 months.

Adverse Events

Pain was the most common adverse event in the surgery group (17% at 1 week compared with 3% in the careful follow-up group). Diplopia occurred in 8% of patients in the surgery group, compared with 1% in the careful follow-up group. One patient developed a central retinal artery occlusion during surgery and had only light perception vision at 6 months. Two surgical patients experienced loss of light perception after surgery that persisted for at least 12 months.

Conclusions

ONDS appears to be of no value to most patients with NAION and may lead to further visual deterioration (55). An unexpectedly high rate of spontaneous improvement in visual acuity (42.7%

improved by three lines or more) was observed at 6 months in the careful follow-up group. ONDS was associated with a lower rate of improvement. At 3, 6, and 12 months of follow-up, patients receiving surgery had a greater risk of losing three or more lines of vision. The most encouraging finding was the high percentage of patients in the careful follow-up group who had visual acuity improvement (42.7%, a value greater than that reported previously) (36,56–58).

Other Measures and Analyses

The IONDT Research Group continued to follow all randomized participants ($n = 258$) to evaluate long-term outcomes. For the primary outcome, the preliminary results demonstrating no benefit for ONDS were confirmed for all follow-up times through 24 months (59). There were no significant differences between the careful follow-up and ONDS groups in mean changes in visual acuity at the 24-month follow-up.

Risk of Fellow Eye Involvement in NAION

NAION with bilateral or sequential involvement has been reported to occur in 10.5% to 73% of patients (26). The patients in the IONDT (both randomized and observed) were followed for 5 years. The primary objective of the follow-up study (60) was to determine the baseline prevalence and cumulative incidence of NAION in the fellow eye in this cohort of 418 patients. Data was explored in the following areas:

- Risk of NAION in the fellow eye;
- Baseline characteristics that are positively or negatively associated with the occurrence of fellow-eye NAION;
- Visual acuity in both the study and fellow eye both short term and long

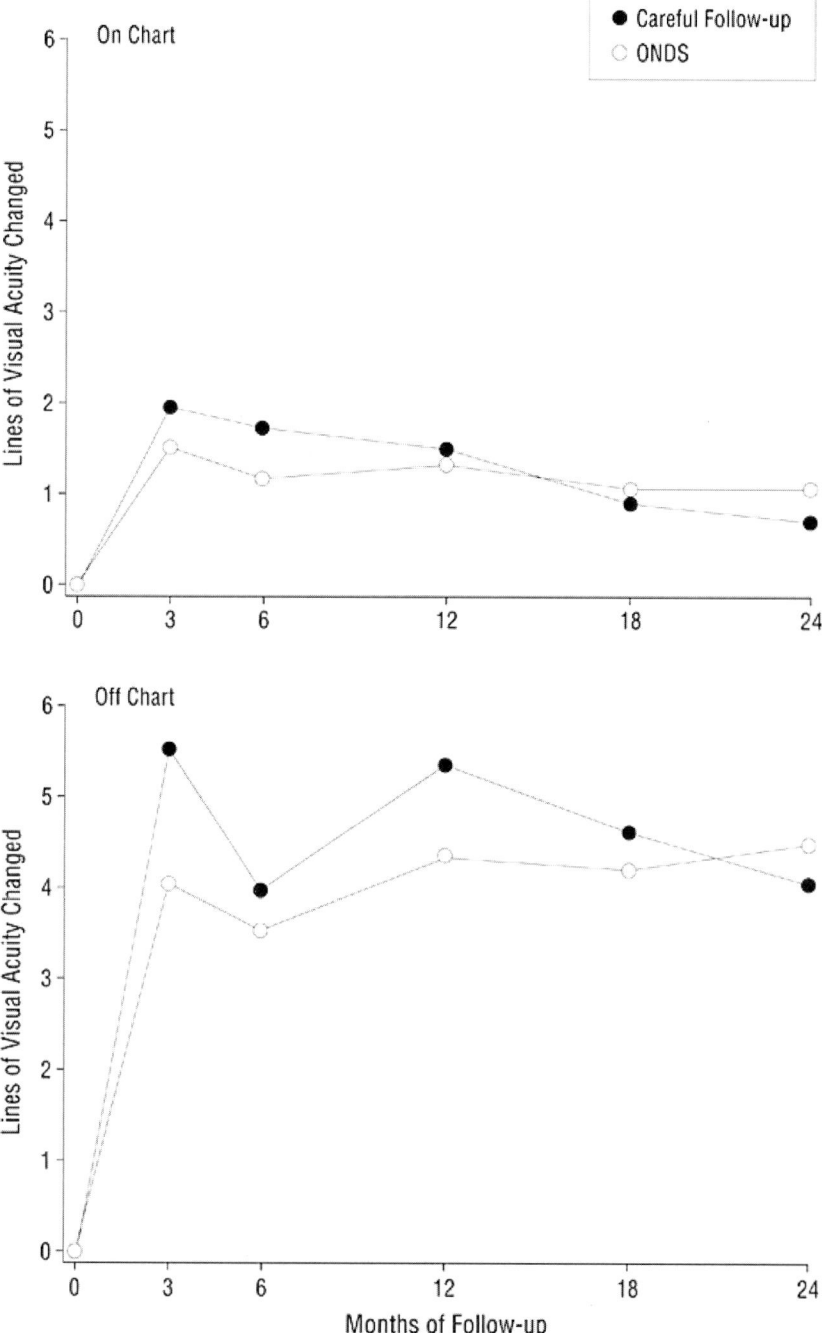

FIGURE 6.3   Top: Mean change from baseline in visual acuity in patients with baseline visual acuity "on chart" (i.e., sufficiently good to be measured with a standardized chart). Results were adjusted for baseline visual acuity, age (<65 years or ≥65 years), and diabetes. Improvement is indicated by positive values. Bottom: Mean change from baseline in visual acuity in patients with baseline visual acuity "off chart" (i.e., count fingers, hand motion, or light perception). Results were adjusted for baseline visual acuity. Improvement is indicated by positive values. (From ref. 59.)

term after NAION occurs in the fellow eye;

- Baseline characteristics associated with the severity of the visual acuity loss in either eye.

Follow-up visits were scheduled at 1 week and at 1, 3, 6, 12, 18, and 24 months after randomization and then at yearly intervals until closeout of the IONDT.

Data on incidence of fellow eye NAION was collected for 326 patients (Fig. 6.4). A cumulative incidence of 14.7% (48/326) and a cumulative prevalence of 30.6% (128/418) of second eye NAION over a median patient follow-up of 5.1 years were observed. Incidence was greater for the randomized compared with the nonrandomized patients (35/20 [(17.4%)] vs. 13/125 [(10.4%)], respectively). The median interval between the study eye NAION (using enrollment date) and occurrence of new NAION in the fellow eye was 1.2 years

(range, 16 days to 6.0 years). Most new cases (22/48, 46%) of fellow eye NAION in all patients occurred during the first 12 months of follow-up. The IONDT found no relationship between age or sex and the risk of second eye NAION.

Having a vascular condition at baseline was weakly associated in univariate analyses with the occurrence of NAION in the fellow eye ($p = 0.06$). A statistically significant association between baseline diabetes and new second eye NAION (24% in patients with diabetes vs. 12% in patients without, $p = 0.02$) was found. No association was found between a history of smoking and the occurrence of new NAION in the fellow eye ($p = 0.55$). Likewise, there was no association between regular aspirin use and incidence of fellow eye NAION.

Thus, in this large cohort of patients with NAION, new NAION in the fellow eye occurred in 14.7% of patients at risk, over a median follow-up of 5.1 years,

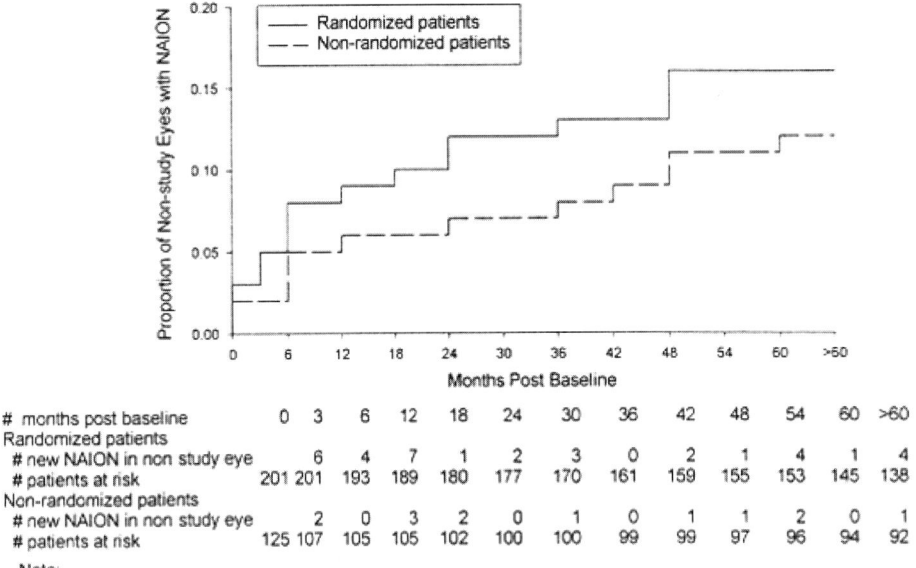

# months post baseline	0	3	6	12	18	24	30	36	42	48	54	60	>60
Randomized patients													
# new NAION in non study eye		6	4	7	1	2	3	0	2	1	4	1	4
# patients at risk	201	201	193	189	180	177	170	161	159	155	153	145	138
Non-randomized patients													
# new NAION in non study eye		2	0	3	2	0	1	0	1	1	2	0	1
# patients at risk	125	107	105	105	102	100	100	99	99	97	96	94	92

Note:
- Persons at risk include all patients not officially withdrawn and without NAION in non-study eye who completed a visit for that time point or a visit beyond that time point. 28 patients in the careful followup group, 28 patients in the surgery group, and 32 patients in the non-randomized group had NAION or optic neuropathy in the non-study eye at baseline and thus are not considered at risk.

FIGURE 6.4 Proportion of nonstudy eyes (fellow eyes) with NAION in patients without NAION or optic neuropathy in the nonstudy eye at baseline by treatment group and months postbaseline. (From ref. 60.)

a lower percentage than previously assumed. Increased incidence of fellow eye NAION is associated with poor visual acuity in the first eye and a history of diabetes mellitus but not with age, sex, smoking history, or aspirin use. Although visual acuities between eyes in patients with bilateral NAION are highly correlated, predicting visual outcome for the second eye in any individual is impossible. Clinicians are encouraged to use data from large prospective studies such as the IONDT when they counsel their patients with NAION.

Future Directions

Therapy to treat acute NAION successfully, reduce the gradual decline in the visual improvement noted in some of these patients, and protect the fellow eye from NAION is not yet available. Neuroprotective agents, reduction of vascular risk factors, aspirin or other antiplatelet medication, and other agents are potential avenues for investigation (26,32,47,60–63). NAION remains a condition that has an unknown etiology and no known means of effective prevention or treatment. Future research must focus on increasing our understanding of the factors leading to the development of NAION and factors associated with prognosis. By achieving better understanding of the pathophysiology and the natural history of the disease, effective prevention and treatment strategies may be developed. However, regardless of any new treatment strategy that is considered, follow-up results of the IONDT underscore the importance of testing any new therapy with a randomized clinical trial, when feasible, before widespread use.

References

1. Percy AK, Nobrega FT, Kurland LT. Optic neuritis and multiple sclerosis. An epidemiologic study. Arch Ophthalmol 1972;87:135–139.

2. Optic Neuritis Study Group. The clinical profile of optic neuritis. Experience of the Optic Neuritis Treatment Trial. Arch Ophthalmol 1991; 109:1673–1678.

3. Optic Neuritis Study Group. ONTT Manual of Procedures. Springfield, VA: National Technical Information Service, 1990. (NTIS accession No. PB90-195728).

4. Beck RW, Cleary PA, Anderson MM Jr, Keltner JL, Shults WT, Kaufman DI, Buckley EG, Corbett JJ, Kupersmith MJ, Miller NR. A randomized, controlled trial of corticosteroids in the treatment of acute optic neuritis. The Optic Neuritis Study Group. N Engl J Med 1992;326:581–588.

5. Cleary PA, Beck RW, Anderson MM Jr, Kenny DJ, Backlund JY, Gilbert PR. Design, methods, and conduct of the Optic Neuritis Treatment Trial. Control Clin Trials 1993;14:123–142.

6. Optic Neuritis Study Group. Visual function more than 10 years after optic neuritis: experience of the Optic Neuritis Treatment Trial. Am J Ophthalmol 2004;137:77–83.

7. Kaufman DI, Trobe JD, Eggenberger ER, Whitaker JN. Practice parameter: the role of corticosteroids in the management of acute monosymptomatic optic neuritis. Report of the Quality Standards Subcommittee of the American Academy of Neurology. Neurology 2000;54:2039–2044.

8. Beck RW, Cleary PA, Backlund JC. The course of visual recovery after optic neuritis. Experience of the Optic Neuritis Treatment Trial. Ophthalmology 1994;101:1771–1778.

9. Beck RW, Cleary PA, Trobe JD, Kaufman DI, Kupersmith MJ, Paty DW, Brown CH. The effect of corticosteroids for acute optic neuritis on the subsequent development of multiple sclerosis. The Optic Neuritis Study Group. N Engl J Med 1993;329:1764–1769.

10. Poser CM, Paty DW, Scheinberg L, McDonald WI, Davis FA, Ebers GC, Johnson KP, Sibley WA, Silberberg DH, Tourtellotte WW. New diagnostic criteria for multiple sclerosis: guidelines for research protocols. Ann Neurol 1983;13: 227–231.

11. Beck RW, Cleary PA. Optic Neuritis Treatment Trial. One-year follow-up results. Arch Ophthalmol 1993;111:773–775.

12. Optic Neuritis Study Group. The 5-year risk of MS after optic neuritis. Experience of the Optic Neuritis Treatment Trial. Neurology 1997; 49:1404–1413.

13. Beck RW. Corticosteroid treatment of optic neuritis: a need to change treatment practices. The Optic Neuritis Study Group. Neurology 1992;42:1133–1135.

14. Beck RW. The Optic Neuritis Treatment Trial. Implications for clinical practice. Optic Neuritis Study Group. Arch Ophthalmol 1992;110:331–332.

15. Keltner JL, Johnson CA, Spurr JO, Beck RW. Baseline visual field profile of optic neuritis. The experience of the Optic Neuritis Treatment Trial. Optic Neuritis Study Group. Arch Ophthalmol 1993;111:231–234.

16. Beck RW, Kupersmith MJ, Cleary PA, Katz B. Fellow eye abnormalities in acute unilateral optic neuritis. Experience of the Optic Neuritis Treatment Trial. Ophthalmology 1993;100:691–697.

17. The Optic Neuritis Study Group. Visual function 5 years after optic neuritis: experience of the Optic Neuritis Treatment Trial. Arch Ophthalmol 1997;115:1545–1552.

18. Beck RW, Arrington J, Murtagh FR, Cleary PA, Kaufman DI. Brain magnetic resonance imaging in acute optic neuritis. Experience of the Optic Neuritis Study Group. Arch Neurol 1993;50:841–846.

19. Optic Neuritis Study Group. High- and low-risk profiles for the development of multiple sclerosis within 10 years after optic neuritis: Experience of the Optic Neuritis Treatment Trial. Arch Ophthalmol 2003;121:944–949.

20. Brex PA, Ciccarelli O, O'Riordan JI, Sailer M, Thompson AJ, Miller DH. A longitudinal study of abnormalities on MRI and disability from multiple sclerosis. N Engl J Med 2002;346:158–164.

21. Jacobs LD, Beck RW, Simon JH, Kinkel RP, Brownscheidle CM, Murray TJ, Simonian NA, Slasor PJ, Sandrock AW. Intramuscular interferon beta-1a therapy initiated during a first demyelinating event in multiple sclerosis. CHAMPS Study Group. N Engl J Med 2000;343:898–904.

22. Goodman BW. Temporal arteritis. Am J Med 1979;67:839–852.

23. Hayreh SS. Anterior Ischemic Optic Neuropathy. New York: Springer-Verlag, 1975.

24. Hayreh SS. Anterior ischemic optic neuropathy. Arch Neurol 1981;38:675–678.

25. Johnson LN, Arnold AC. Incidence of non-arteritic and arteritic anterior ischemic optic neuropathy: population based study in the state of Missouri and Los Angeles county. J Neuro-ophthalmol 1994;14:38–44.

26. Beri M, Klugman MR, Kohler JA et al. Anterior ischemic optic neuropathy. VII. Incidence of bilaterality and various influencing factors. Ophthalmology 1987;94:1020–1028.

27. Boghen DR, Glaser JS. Ischemic optic neuropathy: a clinical profile and natural history. Brain 1975;98:689–708.

28. Repka MX, Savino PJ, Schatz NJ, et al. Clinical profile and long-term implications of anterior ischemic optic neuropathy. Am J Ophthalmol 1983;96:478–483.

29. Hayreh SS, Joos KM, Poddhajsky PA, Long CR. Systemic diseases associated with nonarteritic anterior ischemic optic neuropathy. Am J Ophthalmol 1994;118:766–780.

30. Jacobson DM, Vierkant RA, Belongia EA. Non-arteritic anterior ischemic optic neuropathy: a case-control study of potential risk factors. Arch Ophthalmol 1997;115:1403–1407.

31. Hayreh SS, Podhajsky PA, Zimmerman B. Non-arteritic anterior ischemic optic neuropathy: time of onset of visual loss. Am J Ophthalmol 1997;124:641–647.

32. Landau K, Winterkorn JM, Mailloux LU, Vetter W, Napolitano B. Twenty-four-hour blood pressure monitoring in patients with anterior ischemic optic neuropathy. Arch Ophthalmol 1996;114:570–575.

33. Hayreh SS, Zimmerman MB, Podhajsky P, Alward WL. Nocturnal arterial hypotension and its role in optic nerve head and ocular ischemic disorders. Am J Ophthalmol 1994;117:603–624.

34. Fry CL, Carter JE, Kanter MC, et al. Anterior ischemic optic neuropathy is not associated with carotid artery atherosclerosis. Stroke 1993;24:539–542.

35. Swartz NG, Beck RW, Savino PJ, et al. Pain in anterior ischemic optic neuropathy. J Neuro-ophthalmol 1995;15:9–10.

36. Arnold AC, Hepler RS. Natural history of nonarteritic anterior ischemic optic neuropathy. J Neuroophthalmol 1994;14:66–69.

37. Arnold AC, Badr MA, Hepler RS, Hamilton DR, Lufkin RB. Magnetic resonance imaging of the brain in nonarteritic ischemic optic neuropathy. J Neuroophthalmol 1995;15:158–160.

38. Hayreh SS. Anterior ischemic optic neuropathy. I. Terminology and pathogenesis. Br J Ophthalmol 1974;58:955–963.

39. Galetta SL, Plock GL, Kushner MJ, Wyszynski RE, Brucker AJ. Ocular thrombosis associated with antiphospholipid antibodies. Ann Ophthalmol 1991;23:207–212.

40. Bertram B, Remky A, Arend O, Wolf S, Reim M. Protein C, protein S, and antithrombin III in acute ocular occlusive diseases. Geriatr J Ophthalmol 1995;4:332–335.

41. Rosler DH, Conway MD, Anaya JM, Molina JF, Carr RF, Gharavi AE, Wilson WA. Ischemic optic neuropathy and high-level anticardiolipin antibodies in primary Sjogren's syndrome. Lupus 1995;4:155–157.

42. Hamed LM, Winward KE, Glaser JS, Schatz NJ. Optic neuropathy in uremia. Am J Ophthalmol 1989;108:30–35.

43. Knox DL, Hanneken AM, Hollows FC, Miller NR, Schick HL Jr, Gonzales WL. Uremic optic neuropathy. Arch Ophthalmol 1988;106:50–54.

44. Beck RW, Gamel JW, Willcourt RJ, Berman G. Acute ischemic optic neuropathy in severe pre-eclampsia. Am J Ophthalmol 1980;90:342–346.

45. Katz B. Bilateral sequential migrainous ischemic optic neuropathy. Am J Ophthalmol 1985; 99:489.

46. Arnold AC, Hepler RS, Lieber M, Alexander JM. Hyperbaric oxygen therapy for nonarteritic anterior ischemic optic neuropathy. Am J Ophthalmol 1996;122:535–541.

47. Botelho PJ, Johnson LN, Arnold AC. The effect of aspirin on the visual outcome of nonarteritic anterior ischemic optic neuropathy. Am J Ophthalmol 1996;121:450–451.

48. Sergott RC, Cohen MS, Bosley TM, Savino PJ. Optic nerve decompression may improve the progressive form of nonarteritic ischemic optic neuropathy. Arch Ophthalmol 1989;107: 1743–1754.

49. Kelman SE, Elman MJ. Optic nerve sheath decompression for nonarteritic ischemic optic neuropathy improves multiple visual function measurements. Arch Ophthalmol 1991;109: 667–671.

50. Spoor TC, Wilkinson MJ, Ramocki JM. Optic nerve sheath decompression for the treatment of progressive nonarteritic ischemic optic neuropathy. Am J Ophthalmol 1991;111:724–728.

51. Spoor TC, McHenry JG, Lau-Sickon L. Progressive and static nonarteritic ischemic optic neuropathy treated by optic nerve sheath decompression. Ophthalmology 1993;100:306–311.

52. The Ischemic Optic Neuropathy Decompression Trial (IONDT): design and methods. The IONDT research group. Controlled Clin Trials 1998; 19:276–296.

53. Ware JE Jr, Sherbourne CD. The MOS 36-item short-form health survey (SF-36). I. Conceptual framework and item selection. Med Care 1992;30: 473–483.

54. Westheimer G. Scaling of visual acuity measurements. Arch Ophthalmol 1979;97:327–330.

55. Ischemic Optic Neuropathy Decompression Trial Research Group. Optic nerve decompression surgery for nonarteritic anterior ischemic optic neuropathy (NAION) is not effective and may be harmful. JAMA 1995;273:625–632.

56. Yee RD, Selky AK, Purvin VA. Outcomes of optic nerve sheath decompression for nonarteritic ischemic optic neuropathy. J Neuroophthalmol 1994;14:70–76.

57. Rizzo JF 3rd, Lessell S. Optic neuritis and ischemic optic neuropathy. Overlapping clinical profiles. Arch Ophthalmol 1991;109:1668–1672.

58. Boghen DR, Glaser JS. Ischaemic optic neuropathy. The clinical profile and history. Brain 1975;98:689–708.

59. Ischemic Optic Neuropathy Decompression Trial Research Group. Ischemic optic neuropathy decompression trial. Twenty-four-month update. Arch Ophthalmol 2000;118:793–798.

60. Newman NJ, Scherer R, Langenberg P, et al. for the Ischemic Optic Neuropathy Decompression Trial Research Group. The fellow eye in NAION: report from the ischemic optic neuropathy decompression trial study group. Am J Ophthalmol 2002;134:317–328.

61. Beck RW, Hayreh SS, Podhajsky PA, Tan ES, Moke PS. Aspirin therapy in nonarteritic anterior ischemic optic neuropathy. Am J Ophthalmol 1997;123:212–217.

62. Jacobson DM, Vierkant RA, Belongia EA. Nonarteritic anterior ischemic optic neuropathy: a case-control study of potential risk factors. Arch Ophthalmol 1997;115:1403–1407.

63. Johnson LN, Gould TJ, Krohel GB. Effect of levodpa and carbidopa on recovery of visual function in patients with nonarteritic anterior ischemic optic neuropathy of longer than six months duration. Am J Ophthalmol 1996;121: 77–83.

7

Brain Resuscitation

Edwin Nemoto and Charles C. King, MD

The resuscitation of the brain after systemic circulatory arrest or global anoxic or hypoxic insults is complicated by the interaction of arrest time, duration of cardiopulmonary resuscitation, time to return of spontaneous circulation (ROSC), and comorbidities, all major impact factors affecting long-term outcome. These factors could overshadow and possibly nullify any attempt to attenuate ischemic injury sustained by the brain with therapeutic intervention postresuscitation. Thus, research on the various aspects of cardiopulmonary resuscitation (CPR) on outcome in both animals and clinically have mainly focused on the role of age, resuscitation drugs such as vasopressin and epinephrine, and methods of CPR on survival and outcome after cardiac arrest. Nevertheless, substantial evidence on the pathogenesis of global ischemic brain injury suggests that postresuscitation changes in the brain may be amenable to therapeutic intervention as demonstrated in experimental animals.

Of the large number of pharmacological therapeutic agents explored in brain resuscitation, it is clear that hypothermia is the most effective therapeutic intervention available by far, the efficacy of which was first demonstrated in experimental animals and subsequently in a few clinical studies using moderate hypothermia. In support of this notion, clinically, hypothermia is the mainstay of brain protection against

elective cerebral ischemic insults as may occur during deep hypothermic cardiac arrest in cardiac surgery, although barbiturate suppression may be used as well.

The multifactor effect of hypothermia in brain resuscitation has raised the question as to whether therapies aimed at single factors such as excitotoxic neurotransmitters, inflammatory responses, and metabolic suppression are doomed to failure relative to hypothermia and new multifactor therapeutic interventions may be needed. Animal studies evaluating other therapies should use hypothermia as the standard for comparison before translation to the clinical arena.

Although present understanding of the mechanisms in the pathogenesis of CPR and of ischemic brain injury increases with the elucidation of different mediators such as cytokines, cyclooxygenases, lipid mediators, mitogen activated kinases, and so forth, therapies for brain resuscitation have not developed in parallel and appear to be the major bottleneck in the conduct of clinical research in brain resuscitation.

The impact of factors associated with cardiac arrest and CPR may obfuscate attempts to demonstrate effectiveness of therapeutic intervention for brain resuscitation. Nevertheless, the major problem at this juncture appears to be the lack of potential therapies that would rival the effectiveness of hypothermia in brain resuscitation. Thus, the development of new therapeutic intervention modalities need to be developed before the pace of clinical trials can progress.

EPIDEMIOLOGY AND NATURAL HISTORY

Epidemiology

In the United States, it is estimated that anywhere between 370,000 to 750,000 people suffer cardiac arrest precipitated primarily by an acutely ischemic cardiac myocardium and myocardial infarction associated with arrhythmia of ventricular fibrillation. Of these people, approximately 225,000 or nearly one-half will die before they reach the hospital. The etiology of the arrest is important in that ventricular fibrillation or tachycardia and witnessed cardiac arrest have a hospital discharge rate of 34% compared with other rhythm disturbances with a hospital discharge rate of 6%.

Clearly, factors involved in CPR impact greatly on survival after cardiac arrest. In out-of-hospital cardiac arrest, important predictors for immediate survival are witnessed arrest and ventricular fibrillation as well as the initial arrhythmia and CPR (1). When emergency medicine technicians witness the arrest, ROSC is 49% and hospital discharge is 21% (2). When witnessed by lay people, ROSC was 20.5% and hospital discharge was 4%. When unwitnessed, ROSC was 8.6% and hospital discharge was 1.7%.

In in-hospital cardiac arrest, the most common reasons for cardiac arrest are cardiac arrhythmia, acute respiratory insufficiency, and hypotension (3). The rate of ROSC is 44% overall, with 17% surviving to hospital discharge who were in good neurological outcome with a Cerebral Performance Category (CPC) of 1 (i.e., normal to slightly disabled), on discharge. These observations are relevant to brain resuscitation

after cardiac arrest because they illustrate that there are other factors that influence good recovery such as duration of arrest and CPR time (4) in a dichotomous fashion that could dramatically overshadow the success or failure of therapeutic procedures. Thus, it is important to identify the target patient population.

What is the target population for brain resuscitation? In a study evaluating the efficacy of thiopental in ameliorating brain injury or improving recovery after cardiac arrest (5), the percentage of patients at 1 month with CPCs of 1 (normal/slightly disabled) and 2 (moderately disabled) averaged about 17% and 6%, respectively, whereas the severely disabled group averaged about 3% in both groups. Theoretically, the severely disabled group is the target group. Similarly, in a therapeutic trial with a combination of magnesium and diazepam (6), at 3 months, about 20% were independent and 5% to 7% dependent and none in a vegetative state. The question is with target groups of this size, to what extent are we going to be able to influence recovery from cardiac arrest? With respect to brain resuscitation, the assumption is that good CPR goes hand in hand with good overall and cerebral performance categories.

Natural History

The natural history of cerebral dysfunction as a result of cardiac arrest has long been debated and studied from the early days of subjecting nonhuman primates to circulatory arrest without postarrest intensive care (7) to studies with intensive care (8). Despite the considerable evidence in animal models showing the ability of the brain to recover under controlled conditions from periods of ischemia of 16 (8) and even 60 minutes (9) of complete circulatory arrest, clinically, the predicted tolerance of the brain to complete circulatory arrest remains at 5–6 minutes of arrest time (5). The reasons for this disparity in the tolerable limits of the brain to complete cessation of circulation are unclear. However, as shown by Safar and colleagues (10), the ability to rapidly reperfuse the brain at relatively high perfusion pressure with "diluted blood" appears to play an important role. These studies emphasize the necessity of a rapid and aggressive restoration of cerebral perfusion after cardiac arrest. In the clinical situation, however, periods of cardiac arrest are typically followed by low perfusion pressure with acidotic blood and, likely, high hematocrit with a tendency to clot. Whether this problem of reperfusion might explain the discrepancy between clinically and experimentally tolerable limits of cerebral circulatory arrest remains to be determined. This potential problem with thrombosis after resuscitation has led to studies suggesting the application of thrombolytic therapy after cardiac arrest (11).

In patients suffering cardiac arrest, perhaps the only study using positron emission tomography in the early postarrest period, showed an elevated cerebral blood flow (CBF) and cerebral metabolic rate for oxygen with a low oxygen extraction fraction ratio between postarrest days 1 and 2, and in a patient regaining consciousness, followed by a reduction in CBF and increased oxygen extraction fraction by day (76). In contrast, a patient who did not regain consciousness showed markedly less hyperemia between days 1 and 2 and a marked reduction in CBF and cerebral metabolic rate for oxygen on days 4 and 7 postarrest. Importantly, almost all subjects showed increased oxygen extraction fraction in the later stages postarrest, suggesting that oxygen delivery was limiting relative to oxygen demand corroborating earlier suggestions of increased oxygen demand relative to oxygen delivery (12,13). These results are supported by the observation that postarrest CBF autoregulation is shifted to higher perfusion pressures (14) and suggest that the degree

of reperfusion postarrest appears to be a strong predictor of cerebral recovery from cardiac arrest.

DIAGNOSIS AND SUBPOPULATIONS

The Utstein Style

Before discussing the diagnosis and subpopulations of brain resuscitation after cardiac arrest, it is helpful to introduce the "Utstein style," whose aim is to standardize data gathering and presentation in both clinical and experimental studies on cardiac arrest and cardiopulmonary resuscitation. The Utstein style originated out of a conference in an ancient abbey, "Utstein," in Stavanger, Norway in 1990 where representatives from the American Heart Association, the European Resuscitation Council, the Heart and Stroke Foundation of Canada, and the Australian Resuscitation Council met to establish uniform terms and definitions for out-of-hospital resuscitation, the centerpiece of which was the development of methods for uniform outcome reporting (see http://www.americanheart.org/presenter.jhtml?identifier=3004611). Thereafter, guidelines for in-hospital resuscitation modeled after the out-of-hospital guidelines were developed along with guidelines for uniform reporting of pediatric advanced life support, laboratory CPR research, and data from drowning. These guidelines are comprehensive, providing consensus statements on all aspects related to cardiac arrest CPR.

As first defined in the conference held in 1990, the Utstein style begins from consensus statements on terminology to reporting of data in 22 different categories from the population served to the number of cardiac arrests considered for resuscitation to survival at 1 year, which serves to illustrate the complexity of conducting clinical studies on the effectiveness of therapies aimed at brain resuscitation.

Among the relevant variables to the topic of brain resuscitation and neuroprotection is the assessment of outcome in terms of quality and duration. Short-term survival in expensive neurointensive care units would be of little benefit to society and is therefore important information. The Glasgow-Pittsburgh outcome categories have been the most widely used in evaluating quality of life after successful resuscitation. There are two performance categories, the Overall Performance Categories, reflecting cerebral and noncerebral status, and the Cerebral Performance Categories, reflecting only cerebral performance. Both categories are reliable and easily obtained even by telephone. The time to awakening or return to consciousness is even simpler (see http://www.americanheart.org/presenter.jhtml?identifier=3001853).

Diagnosis

In the assessment of outcome after cardiac arrest, the Utstein style adopted the Glasgow-Pittsburgh Outcome Categories, which feature two 5-point parallel scales: the Cerebral Performance Categories and the Overall Performance Categories (5). The Glasgow-Pittsburgh Overall Performance and Cerebral Performance Categories are presented in detail in Table 7.1.

The Utstein style uses templates for data collection and reporting. The data collection time points during the response to the arrest indicates the specific times that are to be noted (Fig. 7.1).

The characterization of cardiac arrest events begins with the origin of the arrest and the resulting cardiac electrical disturbances that occur. In a study involving 4914 patients, the primary reason for cardiac arrest was myocardial infarction and the primary electrical abnormalities were ventricular fibrillation (45%), asystole (31%), other (14%), pulseless electrical activity (10%), and ventricular tachycardia

(1%). Of the patients who suffer ventricular fibrillation with witnessed arrest, 34% were discharged from the hospital as opposed to 6% with other rhythm disturbances. However, the incidence of ventricular fibrillation has apparently declined (15). For unwitnessed arrests, the corresponding hospital discharge figures are 12% and 1%, respectively. In a smaller study, involving 708 patients (2), EMS witnessed arrest had 49% ROSC and 21% hospital discharge, and when witnessed by lay people, the corresponding figures were 21% and 4.4%. Ventricular fibrillation was highly predictive of outcome. Prehospital arrest in this study had an incidence of 0.95 per 1000 and a survival rate of 6.7%.

In a report of 14,720 cardiac cases from the National Registry of CPR (3), a 17% overall hospital discharge survival rate was observed. Of the survivors, the distribution of CPC 1 and 2 categories represented 85% of the patients, and the moderate (CPC 3) and severe (CPC 4) categories were 26% and 11%, respectively, for a total of 37% of the survivors who may be improved by the application of appropriate therapy (i.e., the target group for therapeutic intervention). However, in the application of therapies that are myocardial depressants, justification for therapeutic intervention can only be made for those patients who do not regain consciousness within 50 minutes of ROSC. For oxygenation as used in the protocol with thiopental therapy (5), the percentage of potentially "salvageable" patients was less. At 10 days postarrest, the percentage of patients in CPC categories 1 and 2 was about 10% for both, whereas the CPC 3 and 4 categories represented about 7% and 15% for a total "salvageable" group of about 20–25%. In a study investigating the efficacy of a combination of magnesium and diazepam in which unconscious cardiac arrest patients after ROSC were entered into the study, at 3 months the percentage of independent survivors was about 23%, whereas

TABLE 7.1   Glasgow-Pittsburgh outcome categorization of brain injury

Cerebral Performance Categories (CPC)	Overall Performance Categories (OPC)
1. *Good cerebral performance:* Conscious, alert, able to work and lead a normal life. Might have minor psychological or neurological deficits (mild dysphasia, nonincapacitating hemiparesis, or minor cranial nerve abnormalities).	1. *Good overall performance:* Healthy, alert, capable of normal life. Good cerebral performance (CPC 1) plus no functional disability from noncerebral organ system abnormalities.
2. *Moderate cerebral disability:* Sufficient cerebral function for part-time work in sheltered environment or independent activities of daily life (dress, travel by public transportation, food preparation). Such patients may have hemiplegia, seizures, ataxia, dysarthria, dysphasia, or permanent memory or mental changes.	2. *Moderate overall disability:* Conscious. Moderate cerebral disability alone (CPC 2) or moderate disability from noncerebral systems dysfunction alone or both. Performs independent activities of daily life (dress, travel, food preparation) or able to work part-time in sheltered environment, but disabled for competitive work.
3. *Severe cerebral disability:* Conscious; patient dependent on others for daily support (in an institution or at home with exceptional family effort) because of impaired brain function. Has at least limited cognition. This category includes a wide range of cerebral abnormalities, from patients who are ambulatory but have severe memory disturbance or dementia precluding independent existence, to those who are paralyzed and can communicate only with their eyes, as in the locked-in syndrome.	3. *Severe overall disability:* Conscious. Severe cerebral disability alone (CPC 3) or severe disability from noncerebral organ system dysfunction alone, or both. Dependent on others for daily support.
4. *Coma/vegetative state:* Not conscious, unaware of surroundings, no cognition. No verbal and/or psychological interaction with environment.	4. Same as CPC 4.
5. *Brain death:* Certified brain dead or dead by traditional criteria.	5. Same as CPC 5.

the dependent group ranged between 4% and 7% (6), a much smaller percentage than in the study on the natural history (3). Neither thiopental nor magnesium/diazepam showed any beneficial effects when applied to this group of patients who could represent the target population. The target population would also have to be further scrutinized based on the specific therapy being tested and the degree to which it can be randomized and that it could lead to potential complications.

It is clear that the variables with the greatest impact on survival or recovery are related to arrest time, CPR time, and time to ROSC. Apart from arrest time, the CPR time and ROSC time are events that could be influenced by appropriate therapeutic intervention. The importance of restoration of cerebral oxygenation is emphasized by the observation that cerebral oxygenation measurements by near-infrared spectroscopy showed that survivors (cerebral saturation, 63%) of cardiac arrest were distinguished from

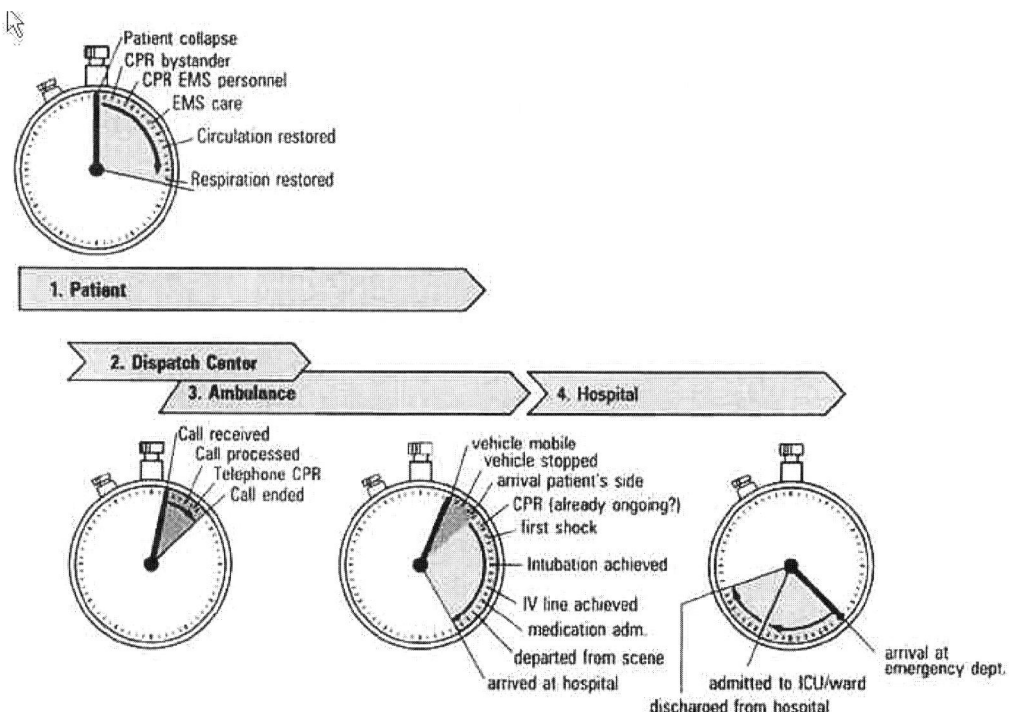

FIGURE 7.1   Utstein style template for recording times during and after cardiac arrest to hospital discharge. (From "Guidelines for Uniform Reporting of Data from Out-of-Hospital Cardiac Arrest: The Utstein style," with permission from The American Heart Association.)

nonsurvivors (cerebral saturation, 46%) with lower cerebral oxygen saturation measurements postarrest measured upon admission into the emergency department (16). A more recent study (17) reported the inability to record an increase in cerebral oxygenation during resuscitation in out-of-hospital cardiac arrest until ROSC and concluded that it is possible that current mechanical methods of CPR provide little or no cerebral oxygenation to the brain during out-of-hospital cardiac arrest, which emphasizes the need to shorten the time to ROSC. Shortening the CPR time, which may be aided by the regimen used to restore spontaneous circulation, and the time to ROSC might be considered steps toward brain resuscitation and survival.

Data collection according to the Utstein style will allow the separation of patients suffering cardiac arrest with successful resuscitation into the different subpopulations that need not be differentiated at the time of resuscitation. Each therapeutic trial will have to decide which subpopulation of cardiac arrest should be included or excluded and the target population identified. Accordingly, for the application of therapeutic interventions such as thiopental, it was decided that rather than put patients who are successfully resuscitated and are responsive to pain at risk with thiopental loading, they were to exclude them. Indeed, it is likely that the beneficial effects of thiopental therapy may have been apparent if applied to these patients as well.

CURRENT DISEASE MANAGEMENT

Management of Cardiac Resuscitation

Guidelines for the management of cardiac arrest have been published by the American Heart Association (see http://circ.ahajournals.org/cgi/content/full/95/8/2172). The management of ventricular fibrillation and pulseless ventricular tachycardia and various drugs useful for cardiac resuscitation are provided.

Management of Brain Resuscitation

Hypothermia

There are no specific guidelines for the application of specific therapies for brain resuscitation except for the application of hypothermia by the American Heart Association (see http://circ.ahajournals.org/cgi/content/full/108/1/118). Important considerations in the application of hypothermia are the patient population, the inclusion and exclusion criteria, and the duration of hypothermia, optimal guidelines that have not necessarily been resolved at this time. The inclusion criteria for the studies differed in some respects.

The Hypothermia After Cardiac Arrest Study Group (18) enrolled 275 patients in a randomized, control, blinded outcome using the following inclusion criteria:

- Witnessed cardiac arrest;
- Ventricular fibrillation or nonperfusing ventricular tachycardia as the initial cardiac rhythm;
- Presumed cardiac origin of the arrest;
- Age 18 to 75;
- Estimated interval of 5 to 15 minutes from the patient's collapse to the first attempt at resuscitation by emergency medical personnel;
- An interval of no more than 60 minutes from collapse to restoration of spontaneous circulation.

Exclusion criteria were as follows:

- Tympanic membrane temperature below 30°C on admission;

- Comatose state before cardiac arrest due to drugs that depress the central nervous system;
- Pregnancy;
- Response to verbal commands after ROSC and before randomization;
- Mean arterial pressure <60 mmHg for more than 30 minutes after ROSC and before randomization;
- Evidence of hypoxemia (arterial oxygen saturation <85%) for more than 15 minutes after ROSC and before randomization;
- A terminal illness preceding the arrest, factors that made participation in follow-up likely;
- Enrollment in another study;
- Cardiac arrest after arrival of emergency medical personnel
- Known preexisting coagulopathy.

Bladder temperature was reduced to 32 to 34°C with external cooling with a goal of achieving the target temperature within 4 hours after ROSC with sedation with midazolam and fentanyl and muscle paralysis with pancuronium. The temperature was maintained for 24 hours followed by passive rewarming, which lasted for about 8 hours. Outcome (CPC 1 and 2) in the hypothermia group was significantly ($p < 0.009$) better than the normothermic group. Complications between the hypothermia and normothermia groups were similar.

In a feasibility and safety study (19) involving nine patients, the inclusion criteria were as follows:

- Documented out-of-hospital cardiac arrest;
- ROSC with systolic blood pressure <90 within 60 minutes of initiation of ROSC;
- Enrollment and initiation of hypothermia within 90 minutes of initiation of advanced cardiac life support (ACLS);
- Age 18 to 75;
- Comatose upon enrollment (Glasgow Coma Scale <8)

- Informed consent obtained.

Exclusion criteria were as follows:

- Significant cardiac dysrhythmias or cardiac instability;
- Electrocardiographic evidence of continuing acute myocardial ischemia;
- Evidence of sepsis;
- Vasoactive drugs required to maintain adequate perfusion;
- Cardiogenic shock (systolic blood pressure <90)
- Coagulopathy or thrombocytopenia
- QT interval >470 ms
- In-hospital cardiac arrest (CA);
- Any condition precluding treatment in the opinion of the primary physician.

The advisory statement by the Advanced Life Support Task Force of the International Liaison Committee on Resuscitation made the following recommendations for hypothermia after cardiac arrest (20):

- Unconscious adult patients with spontaneous circulation after out-of-hospital cardiac arrest should be cooled to 32°C to 34°C for 12 to 24 hours when the initial rhythm was ventricular fibrillation;
- Such cooling may also be beneficial for other rhythms or for in-hospital cardiac arrest.

Bernard and colleagues (2) studied 77 patients randomly assigned to hypothermia to 33°C within 2 hours after ROSC with 43 patients subjected to hypothermia and 34 patients in the normothermic group. Twenty-one of the 43 patients subjected to hypothermia had a good outcome compared with 9 of the 34 normothermic patients who had a good outcome. Adjusting for baseline differences, the odds ratio for a good outcome with hypothermia compared with normothermia was 5.25

($p = 0.01$). Their inclusion and exclusion criteria were as follows:

Inclusion criteria
- Initial cardiac rhythm of ventricular fibrillation at time of ambulance arrival;
- Successful ROSC;
- Persistent coma after ROSC.

Exclusion criteria
- Age < 18 years for men, < 50 years for women (for pregnancy concerns);
- Cardiogenic shock (blood pressure < 90 mmHg despite epinephrine infusion);
- Possible causes of coma other than cardiac arrest;
- No intensive care unit bed available.

Management Concerns

Although the application of hypothermia is being promoted, investigators are becoming increasingly concerned about elevated temperatures after cardiac arrest (22), where a lower temperature upon admission and a higher maximum temperature by 1°C differentiated patients making good and unfavorable neurological recovery. Although it is well appreciated that hyperthermia is a consequence of cerebral ischemia in stroke (23) and head injury (24), it is not well appreciated that cerebral ischemia itself causes an increase in brain temperature because the blood perfusing the brain is normally cooling the brain. If metabolism continues during ischemia, brain tissue temperature increases of up to 2 to 3°C may occur (25).

The overall response to cerebral injury whether due to cardiac arrest, stroke, or head injury elicits activation of the pituitary adrenal axis and the stress response. The stress response is intended to counter the insult on the brain and maintain continued perfusion of the brain. In doing so, however, it may aggravate the injury to the brain through direct effects of blood-borne catecholamines on the brain and its blood vessels (26). Thus, suppression of the adrenergic response should be instituted.

It should also be noted that the induction of hypothermia by external cooling of the body elicits a severe stress response (27) and is a well-known stress inducer, which also has the potential of counteracting the beneficial effects of hypothermia in attenuating ischemic brain damage after cardiac arrest. Finally, it is also important to maintain arterial blood pressure on the high side of normal to ensure adequate perfusion of the brain post–cardiac arrest when the perfusion pressure is needed to maintain adequate cerebral perfusion (14). CBF autoregulation also appears to be disrupted postarrest.

ISSUES OF CLINICAL THERAPEUTIC PROTOCOL DEVELOPMENT

Dosing Regimen

The dosing regimen should be designed according to the specific therapeutic agent being applied, the target of the injury process addressed, and the dosing regimen used in the preclinical studies that demonstrated a resuscitative effect. In the application of thiopental therapy postresuscitation, the drug was administered as early as possible after ROSC using a dose of 30 mg/kg within the first 30 minutes using vasopressors as needed to maintain arterial blood pressure within normal levels (5). The intention in the application of thiopental was rapid saturation of the brain with the initial ROSC and loading during the early postresuscitative phase as applied in the nonhuman primate model.

In the application of hypothermia, the temperature was guided by the animal studies demonstrating efficacy (28) using moderate hypothermia with temperatures in the range of 32°C to 34°C. On the other hand, the optimal duration of hypothermia and the method of rewarming were unknown at the time of implementation. It is almost a certainty that the application of cerebral resuscitation drugs after cardiac arrest would attempt to apply the therapeutic agent early in the postresuscitative phase.

Recommendations for Consideration in Clinical Protocol Development

In the application of any therapeutic regimen in any clinical trial, the following should be considered:

- The relative efficacy of the proposed therapeutic intervention relative to the efficacy of moderate hypothermia and the animal studies (preferably non-human primate) defining optimal parameters for application;
- The target patient population within the group of patients suffering cardiac arrest with survival and the power of the study with the predicted outcomes based on the size of the target population;
- The intended pathological process being targeted (i.e., inflammatory response, delayed excitotoxic injury, cerebral hypoperfusion, free radicals, etc.) and the relevant time window postarrest;
- The potentially harmful effects of the drug being administered in compromising the intensive care of the patient and the perfusion of the brain;
- The duration of the application of the drug;
- The variables that are used to monitor the appropriate application of the drug, that is, a physiological response rather than a dose per milligram application because of interindividual variability in drug sensitivity;

- The endpoints both interim and final that will be used to judge the efficacy of the drug;
- A Data and Safety Monitoring Board specifying stopping rules;
- Identification of all key personnel and roles in the project.

OUTCOMES OF RESUSCITATION

In addition to the data collected as illustrated in Figure 7.1, the Utstein style guidelines published by the American Heart Association recommend recording the following variables in the documentation of the outcome after resuscitation in addition to the Overall Performance and Cerebral Performance Categories illustrated in Table 7.1

- Site of cardiac arrest (core);
- Prearrest clinical status (supplementary): Overall Performance Category and Cerebral Performance Category;
- Witnessed arrest: yes or no;
- Precipitating event (supplementary) (determined as best possible at the scene);
- Clinical status of patient when ambulance arrives (core): breathing (yes/no);
- Initial recorded rhythm (core);
- Treatment (core);
- Final patient status at the scene (core);
- Status on arrival at emergency department (supplementary);
- Status after treatment in the emergency department;
- Discharged alive (core);
- Discharge destination (supplementary);
- Alive at 1 year (yes/no) (core). Supplementary data include the best Cerebral Performance Category achieved.

Other outcome scales that could be used to judge the quality of life are the modified Rankin scale, the Barthel Index, and the Quality of Life scales. The modified Rankin Scale is a one-item, 5-point scale that rates

disability and need for assistance. It is the most commonly used outcome classification scale for disabilities and handicaps and has demonstrated adequate interobserver reliability and validity (29). The Barthel index (30) (20-point version) is used with a dichotomous category of scores = 16 versus 17–20 as an outcome variable. The Barthel Index is a self-report measure of functional disability, focused on bodily oriented personal care. Family members are interviewed for individuals unable to provide a self-report. This index has been extensively used in clinical research, has high interrater reliability, and correlates highly with performance functional measures (31).

GOLD AND SILVER STANDARDS

In a statement on the "Gold Standards for Outcome Comparisons" in cardiac arrest and resuscitation guidelines, the American Heart Association stated that "The task force recognizes that no individual hospital will have enough patients to support meaningful interhospital outcome comparisons." Therefore, no single gold standard recommendation for outcome comparisons was made. Nevertheless, researchers should provide the survival rates listed in template boxes, 10 (discharged alive), 11 (died within 1 year of discharge), and 12 (alive at 1 year).

CONCLUSION

Despite decades of related research, experimental and clinical brain resuscitation research is in relative infancy. The complications of brain resuscitation research are partly due to the impact of factors that are more directly associated with cardiac arrest, such as arrest time, CPR time, and time to ROSC, which may overshadow the effects of brain-directed therapies in the postresuscitative phase. Furthermore, even though present knowledge on the pathophysiological and metabolic processes involved in the pathogenesis of global ischemic brain injury is being elucidated at a rapid pace, none has stood out as the primary factor in the myriad of processes contributing to the severity of ischemic brain injury. The only proven therapy for brain resuscitation that has been experimentally and clinically proven is moderate hypothermia, and even that without clearcut guidelines as to optimal temperature, onset, duration, and rate of rewarming. It may be that the inability of drugs directed to specific processes in the pathogenesis of ischemic brain injury in the postresuscitative phase is due to the fact that they target isolated processes in contrast to the blanket multifactor effects of hypothermia. Presently, the future of efforts in brain resuscitation depends on the development of therapies with hypothermia as the standard of comparison.

References

1. Engdahl J, Homberg M, Karlson BW, Luepker R, Herlitz J. The epidemiology of out-of-hospital "sudden" cardiac arrest. Resuscitation 2002;52:235–245.
2. Kette F, Sbrojavacca R, Rellini G, Tosolini G, Capasso M, Arcidiacono D, Bernardi G, Fittitta P. Epidemiology and survival rate of out-of-hospital cardiac arrest in north-east Italy: the F.A.C.S. study. Resuscitation 1998;36:153–159.
3. Perbedy MA, Kaye W, Ornato JP, Larkin GL, Nadkarni V, Mancini ME, Berg RA, Nichol G, Lane-Trultt T. Cardiopulmonary resuscitation of adults in the hospital: A report of 14720 cardiac arrests from the National Registry of Cardiopulmonary Resuscitation. Resuscitation 2003;58:294–308.
4. Abramson N, Safar P, Detre K, Kelsey SF, Monroe J, Reinmuth O, Snyder J. Brain Resuscitation Clinical Trial I Study Group: Steering Committee. Neurologic recovery after cardiac arrest: effect of duration of ischemia. Crit Care Med 1885; 13:930–931.
5. Brain Resuscitation Clinical Trial I Study Group. Randomized clinical study of thiopental loading in comatose survivors of cardiac arrest. N Eng J Med 1886;314:397–403.
6. Longstreth WT, Fahrenbruch CE, Olsufka N, Walsh TR, Copass MK, Cobb LA. Neurology 2002;59:506–514.
7. Miller JR, Myers RE. Neurological effects of systemic circulatory arrest in the monkey. Neurology 1970;20:715–724.
8. Nemoto EM, Bleyaert AL, Stezoski SW, Moossy J, Safar P. Global brain ischemia: a reproducible monkey model. Stroke 1977;8:558–564.
9. Hossmann K-A, Zimmermann,V. Resuscitation of the monkey brain after 1 h complete ischemia. I. Physiological and morphological observations. Brain Res 1974;81:59–74.
10. Safar P, Stezoski SW, Nemoto EM. Amelioration of brain damage after 12 minutes cardiac arrest in dogs. Arch Neurol 1976;33:91–95.
11. Bottinger BW, Martin EW. Thrombolytic therapy during cardiopulmonary resuscitation and the role of coagulation activation after cardiac arrest. Curr Opin Crit Care 2001;7:176–183.
12. Hossmann KA, Lechtape-Gruter HL, Hossmann V. The role of cerebral blood flow for the recovery of the brain after prolonged ischemia. Z Neurol 1973;204:281–299.
13. Nemoto EM, Snyder JV, Carroll RG. Global ischemia in dogs: cerebrovascular CO_2 reactivity and autoregulation. Stroke 1975;6:415–431.
14. Sundgreen C, Larsen FS, Herzog TM, Knudsen GM, Boesgaard S, Aldershvile J. Autoregulation of cerebral blood flow in patients resuscitated from cardiac arrest. Stroke 2001;32: 128–132.
15. Cobb LA, Fahrenbruch CE, Oluska M, Copass MK. Changing incidence of out-of-hospital ventricular fibrillation. JAMA 2002;18:3008–3013.
16. Mullner M, Sterz F, Binder M, Hirschl MM, Janata K, Laggner AN. Near infared spectroscopy during and after cardiac arrest-preliminary results. Clin Intens Care 1995;6:107–111.
17. Newman DH, Freed J, Callaway CW. Cerebral oximetry and ventilation changes in out-of-hospital cardiac arrest. Ann Emerg Med 2002; 40:S23 (Abstract).
18. The Hypothermia After Cardiac Arrest Hypothermia Study Group. Mild therapeutic hypothermia to improve the neurologic outcome after cardiac arrest. N Engl J Med 2002;346: 549–556.
19. Felberg RA, Krieger DW, Chuang R, Persse DE, Burgin WS, Hickenbottom SL, Morgenstern LB, Rosales O, Grotta JC. Hypothermia after cardiac arrest. Circulation 2001;104:1799–1804.
20. Nolan JP, Morley PT, Vanden Hoek TL, Hickey RW. Therapeutic hypothermia after cardiac arrest. Circulation 2003;108:118–121.
21. Bernard SA, Gray TW, Buist MD, Jones BM, Silvester W, Gutteridge G, Smith K. Treatment of comatose survivors of out-of-hospital cardiac arrest with induced hypothermia. N Engl J Med 2002;21:557–563.
22. Zeiner A, Holzer M, Sterz F, Schorkhuber W, Eisenburger P, Havel C, Kliegel A, Laggner A. Hyperthermia after cardiac arrest is associated

with an unfavorable neurologic outcome. Arch Intern Med 2001;161:2007–2012.

23. Castillo J, Davalos A, Marrugat J, Noya M. Timing for fever-related brain damage in acute ischemic stroke. Stroke 1998;29:2455–2460.

24. Thompson HJ, Pinto-Martin J, Bullock MR. Neurogenic fever after traumatic brain injury: an epidemiological study. J Neurol Neurosurg Psych. 2003;74:614–619.

25. Nemoto EM, Jungreis CA, Larnard D, Kuwabara H, Horowitz M, Kassam A. Hyperthermia and hypermetabolism in focal cerebral ischemia. Adv Exp Med 2004.

26. MacKenzie ET, McCulloch J, Harper AM. Influence of endogenous noradrenaline on local cerebral blood flow and metabolism. Am J Physiol 1976;274:149–156.

27. Frank SM, Satipunwaycha P, Bruce SR, Herscovitch P, Goldstein DS. Increased myocardial perfusion and sympathoadrenal activation during mild core hypothermia in awake humans. Clin Sci 2003;104:503–508.

28. Ebymeyer U, Safar P, Radovsky A, Obrist W, Pomeranz AlH. Moderate hypothermia for 48 hours after temporary epidural brain compression in a canine outcome model. J Neurotrauma 1998;15:323–326.

29. Wolfe CDA, Taub NA, Woodrow EJ, Burney PGJ. Assessment scales of disability and handicap for stroke patients. Stroke 1991;22:1242–1244.

30. Wade DT, Collin C. The Barthel ADL index: A standard measure of physical disability? Int Disabil Stud 1988;11:89–92.

31. Shinar D, Gross C, Bronstein KS, Licata-Gehr EE, Eden DT, Cabrera AR, Fishman IG, Roth AA, Barwick JA, Kunitz SC. Reliability of the activities of daily living scale and its use in telephone interview. Arch Phys Med Rehabil 1987;68:723–728.

8

Clinical Trials in Brain Injury After Cardiac Arrest

Romergryko G. Geocadin and Daniel F. Hanley

EPIDEMIOLOGY

Cardiac arrest is common a medical problem with about 200,000 to 375,000 cases annually (1). Historically, cardiopulmonary resuscitation (CPR) restores spontaneous circulation in at least 70,000 patients a year in the United States (2). About 37% of patients survive to discharge after in-hospital cardiac arrest, whereas only 9% of out-of-hospital cardiac arrest patients survive to discharge (3). Approximately half of the survivors experience persistent brain dysfunction (4), and only 3% to 10% are able to resume their former lifestyle (4,5).

Efforts to improve CPR protocols continue to be undertaken. In 2000, Advanced Cardiac Life Support was modified to include vasopressin by the American Heart Association as a treatment option (6). The use of vasopressin improved survival after out-of-hospital cardiac arrest but demonstrated no added benefit in neurological outcome (7). Patient survival from out-of-hospital cardiac arrest has increased with efforts such as faster response time, the use of early cardiac defibrillation, and increased public awareness of basic life support skills (8,9). An increase in cardiac arrest victims who survive with severe neurological injury has been noted (10).

RESEARCH AND PRACTICE INITIATIVES

The predominance of unfavorable functional outcome despite successful resuscitation has resulted in the American Heart Association highlighting brain injury after cardiac arrest as an important concern. In its 2000 guidelines for cardiopulmonary resuscitation and emergency cardiovascular care, the American Heart Association stated the following: "The cerebral cortex, the tissue most susceptible to hypoxia, is irreversibly damaged, resulting in death or severe neurological damage. The need to preserve cerebral viability must be stressed in research endeavors and in practical interventions. The term cardiopulmonary-cerebral resuscitation has been used to further emphasize this need" (6).

A conference in 2000, organized by the leaders of international scientific community, focused on research initiatives toward major improvements in clinical outcomes after CPR and after resuscitation from serious traumatic injury (11). The Post-Resuscitative and Initial Utility in Life Saving Efforts (PULSE) initiative was formed (11,12) to focus on scientific research that would yield major advances in lifesaving care, including measurable increases in survival and functional recovery. The PULSE initiative urged the prioritization of resuscitation research and appropriate funding of research by federal and voluntary agencies.

The PULSE Leadership Group collated, distilled, and prioritized research initiatives. It carefully identified and classified subjects for priority attention and opportunities for implementation. Five domains of resuscitation science were identified: Mechanisms, Pharmacology, Translational Studies, Bioengineering, and Clinical Evaluative Research (2) (Table 8.1).

The push to improve neurological functional outcome and survival, especially in comatose patients after cardiac arrest, has come a long way from the development of the modern resuscitation with the mouth to mouth and mouth to airway methods of artificial respiration by Safar and colleagues (13) in 1958 at the Baltimore City Hospital and the closed cardiac massage with external defibrillation by Koewenhoven, Jude, and Knickerbocker (14,15) in the early 1960s at the Johns Hopkins Hospital. Several well-organized clinical trials, undertaken over three decades, have provided critical insights on pathophysiology, clinical design, and outcome evaluation. Several academic, professional, industrial, and governmental initiatives have steered the development of therapies for this devastating injury. Beneficial results in two studies using moderate hypothermia at 32°C to 34°C for 12 to 24 hours (16,17) indicated that this injury

TABLE 8.1   PULSE initiatives (11,12)

Strategies and Priorities for Resuscitation Research

Mechanisms
1. Hibernation physiology in settings of ischemia and reperfusion
2. Controlled reperfusion

Pharmacology
1. Improved understanding of currently used and newly proposed agents in the management of cardiac arrests
2. Inducing blood coagulation at the site of severe hemorrhage
3. Prevention of diffuse coagulopathies
4. Minimizing endothelial injury
5. Minimizing parenchymal injury

Translational research
1. Hypothermia
2. Controlled reperfusion
3. Mechanisms of generating greater blood flows during CPR
4. Animal models
5. Induction of hypometabolic states together with the option of reducing the effects of free radicals

Bioengineering
1. New biosensors for detection of critical limitations of blood flows
2. Methods for inducing hypothermia
3. Devices for remote notification and improved CPR and trauma care
4. Vascular access
5. New mechanical devices and methods for securing maximal forward flow during cardiac arrest
6. Gene product sensors
7. Simulation and telemedicine technologies

Clinical evaluative research
1. Regional, national, and international registries on trauma and CPR
2. Clinical trial networks

is amenable to therapy. These findings have led to the recommendation of the use of moderate hypothermia in patients who are unconscious after resuscitation from cardiac arrest (18). This chapter provides an overview of the development of cardiac resuscitation as it impacts on the brain and the development of brain-directed therapies for brain injury after cardiac arrest.

PATHOPHYSIOLOGY

The field of resuscitation medicine, especially that which involves CPR, differs from more traditional clinical disciplines because impairment involves the entire body after ischemic or hypoxic injury. The inciting events that lead to arrest may be varied, such as primary ventricular fibrillation or asphyxiation. But with the onset of circulatory failure or insufficiency of circulation, common pathophysiological responses are set in motion, leading to multiorgan dysfunction with complex interorgan interactions. In settings of cardiac arrest that primarily center on the heart, the other predominant organ of injury is the brain.

Neuronal Injury and Death

The complex pathophysiological mechanisms leading to cerebral ischemic injury after global ischemia and energy failure from cardiac arrest are just beginning to be understood. These processes are presented at the various stages as they occur over the course of injury and recovery. The global cerebral ischemia that results from cardiopulmonary arrest is complete but temporary if successful resuscitation results in return of spontaneous circulation (ROSC).

During the period of total circulatory arrest, energy failure with loss of ATP production occurs, which then leads to malfunctioning membrane pumps (Na^+-K^+ pumps). This leads to the release of glutamate, causing excitotoxic injury (19) that is mediated largely through

N-methyl-D-aspartate receptors (20). Other neurotransmitters that modulate glutamate function, such as glycine and γ-aminobutyric acid, are also affected (21).

Glutamate release increases intracellular calcium, leading to second messenger activation, which amplifies injury by changing permeability to allow additional calcium entry and further glutamate release from storage pools (22). Glutamate release also gives rise to oxygen free radical production, which in turn results in the further release of glutamate (22,23). During reperfusion, delayed excitotoxicity can be further enhanced by providing oxygen to cells for energy metabolism and by serving as a substrate for several enzymatic oxidation reactions that produce oxygen free radicals (24). These reactive oxygen species are known to cause damage that includes lipid peroxidation, protein oxidation, and DNA oxidation, all of which contribute to cell death (25). The release of these neurotransmitters has been shown to be affected by manipulation of temperature. This observation has been the basis for the use of hypothermia as a therapeutic intervention.

Cerebral Reperfusion Failure

During the immediate postresuscitation period, a state of global hypoperfusion for up to 12 hours follows in spite of maintaining normal blood pressures (19). This is believed to be due to impairment of CO_2 reactivity and increased vascular tone of the cerebrovascular system (26). Endothelin type A receptors mediate this response (27), and endothelin-1, a powerful vasoconstrictor, is the endogenous ligand to these receptors (28,29). Nitric oxide, on the other hand, is a vasodilator. It is the balance between the effects of nitric oxide and endothelin-1 that regulates cerebral blood flow (28). Excessive release of endothelin-1 has been implicated in the pathophysiology of delayed hypoperfusion after cardiac arrest (30). This hypoperfusion tends to be heterogeneous and can cause regional injury (31). It is possible to overcome the immediate postarrest no-reflow phenomenon if reperfusion pressure can be maintained at or above normal levels (32). Increasing perfusion by induced hypertensive-hemodilution therapy may be able to partially reverse this state of hypoperfusion (33). Treatment with an endothelin type A receptor antagonist has also been shown to improve functional outcomes in experimental models of cardiac arrest (28,29).

Cerebral and Extracerebral Injury

Total circulatory failure with cardiac arrest injures both brain and extracerebral organs. The injury leads to complex adaptive and maladaptive responses at the level of cells, tissues, organs, and ultimately the organism. The primary injury in specific organs leads to secondary effects on other organs (3), which can worsen the brain and heart injury. The more common example of these injuries includes the disruption of the gastrointestinal barrier caused by ischemia, leading to bacterial toxin leakage into the bloodstream (34,35). Excesses of lactic acidosis production can be caused by ischemia to the liver and skeletal muscle, which may lead to altered systemic acid-base conditions. The ischemic injury can also disrupt blood vessels, which can lead to release of mediators, which potentially injure otherwise unaffected tissues and cells (36,37). This complex process worsens cardiorespiratory and neurological malfunction, leading to adverse outcomes, but the specific mechanistic processes are still unclear. This complex interplay between the heart, brain, vasculature, gastrointestinal tract, muscle, inflammatory system, coagulation system, and other organs necessitates a multidisciplinary team who can address multiorgan-integrated pathophysiology of injury (38). An understanding of the cellular and molecular pathophysiological mechanisms of ischemic injury can then be translated into targeted therapies to

improve outcome. Such treatments may include induced moderate hypothermia, anti-inflammatory agents, free radical scavengers, thrombolytic agents, and genetic manipulation of gene expression.

Duration of Ischemia and Cerebral Injury

The duration of cerebral ischemia in relation to neurological outcome is well documented, and the precise duration of cardiac arrest represents the primary insult and is the single most important factor in determining the severity of the injury (39,40). The good and poor outcome threshold seems to be at 6 minutes for cardiac arrest and about 30 minutes for CPR (Table 8.2).

The duration and severity of the ischemia, however, is almost always unknown and rarely easy to estimate; therefore, the extent of neurological damage must be assessed by other means. Specifically, the time of anoxia was found to differ significantly in patients with favorable (4.1 minutes) versus unfavorable (8.0 minutes) outcomes. CPR time with an average of 17 minutes was more likely to result in a favorable outcome than CPR time of 34.5 minutes. Anoxic times of 7.5 minutes were more likely in patients who ended up dying of cardiac causes as opposed to times of 11.6 minutes, in which patients were more likely to ultimately die of neurological cause (40).

It has been shown that the degree of neuronal injury and loss is largely

TABLE 8.2   Arrest time and CPR post hoc analysis from the Brain Resuscitation Clinical Trial (BRCT) I (39)

Arrest time	CPR time	Patients	Favorable outcome
<6 min	<30 min	$n = 158$	50%
>6 min	<5 min	$n = 70$	50%
>6 min	6–15 min	$n = 32$	19%
<6 min	>30 min	$n = 30$	3%
>6 min	>15 min	$n = 34$	0%

dependent on the duration and severity of cerebral ischemia during cardiac arrest (41–45). Along with the bedside neurological examination, the information on the duration of cardiac arrest and CPR time (if both are available) is important in the estimation of the acute brain injury. A significant problem in this assessment stems from the fact that cardiac arrest times are known in only 50–70% of those out-of-hospital cardiac arrests. In these cases, neurodiagnostic tests such as the electroencephalogram or evoked potentials, with some limitations, have provided some assistance in the estimation of injury.

Selective Vulnerability and Clinical Manifestations

The pathology of global cerebral ischemia has been well described (46), but the mechanisms leading to these injuries are as yet not fully understood, but certain areas of brain are clearly more vulnerable. Global cerebral ischemic injury affects the brain in a heterogeneous manner. The most vulnerable area includes the cortex, especially the CA-1 region of the hippocampus

(47). This area plays an important role in memory. Selective injury is also noted in different layers of the cortex, with layer 3 as the most sensitive, followed by layers 5 and 6. The cortex is more susceptible to injury compared with the deeper structures, such as the brainstem. With cortical injury, neurological impairment in cognition and memory and seizure disorders may be observed. The thalamus and basal ganglia are other structures that are susceptible to injury (47,48). Among the complex functions of the thalamus, it is probably it role in arousal and consciousness that is most critical during the recovery phase from global ischemia. The return of baseline interaction of the cortex and thalamus is critical in recovery from coma after cardiac arrest (49). Other areas that are prone to injury are the basal ganglia and cerebellum, which may account for problems with movement and coordination. The clinical spectrum of neurological injury and recovery is provided in Figure 8.1.

With the drastic reduction in systemic perfusion, areas perfused most distally are also affected more. This is the mechanism

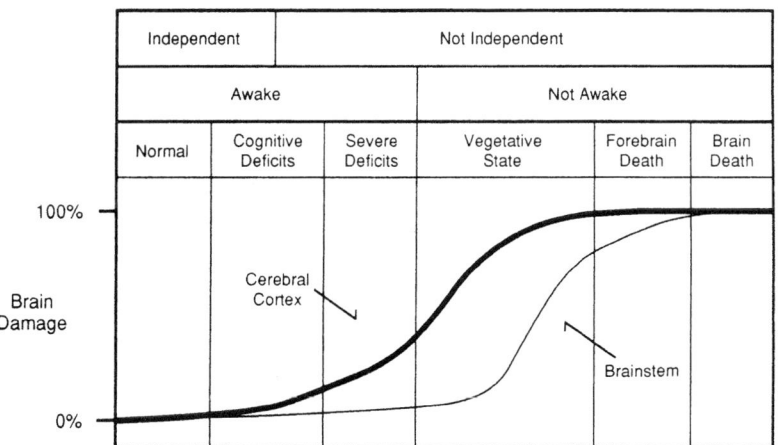

FIGURE 8.1   Scheme describing the spectrum of neurological injury and recovery after cardiac arrest. Used with permission from reference 65.

for the watershed infarcts occasionally described with after severe hypotension or resuscitation from cardiac arrest. These infarcts are mostly observed in the watershed areas of the middle cerebral, the posterior cerebral, and anterior cerebral arteries. This can affect the motor neuron activating the proximal arm, leading to the "man in the barrel syndrome" (50).

Primary Brain-Directed Therapies After Cardiac Arrest

The expanse of controlled trials primarily targeting brain injury for global ischemia related to cardiac arrest started with the Brain Resuscitation Clinical Trials (BRCTs) in 1986 (59) to the more recent use of moderate hypothermia by the Hypothermia After Cardiac Arrest (HACA) Study Group (16) and Australian groups (17) in 2002. Many other controlled clinical trials were undertaken in the intervening period targeting numerous factors in the injury cascade of brain injury from cardiac arrest. The clinical trials are presented in chronological order. A list of drug and interventions used in clinical trials is provided in Table 8.3, and

TABLE 8.3   Published clinical studies in brain injury and cardiac arrest

---

Primary brain-directed drugs or interventions
  1. Thiopental
  2. Glucocorticoids
  3. Lidoflazine
  4. Nondextrose solutions
  5. Nimodipine
  6. Magnesium
  7. Diazepam
  8. Moderate hypothermia (external cooling)
  9. Flunarizine* (study design only) (51)
 10. Phenytoin† (uncontrolled) (52)

CPR-related interventions with neurological or functional recovery endpoints
  1. High-dose epinephrine (53–55)
  2. Active compression-decompression CPR (38,56–57)
  3. Vasopressin (7,58)

---

the first eight trials (seven randomized controlled studies and one noncontrolled study) (16,59–65) are discussed in detail in the succeeding sections. Additional agents are included such as flunarizine (51) as an agent for a clinical trial design but without subsequent outcome data, and phenytoin (52) in a small uncontrolled trial showing no benefit. Several strategies undertaken to minimize injury by improving CPR-related interventions with neurological or functional recovery endpoints as also listed (7,38,53–58). All these strategies provided no significant neurological or functional outcome benefit.

Drugs and Interventions

The overwhelming basis for the use of the experimental agents came from animal experimentation and observed benefits in preliminary clinical trials or in closely related ischemic injury of the brain. The neuroprotective mechanisms are all in response to the injury cascade that is observed after global ischemia. The BRCT I used thiopental (59) with glucocorticoids (61) as an add-on therapy, and BRCT II used the calcium channel blocker lidoflazine (60). Thiopental was successfully tested in a primate model of global ischemia (66) and was followed closely by successful pilot studies in humans resuscitated from cardiac arrest (67). Thiopental was used because of its "ability to reduce cerebral metabolism, edema formation, intracranial pressure, seizure activity and damage by focal and incomplete ischemia" (59). The thiopental study (BRCT I) allowed for the use of glucocorticoids at the discretion of the investigators in the standard treatment group. Practice differences in the use of steroids after cardiac arrest prevented the BRCT group from fully controlling this therapy. Glucocorticoids were studied because of their "benefit in the treatment of perifocal vasogenic edema in of an intrinsic mass lesion," but "they have not been proven to be beneficial in models of intracellular cytotoxic edema after brain ischemia" (61). The BRCT II acknowledged the important role of calcium in mediating neuronal death. So the agent lidoflazine, a calcium channel blocker, was used because it has "been shown to lessen cerebral infarct size, improve cerebral blood flow and improve outcome in animal models of focal cerebral ischemia and in some models of global ischemia" (60).

The successive studies underwent similar development. The neuroprotective effect of calcium antagonism from brain injury after cardiac arrest was further tested by Roine and colleagues (62) by using the drug nimodipine. The randomized study using nimodipine was preceded by an open pilot study on the safety and efficacy of nimodipine in patients after out-of-hospital ventricular cardiac arrest (68). Nimodipine was earlier shown to reduce death or severe ischemic deficits in subarachnoid hemorrhage (69). At about this time, Longstreth and colleagues in Seattle observed that hyperglycemia was associated with poor recovery after cardiac arrest (70). With other studies suggesting that hyperglycemia may be detrimental to an injured brain, they studied the effect of glucose free solutions during the resuscitation process (64).

The search for effective therapies moved on to magnesium, with its antiarrhythmic effects and ability to block excitatory neurotransmitters (63). This led to the undertaking of two clinical trials in this area. The first study was done by Thel and others in 1997 (63) followed by another study by the group of Longstreth using magnesium in association with diazepam (65). The inhibitory action of diazepam was hoped to reduce the neuroexcitotoxic injury after cardiac arrest.

With the critical role of temperature on cerebral metabolism and the numerous studies showing worsening injury with hyperthermia after cerebral ischemia and beneficial effects of hypothermia in preclinical studies, two randomized

clinical trials using moderate hypothermia (32–34°C) for 12 to 24 hours were undertaken (16,17). These clinical trials were preceded by extensive preclinical studies showing functional and survival benefits in experimental studies in rodents (71,72) and dog (73,74). Several pilot studies (75,76) on human subjects using moderate hypothermia were undertaken before the two successful randomized control studies. The precise mechanism of hypothermia leading to improved survival and functional recovery is not well understood. Numerous mechanisms have been suggested (77). These include the effect of hypothermia on cerebral blood flow and metabolism by retarding the initial rate of ATP depletion (78,79) and on the reduction of excitotoxic neurotransmitter release (80), alteration of intracellular messengers and mediators activity (81), blood–brain barrier breakdown (82), reduction of inflammatory response (83), and alteration in gene expression and protein synthesis (84,85). Although the brain was the primary beneficiary of hypothermia therapy, other organ systems may have also benefited from this treatment. This additional benefit may play an important role in reducing extracerebral injury that helped to promote overall survival in patients treated.

Clinical Study Design

The progress of research in this area led to the development of numerous and varied research protocols and outcome measures. All eight clinical trials were prospective, randomized, and placebo controlled. One study using glucocorticoid undertaken as a companion study of the thiopental in BRCT I was uncontrolled. With the time dependency of the injury and urgency of intervention, all eight controlled clinical trials were undertaken with waiver of informed consent or deferment of consent. The major inclusion criterion of all these trials was the unresponsiveness or coma state of patients after resuscitation from cardiac arrest. As investigators of clinical trials learn from previous trials, the designs of succeeding trials are modified. At the start of the BRCT I and II, all comatose patients after ROSC had no differentiation on enrollment based on initial cardiac rhythm, the duration of cardiac arrest and resuscitation, and place of arrest (either as in-hospital or as out-of-hospital cardiac arrest). Several important outcome indicators were learned from the BRCT trials. After the BRCT I trials, a consensus meeting was undertaken under the auspices of the American Heart Association and the European Resuscitation Council to facilitate research and reduce the variability in the methodologies used. This led to the development of the "Utstein style guidelines" (Table 8.4) that provides recommendations for the reviewing, reporting, and conducting research in- and out-of-hospital cardiac arrest in 1991 (86) and later in-hospital resuscitation in 1997 (87).

As the impact of global ischemic injury to survival and outcome became better understood, many factors in the original BRCT clinical trial were modified. One early major change was the differentiation of patients by place of arrest and resuscitation. The place of cardiac arrest indicated the state of health of the patient, with more comorbidities associated with in-hospital cardiac arrest than out-of-hospital arrest. With the exception on the study by Thel and colleagues (63), succeeding trials focused only on out-of-hospital cardiac arrest. The severity of injury as defined by the duration of cardiac arrest and CPR was also associated with outcome. This realization led some studies to focus on the less extreme injury with the exclusion of prolonged resuscitation (>30 minutes), as in the study by Roine and colleagues (62). A more selective criteria was used by the HACA study, which only included patients with collapse to CPR time 5–15 minutes and CPR time <30 minutes (16). The initial cardiac rhythms have also been used as

TABLE 8.4 Recommended guidelines for uniform reporting of data from out-of-hospital cardiac arrest: the Utstein style (used with permission from reference 4)

Recommended data
1. Population served
2. Confirmed cardiac arrests considered for resuscitation
3. Resuscitations not attempted
4. Resuscitations attempted
5. Cardiac etiology
6. Noncardiac etiology
7. Arrest witnessed
8. Arrest not witnessed
9. Arrests after arrival of emergency personnel
10. Initial rhythm ventricular fibrillation
11. Initial rhythm ventricular tachycardia
12. Initial rhythm asystole
13. Other initial rhythms
14. Determine presence of bystander CPR
15. Any return of spontaneous circulation
16. Never achieved return of spontaneous circulation
17. Efforts ceased: (a) patient died in the field or (if transported) (b) in the emergency department
18. Admission to intensive care unit/ward
19. Patient died in hospital
    a. Total
    b. Within 24 hours
20. Discharged alive
21. Death within 1 year of discharge
22. Alive at 1 year

Time events to be recorded
1. Time of collapse/time of recognition
2. Time first emergency response vehicle is mobile
3. Time vehicle stops (core) time of scene arrival
4. Time of first CPR attempts (core)
5. Time of first defibrillatory shock (core)
6. Time of return of spontaneous circulation (core)
7. Time intubation achieved
8. Time intravenous access achieved and time medications administered
9. Time CPR abandoned/death (core)
10. Departure from scene and arrival at emergency department

inclusion criteria. Relative to other initial cardiac rhythms, ventricular fibrillation or nonperfusing ventricular tachycardia are associated with better outcome. The studies by HACA (16), Bernard and colleagues (17), and Roine and colleagues (62) used ventricular fibrillation and ventricular tachycardia as inclusion criteria.

Outcome Measures

The BRCT trials (41) developed a functional outcome measure as defined by the 5-point scale of the Glasgow-Pittsburgh Outcome Categories (Table 8.5). This scale was divided into favorable outcome, which was defined as Cerebral Performance Category (CPC) 1 and 2, and unfavorable outcome, defined as with CPC 3, 4, and 5. For the BRCT trial, the outcome was determined at 48 hours, 72 hours, 10 days, and months 1, 3, 6, and 12. The use of Glasgow-Pittsburgh Outcome Categories as the outcome measure has been recommended as the Utstein style guidelines of the American Heart Association, the European Council on Resuscitation, and other resuscitation research organizations (86,87) (Table 8.5).

Other than the BRCT series, only the study on moderate hypothermia by the HACA study group used these guidelines and the CPC as their primary outcome measure at 6 months. The other studies developed their own functional outcome measure. The Roine study (62) used the Glasgow Outcome Score at 1 year and a secondary endpoint of the Glasgow Coma Scale at 24 hours and 1 week. Interval functional evaluations were done at 3 and 12 months using the Mini-mental Examination, Barthel index, and Katz ADL Scale. The two studies (nonglucose containing solutions and magnesium-diazepam studies) by Longstreth and coworkers (64,65)

TABLE 8.5 The Glasgow-Pittsburgh Outcome Categories: Cerebral Performance Categories and Overall Performance Categories (41)

Category 1: conscious and normal, without disability
Category 2: conscious with moderate disability
Category 3: conscious with severe disability
Category 4: comatose or vegetative state
Category 5: death

The Utstein Consensus Conference participants recommend use of the Glasgow-Pittsburgh Outcome Categories to record prearrest status, status at the time of discharge, and status after 1-year survival.

used the awakening time, defined as the time from unresponsiveness after resuscitation to time of comprehensible speech or after commands. The Thel study (63) used the ROSC as the primary outcome and survival and Glasgow Coma Scale to discharge as the secondary outcome measure. The Bernard study (17) used the place of discharge as a surrogate for functional outcome as the primary outcome measure.

Despite the detailed and extensive recommendation for Utstein style uniform reporting in clinical trials on out-of-hospital cardiac arrest, a study undertaken in 1999 showed that of the 102 studies where Utstein style was indicated, only 41 (40%) used the Utstein style and 61 (60%) did not (88). Several reasons have been provided for this failure, such as the lack of enforcement by journal editors, uninformed researchers, and authors and the fact that guidelines require change in practice and the variables required are impossible to gather (88,89).

Survival and Functional Outcome

The BRCT was organized to conduct clinical trials with the primary goal of ameliorating brain injury after resuscitation from cardiac arrest. The first randomized clinical trial undertaken by this group compared standard of care to a barbiturate, thiopental, given as an intravenous loading dose of 30 mg/kg at 10 to 50 minutes after ROSC (59). An equal number of patients were randomized between the thiopental arm ($n = 131$) and standard of care ($n = 131$). The primary outcome as measured by a treatment-blinded investigator using the 5-point CPC scale did not show a difference between treated and control groups. The thiopental group had an overall good outcome (CPC 1, 2) of 20% and the control group, 15%. Mortality was similar in both study arms, with 70% in the thiopental-treated group and 80% in the control group. An important factor that may have contributed to the failure of thiopental was the excess in hypotension episodes (60%) compared with in the control group (29%).

The anti-inflammatory effects of glucocorticoids were believed to be beneficial in the treatment of patients after cardiac arrest. As such it was widely used during the time of the thiopental study that a study group allowed the different sites to administer glucocorticoids at the discretion of the centers participating (61). The glucocorticoid treatment arm was not controlled, and the data were analyzed retrospectively. The inclusion criteria and the primary outcome measure were identical to the thiopental study. The choice of the glucocorticoid varied (dexamethasone, methylprednisolone, etc.) depending on the treatment preference of a center and was administered during the first 8 hours post-ROSC. The glucocorticoid treatments were grouped into no dose, low dose (equivalent to 1–20 mg of dexamethasone), medium dose (equivalent to 21–70 mg of dexamethasone), and high dose (equivalent to >70 mg of dexamethasone). The good outcome (CPC 1 and 2) in the different treatment groups was similar and ranged from 34.0% to 37.3%. Mortality was also similar in all groups, from 71.6% in low dose, 74.1% in medium dose, to 81.8% in high dose and 85.7% in glucocorticoid nontreated group. The uncontrolled and nonstandardized study design may have contributed much to it failure (61).

The BRCT II (61) study using intravenous lidoflazine at 1 mg/kg as soon as possible (no later than 30 minutes after ROSC) after resuscitation and then 0.25 mg/kg at 8 hours and 16 hours was compared with placebo. A total of 520 patients was enrolled. Four were lost to follow-up. The lidoflazine group had 257 patients and the placebo arm had 259 patients. The 6-month functional outcome by CPC showed no difference with good outcome in 15% of lidoflazine-treated patients compared with 13% in the placebo-treated patients. The 6-month mortality was also similar at 82% with lidoflazine treatment and 83% with placebo (60).

In the Roine study (62), nimodipine as bolus of 10 μg/kg followed by 0.5 μg/kg over 24 hours was compared with placebo. The study agent was given immediately after resuscitation, which was usually at the site of the resuscitation. In the study, the nimodipine arm had 75 patients and the placebo arm had 80 patients. The 1-year good outcome score for the nimodipine-treated patients was 29% compared with placebo, at 24%. The 1-year mortality in the nimodipine-treated group was 60% and in the placebo group was 64% (62).

The study by Longstreth and colleagues (64) comparing the effect of glucose, non-glucose, and glucose-containing solutions during resuscitation was undertaken primarily by the Emergency Medical Services in the process of resuscitation. The use of 0.5 normal saline solution was compared with a solution of 5% dextrose in water. A total of 858 patients was evaluated for the study, and 748 were randomized. The arm using dextrose water treated 371 patients, and the arm using nonglucose solution treated 377 patients. Of the 748 patients randomized, 291 patients were admitted to the hospital. Of those admitted, 141 patients received the dextrose solution and 150 received the nondextrose solution. With the patient awakening as the functional endpoint, no difference was observed between the dextrose solution arm (16.7%) and the nondextrose arm (14.6%). Further comparison of the two groups showed similar recovery proportions were noted in those who awakened, those who were discharged, and those with moderate deficits. Significantly more patients with mild deficits were noted in those treated with dextrose solution (27.7%) than with nondextrose solution (14.7%). The patients surviving to be discharged after treatment were not different, with 15.1% in the dextrose solution arm and 13.3% in the nondextrose solution arm.

In a randomized, double-blind, placebo-control study, Thel and colleagues (63) compared the efficacy of empiric magnesium supplementation to placebo in patients resuscitated from in-hospital cardiac arrest on successful ROSC and survival to discharge. The magnesium administered was 2 g intravenous bolus followed by 8 g over 24 hours. Based on the primary endpoints, the efficacy of magnesium and placebo was not different, with ROSC 54% in magnesium compared with 60% in placebo treatment, 43% in the magnesium arm compared with 50% in placebo survived to 24 hours, and identical proportions (21%) who survived to discharge (63).

The comparative effectiveness of magnesium, diazepam, magnesium + diazepam, and placebo in patients awakening from CPR from OHCA was studied by Longstreth and colleagues (65). The study agents were administered intravenously upon ROSC at the following doses: magnesium 2 g, diazepam 10 mg, magnesium 2 g + diazepam 10 mg, and placebo. No significant difference was noted in the proportions of patient awakening at 3 months primary outcome (magnesium, 46.7%; diazepam, 30.7%; magnesium + diazepam, 29.3% and placebo, 37.3%). The mortality at 3 months was likewise not different in the four groups: magnesium 62.7%, diazepam 76.0%, magnesium + diazepam 77.3%, and placebo 68% (65).

For the first of the two hypothermia trials, moderate hypothermia was induced immediately at the site of resuscitation and continued in the hospital by external cooling to a target temperature of 33°C for 12 hours compared with a normothermia control subject at 37°C for the same period (17). The hypothermia arm had 43 patients and the normothermia arm had 34 patients. With the primary outcome as good discharge to home or rehabilitation facility and poor outcome as death or discharge to long-term nursing facility, the

hypothermia arm had 49% good outcome compared with 26% in the normothermia arm ($p = 0.046$). The overall mortality was not significantly different at 51% for the hypothermia arm and 68% for the normothermia arm. The other hypothermia study cooled the patients externally as well to temperatures of 32°C to 34°C for 24 hours and comparison was made to normothermia control subjects at 37°C for the same period (16). The hypothermia arm had 137 patients and the normothermia arm had 138 patients. This is the only study that showed a significant difference in both mortality and functional outcome. Using the Utstein style CPC scale, 55% of patients treated with hypothermia had a favorable outcome (CPC 1, 2) compared with 39% in the normothermia control group (risk ratio 1.40; 95% confidence interval [CI], 1.08–1.81). The hypothermia arm also had a significantly lower 6-month mortality at 41% compared with 55% of the normothermia control group (risk ratio, 0.74; 95% CI, 0.58 to 0.95).

All studies made a careful effort to indicate the effect of their intervention on mortality as compared with control. All the studies also described the occurrence of complication in relation to the drug or the intervention provided. Only two complications were reported to occur significantly greater than controls. First is the excess occurrence of hypotension with the administration of thiopental (60%) compared with control (29%) ($p < 0.001$), leading to higher requirement of a vasopressor in the thiopental group in the BRCT I study (59). In the Roine study (62), the group treated with nimodipine was also noted to have more hypotensive episodes requiring dopamine infusion (73%) compared with the control group (49%) ($p = 0.003$). Several other adverse events worth noting but did not reach statistical significance are the infection rates with glucorticoid treatment in BRCT I (61) and the problems associated with hypothermia (bleeding, pneumonia, sepsis, and arrhythmia) in HACA study (16).

SECONDARY BRAIN INJURIES AFTER CARDIAC ARREST

Despite the success of hypothermia, the progression of secondary brain injury becomes very important in the course of the recovery process of patients. It is important therefore that a preventive strategy that may prevent or attenuate secondary brain injury is also presented. No randomized controlled trials have been undertaken in relation to the secondary brain injuries after cardiac arrest. The approaches and interventions provided below are from uncontrolled studies and observational studies in brain injury after cardiac arrest and other similar ischemic brain injuries.

Cerebral Edema and Intracranial Pressure Elevation

Global cerebral ischemia leads to the development of cytotoxic edema. Up to 47% of patients resuscitated from out of hospital cardiac arrest (OHCA) showed cerebral edema on head computed tomography (CT) at day 3 (90). In another study, more patients (92%) with cerebral edema on head CT were noted in those with primary respiratory arrest. In a separate study on cerebral edema as seen the degree of obliteration gray matter–white matter demarcation by brain CT, Torbey *et al.* (91) found that the progressive loss of gray matter–white matter demarcation as a reflection of brain injury was associated with poor outcome. In these reports, brain edema is a marker of brain injury and associated with poor neurological outcomes. Several small studies have attempted to define the occurrence of intracranial pressure elevation after resuscitation from cardiac arrest. A study showed that intracranial pressure elevation was associated

with delayed hyperemia by transcranial Doppler ultrasound (92). Acute hyperventilation and mannitol therapy may be used at the time of intracranial pressure elevation. These therapies have been used successfully in other pathologies, but their use in edema related to global ischemia has not been well described. Most importantly, the use of steroids does not provide benefit and can lead to adverse outcomes (61).

Cerebral Perfusion

Using transcranial Doppler ultrasound and norepinephrine infusion testing, the cerebral blood flow autoregulation during the first 24 hours after cardiac arrest (18 patients and 6 healthy volunteers) was found to be impaired in 8 of 18 or right-shifted in 5 of 10 post–cardiac arrest patients. In another study of human cardiac arrest survivors ($n = 136$), good functional neurological recovery was independently and positively associated with arterial blood pressure during the first 2 hours after human cardiac arrest. The investigators also observed that hyperacute hypertensive reperfusion (first minutes after ROSC) did not provide additional benefit (93).

Hyperglycemia

In global ischemia from cardiac arrest (64,94), elevated serum glucose has been associated with unfavorable outcome. Serum glucose elevation is believed to be a marker of the severity of injury. The direct effect of elevated glucose on neurological injury is post–cardiac arrest patients and the definite benefit in terms of neurological outcome is not well defined. In a series of 145 nondiabetic patients evaluated after witnessed ventricular fibrillation cardiac arrest, a strong association between high median blood glucose levels over 24 hours and poor neurological outcome was found ($rs = -0.2$, $p = 0.015$) (95).

Because some animal and uncontrolled human studies indicate that hyperglycemia may be harmful, avoiding early hyperglycemia in patients with nonlacunar stroke and global ischemia has been advocated (96). Similarly, in an effort to avoid potential neurological injury, American Heart Association guidelines state that during CPR, drugs should be given in non–glucose-containing solutions. Controlled clinical trials on the regulation of hyperglycemia in focal and global ischemia are not currently available. In the absence of controlled human trials showing the benefit of glucose control in strokes and global cerebral ischemia, some insights may be taken from the general critical care literature. In a general intensive care unit with patients having primarily systemic pathology, a randomized controlled study showed that the tight control of serum glucose (80 and 110 mg/dl) was able to reduce overall mortality by about 50% (97). Glucose monitoring in the intensive care unit is a routine procedure and adapting a tight control may prove beneficial to post–cardiac arrest patients.

Hyperthermia

Increasing body temperature after cardiac arrest has been associated with poor neurological outcome. In a study of patients with cardiac arrest presenting in the emergency department, a body temperature higher than 37°C within 48 hours after resuscitation was associated with unfavorable functional neurological recovery after. The investigators also concluded that for every degree celsius higher than 37°C, an increased association with severe disability, coma, or a persistent vegetative state was found (98). In another study (99), body temperature of 39°C or higher after resuscitation from cardiac arrest was associated with poor outcome and brain death.

The mechanism by which hyperthermia worsens neurological outcome is still unclear. Several studies have suggested the role of infection leading to hyperthermia, such as intestinal ischemia and the systemic translocation of bacteria or toxins (100), and pulmonary aspiration because of

lack of airway protective reflexes in coma (101). However, in a significant portion of patients, an infectious source is not identified. Hyperthermia may be related to injuries affecting the anterior hypothalamus (102,103). But its precise impact during global ischemic injury is also unclear. In animal models, the cellular mechanisms appear to affect the neurons' response to ischemic damage. Hyperthermia accentuates the release of neurotransmitters and free radical activity and promotes excitotoxic injury in global ischemia (104–106). The adverse effects of hyperthermia highlight the need for hypothermia as well as provide some possible mechanisms for it beneficial effect.

Status Epilepticus/Myoclonus

Postanoxic myoclonus was previously regarded as a predictor of poor outcome (107,108). But a report of survivors with postanoxic myoclonus indicated that the myoclonus may improve as neurological status improves (109). Postanoxic myoclonus tends to occur after respiratory causes of arrest and often persists after other signs of neurological damage have improved. However, a state of status myoclonus, defined as more than 30 minutes of myoclonic activity (107), which tends to occur after cardiac causes of arrest and is often associated with burst suppression on electroencephalogram, is considered an indicator of extremely poor prognosis, and treatment with antiepileptics tends not to influence short- or long-term outcome (107).

Associated Risk of Brain Hemorrhage After Systemic Thrombolysis

About 40% of out-of-hospital cardiac arrest has underlying myocardial infarction. Thrombolysis as a therapy for acute ischemia has been a concern because of the potential increased risk of intracranial bleeding. Several uncontrolled trials found thrombolysis relatively safe for the brain. A retrospective analysis was performed of 68 patients who received systemic thrombolytics (18 with strepto-kinase, 28 with alteplase, and 22 with reteplase) after resuscitation from cardiac arrest for presumed acute myocardial in-farction. Cardiac reperfusion was achieved in 71% of the patients treated. Hemor-rhagic complications included intracranial hemorrhage (one patient), gastrointes-tinal bleeding (two patients), bleeding from the puncture site (one patient), and epistaxis (one patient). Sixty-three patients (93%) were admitted alive to the hos-pital, with 36 subsequently surviving to discharge (110). In another study of 303 acute myocardial infarction patients, 67 patients were administered systemic thrombolysis (group I) and 236 patients were managed without thrombolysis. Systemic thrombolysis showed cardiac benefit with less mechanical ventilation ($p < 0.00001$) and fewer cardiopulmo-nary resuscitation attempts in group I ($p < 0.0001$). No fatal hemorrhagic com-plications in either group occurred (111). Given these reports, the incidence of intra-cerebral hemorrhage is small and treat-ment of acute myocardial infarction with systemic thrombolysis is justifiable.

FUTURE DIRECTIONS

The most important consideration in the development of therapies to improve survival and function after brain injury from global cerebral ischemia is the fact that the injury can be ameliorated, and this amelioration has been translated to improved survival and better function of survivors. The foundation to further advance research leading to yet a better outcome of brain injury after cardiac arrest has been laid with the clinicopathological aspects learned from clinical trials as exemplified by the Utstein recommen-dations on the conduct of research in this area and the direction and focus of further

research has been defined by the PULSE initiatives. The global nature of the injury, involving brain and systemic organs, makes the problem relevant to a host of medical specialties and field of study. Safar and colleagues (112) summarized the key aspects of permanent brain damage after cardiac arrest and resuscita-tion to be determined predominantly by three factors: arrest (no-flow) time, CPR (low-flow) time, and temperature.

The future direction can be gleaned from priorities set by the PULSE initia-tive. For this discussion we adopt the five domains of resuscitation science from the PULSE initiative.

Mechanisms

To further develop therapeutic options, a need for better understanding the basic sciences fundamental to resuscitation medicine persists. The PULSE initiative has focused on two areas, hibernation physiology and controlled reperfusion. The development of transient therapeutic hibernation wherein injury to human organs would be prevented during ische-mic states may have a profound benefit in vital organ perfusion. The prevention of reperfusion injury also needs a focus with concept "controlled reperfusion" to maximize cell recovery. Specific to the neurological aspects, understanding mechanisms of neuronal injury and recov-ery needs to be pursued.

Pharmacology

With the failure of all pharmacological interventions in all previous trials, the developments of molecular medicine, genomics, and proteinomics promise the development of novel drugs. The develop-ment in this area needs to focus not only on the primary brain injury, but also to encompass the systemic injuries sustained with cardiac arrest. There is also a need to understand the therapies currently under-taken in the management of cardiac arrests itself. The PULSE initiative has highlighted minimizing endothelial injury, which may prevent capillary plugging, and loss of vascular tone may account for exacerbation of injury of remote organs or cells.

Translational Research

This field of research provides the critical link between mechanisms and interventions that develop out of basic science laboratories and that are advanced to preclinical testing. The PULSE initiative has placed the highest priorities on hypothermia, which at the time of its drafting in 2002 was to improve outcome after CPR and cardiac arrest. This expectation has been realized, but the success of the hypothermia is limited, especially in its complexity as therapies and all the potential complications that it can cause. Several centers have undertaken research on mechanical devices that promote forward blood flow, including devices that are portable and adaptable for out-of-hospital uses as part of this translational thrust. The success of translational research hinges on the development of clinically realistic animal models (small and large animal) of cardiac arrest. The design of these animal models must also have clinical functional outcomes that reflect the clinical problem.

Bioengineering

Advances in bioengineering are needed, especially in areas directly related to cardiac arrest and CPR with the development of defibrillators, electrical and mechanical monitors of cardiopulmonary function, airway devices, and vascular access technologies. The PULSE initiative has provided a focus on the development new biosensors for detection of critical limitations of blood flows, methods for inducing hypothermia, new mechanical devices and methods for securing maximal forward flow during cardiac arrest, gene product sensors, and devices for remote notification and improved CPR and trauma care.

In response to the thrust of developing brain-directed therapies, a need to develop rapidly deployable and easily interpretable brain injury indicator and monitor is essential. This technology must be able to detect real-time changes in the brain as it responds to injury, recovery, drugs, or other interventions. Novel neurophysiological analysis methods can be developed for this use. Other areas such of development may include bedside static and functional brain imaging.

CLINICAL EDUCATION AND TRIAL NETWORKS

The complexity of cardiac arrest requires the development of sophisticated human patient simulators for training and testing to allow a reproducible learning environment with consistent standards. Together with the educational thrust is the necessity to develop regional, national, and international registries on trauma and CPR and for the evaluation of promising new interventions derived from translational research and phase II clinical trials.

References

1. Thel M, O'Connor C. Cardiopulmonary resuscitation: historical perspective to recent investigations. Am Heart J 1999;137:39–48.
2. O'Neil BJ, Krause GS, Grossman LI, Gruenberger G, Rafols J, DeGraciaz EAD, Neuwar R, Tiffany B, White B. Global brain ischemia and reperfusion by cardiac arrest and resuscitation. In: Paradis N, Halpern H, Nowak R, eds. Cardiac Arrest. The Science and Practice of Resuscitation Medicine. Baltimore: Williams and Wilkins, 1996.
3. Herlitz J, Andersson E, Bang A, Engdahl J, Holmberg M, Lindqvist J, Karlson BW, Waagstein L. Experiences from treatment of out-of-hospital cardiac arrest during 17 years in Goteborg. Eur Heart J 2000;21:1251–1258.
4. Pusswald G, Fertl E, Faltl M, Auff E. Neurological rehabilitation of severely disabled cardiac arrest survivors. Part II. Life situation of patients and families after treatment. Resuscitation 2000; 47:241–248.

5. Krause GS, Kumar K, White BC, Aust SD, Wiegenstein JG. Ischemia, resuscitation, and reperfusion: mechanisms of tissue injury and prospects for protection. Am Heart J 1986;111:768–780.

6. American Heart Association. Guidelines 2000 or Cardiopulmonary Resuscitation and Emergency Cardiovascular Care. 2000.

7. Lindner KH, Dirks B, Strohmenger HU, Prengel AW, Lindner IM, Lurie KG. Randomised comparison of epinephrine and vasopressin in patients with out-of-hospital ventricular fibrillation. Lancet 1997;349:535–537.

8. Ekstrom, L, Herlitz J, Wennerblom B, Axelsson A, Bang A, Holmberg S. Survival after cardiac arrest outside hospital over a 12-year period in Gothenburg. Resuscitation 1994;27:181–187.

9. Stiell IG, Wells GA, Field BJ, Spaite DW, De Maio VJ, Ward R, Munkley DP, Lyver MB, Luinstra LG, Campeau T, Maloney J, Dagnone E. Improved out-of-hospital cardiac arrest survival through the inexpensive optimization of an existing defibrillation program: OPALS study phase II. Ontario Prehospital Advanced Life Support. JAMA 1999;281:1175–1181.

10. Grubb NR. Managing out-of-hospital cardiac arrest survivors. 1. Neurological perspective. Heart 2001;85:6–8.

11. Becker LB, Weisfeldt ML, Weil MH, Budinger T, Carrico J, Kern K, Nichol G, Shechter I, Traystman R, Webb C, Wiedemann H, Wise R, Sopko G. The PULSE initiative: scientific priorities and strategic planning for resuscitation research and life saving therapies. Circulation 2002;105:2562–2570.

12. Weil MH, Becker L, Budinger T, Kern K, Nichol G, Shechter I, Traystman R, Wiedemann H, Wise R, Weisfeldt M, Sopko G. Workshop Executive Summary Report: post-resuscitative and initial Utility in Life Saving Efforts (PULSE): June 29–30, 2000; Lansdowne Resort and Conference Center; Leesburg, VA. Circulation 2001;103:1182–1184.

13. Safar P, Escarraga LA, Elam JO. A comparison of the mouth-to-mouth and mouth-to-airway methods of artificial respiration with the chest-pressure arm-lift methods. N Engl J Med 1958; 258:671–677.

14. Kouwenhoven WB, Jude JR, Knickerbocker GG. Closed-chest cardiac massage. JAMA 1960;173: 1064–1067.

15. Jude JR, Kouwenhoven WB, Knickerbocker GG. An experimental and clinical study of a portable external cardiac defibrillator. Surg Forum 1962;13:185–187.

16. Mild therapeutic hypothermia to improve the neurologic outcome after cardiac arrest. N Engl J Med 2002;346:549–556.

17. Bernard SA, Gray TW, Buist MD, Jones BM, Silvester W, Gutteridge G, Smith K. Treatment of comatose survivors of out-of-hospital cardiac arrest with induced hypothermia. N Engl J Med 2002;346:557–563.

18. Nolan JP, Morley PT, Vanden Hoek TL, Hickey RW, Kloeck WG, Billi J, Bottiger BW, Okada K, Reyes C, Shuster M, Steen PA, Weil MH, Wenzel V, Carli P, Atkins D. Therapeutic hypothermia after cardiac arrest: an advisory statement by the advanced life support task force of the International Liaison Committee on Resuscitation. Circulation 2003;108:118–121.

19. Vaagenes P, Ginsberg M, Ebmeyer U, Ernster L, Fischer M, Gisvold SE, Gurvitch A, Hossmann KA, Nemoto EM, Radovsky A, Severinghaus JW, Safar P, Schlichtig R, Sterz F, Tonnessen T, White RJ, Xiao F, Zhou Y. Cerebral resuscitation from cardiac arrest: pathophysiologic mechanisms. Crit Care Med 1996;24:S57–S68.

20. Lipton S, Rosenberg PA. Excitatory amino acids as a final common pathway for neurologic disorders. N Engl J Med 1994;330:613–622.

21. Globus MY, Ginsberg MD, Busto R. Excitotoxic index–a biochemical marker of selective vulnerability. Neurosci Lett 1991;127:39–42.

22. Choi DW. Excitotoxic cell death. J Neurobiol 1992;23:1261–1276.

23. Traystman RJ, Kirsch JR, Koehler RC. Oxygen radical mechanisms of brain injury following ischemia and reperfusion. J Appl Physiol 1991;71:1185–1195.

24. Chan PH. Role of oxidants in ischemic brain damage. Stroke 1996;27:1124-1129.

25. Chan PH. Reactive oxygen radicals in signaling and damage in the ischemic brain. J Cereb Blood Flow Metab 2001;21:2–14.

26. Takagi S, Cocito L, Hossmann KA. Blood recirculation and pharmacological responsiveness of the cerebral vasculature following prolonged ischemia of cat brain. Stroke 1977;8:707–712.

27. Spatz M, Yasuma Y, Strasser A, McCarron RM. Cerebral postischemic hypoperfusion is mediated by ETA receptors. Brain Res 1996;726: 242–246.

28. Krep H, Brinker G, Pillekamp F, Hossmann KA. Treatment with an endothelin type A receptor-antagonist after cardiac arrest and resuscitation improves cerebral hemodynamic and functional recovery in rats. Crit Care Med 2000;28:2866–2872.

29. Krep H, Brinker G, Schwindt W, Hossmann KA. Endothelin type A-antagonist improves long-term neurological recovery after cardiac arrest in rats. Crit Care Med 2000;28:2873–2880.

30. Barone FC, Globus MY, Price WJ, White RF, Storer BL, Feuerstein GZ, Busto R, Ohlstein EH. Endothelin levels increase in rat focal and global ischemia. J Cereb Blood Flow Metab 1994; 14:337–342.

31. Sterz F, Leonov Y, Safar P, Johnson D, Oku K, Tisherman SA, Latchaw R, Obrist W, Stezoski SW, Hecht S, et al. Multifocal cerebral blood flow by Xe-CT and global cerebral metabolism after prolonged cardiac arrest in dogs. Reperfusion with open-chest CPR or cardiopulmonary bypass. Resuscitation 1992;24:27–47.

32. Ames A 3rd, Wright RL, Kowada M, Thurston JM, Majno G. Cerebral ischemia. II. The no-reflow phenomenon. Am J Pathol 1968;52:437–453.

33. Leonov Y, Sterz F, Safar P, Johnson DW, Tisherman SA, Oku K. Hypertension with hemodilution prevents multifocal cerebral hypoperfusion after cardiac arrest in dogs. Stroke 1992;23:45–53.

34. Kong SE, Blennerhassett LR, Heel KA, McCauley RD, Hall JC. Ischaemia-reperfusion injury to the intestine. Aust N Z J Surg 1998;68: 554–561.

35. Stechmiller JK, Treloar D, Allen N. Gut dysfunction in critically ill patients: a review of the literature. Am J Crit Care 1997;6:204–209.

36. Carden DL, Granger DN. Pathophysiology of ischaemia-reperfusion injury. J Pathol 2000;190: 255–266.

37. Pittard AJ, Hawkins WJ, Webster NR. The role of the microcirculation in the multi-organ dysfunction syndrome. Clin Intensive Care 1994;5:186–190.

38. Mauer D, Schneider T, Dick W, Withelm A, Elich D, Mauer M. Active compression-decompression resuscitation: a prospective, randomized study in a two-tiered EMS system with physicians in the field. Resuscitation 1996;33:125–134.

39. Abramson NS, Safar P, Detre KM, Kelsey SF, Monroe J, Reinmuth O, Snyder JV. Neurologic recovery after cardiac arrest: effect of duration of ischemia. Brain Resuscitation Clinical Trial I Study Group. Crit Care Med 1985;13:930–931.

40. Berek K, Jeschow M, Aichner F. The prognostication of cerebral hypoxia after out of hospital cardiac arrest in adults. Eur Neurol 1997;37: 135–145.

41. Brain Resuscitation Clinical Trial I Study Group. A randomized clinical study of cardiopulmonary-cerebral resuscitation: design, methods, and patient characteristics. Am J Emerg Med 1986; 4:72–86.

42. Plum F, Posener J. The Diagnosis of Coma and Stupor, 3rd ed. Philadelphia: F. A. Davis Company, 1982.

43. Levy D, Bate D, Carrona J, et al.. Prognosis in nontraumatic coma. Ann Intern Med 1981;94: 293–301.

44. Levy DE, Caronna JJ, Singer BH, Lapinski RH, Frydman H, Plum F. Predicting outcome from hypoxic-ischemic coma. JAMA 1985;253: 1420–1426.

45. Rogove HJ, Safar P, Sutton-Tyrrell K, Abramson NS. Old age does not negate good cerebral outcome after cardiopulmonary resuscitation: analyses from the brain resuscitation clinical trials. The Brain Resuscitation Clinical Trial I and II Study Groups. Crit Care Med 1995;23: 18–25.

46. Auer R, Benvinste H, Hypoxia and Related Conditions. New York: Oxford University Press, 1997.

47. Wijdicks EF, Campeau NG, Miller GM. MR imaging in comatose survivors of cardiac resuscitation. AJNR Am J Neuroradiol 2001;22: 1561–1565.

48. Fujioka M, Okuchi K, Sakaki T, Hiramatsu K, Miyamoto S, Iwasaki S. Specific changes in human brain following reperfusion after cardiac arrest. Stroke 1994;25:2091–2095.

49. Steriade M. Corticothalamic resonance, states of vigilance and mentation. Neuroscience 2000;101: 243–276.

50. Sage JI, Van Uitert RL. Man-in-the-barrel syndrome. Neurology 1986;36:1102–1103.

51. Schroder R. Flunarizine i.v. after cardiac arrest (Fluna-study): study design and organisational aspects of a double-blind, placebo-controlled randomized study. FLUNA Study Group Berlin (corrected). Resuscitation 1989;17(suppl): S121–S127; discussion S199–S206.

52. Aldrete JA, Romo-Salas F, Mazzia VD, Tan SL. Phenytoin for brain resuscitation after cardiac arrest: an uncontrolled clinical trial. Crit Care Med 1981;9:474–477.

53. Stiell IG, Hebert PC, Weitzman BN, Wells GA, Raman S, Stark RM, Higginson LA, Ahuja J, Dickinson GE. High-dose epinephrine in adult cardiac arrest. N Engl J Med 1992;327:1045–1050.

54. Gueugniaud PY, Mols P, Goldstein P, Pham E, Dubien PY, Deweerdt C, Vergnion M, Petit P, Carli P. A comparison of repeated high doses and repeated standard doses of epinephrine for cardiac arrest outside the hospital. European Epinephrine Study Group. N Engl J Med 1998;339:1595–1601.

55. Brown CG, Martin DR, Pepe PE, Stueven H, Cummins RO, Gonzalez E, Jastremski M. A comparison of standard-dose and high-dose epinephrine in cardiac arrest outside the hospital. The Multicenter High-Dose Epinephrine Study Group. N Engl J Med 1992;327:1051–1055.

56. Schwab TM, Callaham ML, Madsen CD, Utecht TA. A randomized clinical trial of active compression-decompression CPR vs standard CPR in out-of-hospital cardiac arrest in two cities. JAMA 1995;273:1261–1268.

57. Plaisance P, Lurie KG, Vicaut E, Adnet F, Petit JL, Epain D, Ecollan P, Gruat R, Cavagna P, Biens J, Payen D. A comparison of standard cardiopulmonary resuscitation and active compression-decompression resuscitation for out-of-hospital

cardiac arrest. French Active Compression-Decompression Cardiopulmonary Resuscitation Study Group. N Engl J Med 1999;341:569–575.

58. Wenzel V, Krismer AC, Arntz HR, Sitter H, Stadlbauer KH, Lindner KH. A comparison of vasopressin and epinephrine for out-of-hospital cardiopulmonary resuscitation. N Engl J Med 2004;350:105–113.

59. Brain Resuscitation Clinical Trial I Study Group. Randomized clinical study of thiopental loading in comatose survivors of cardiac arrest. N Engl J Med 1986;314:397–403.

60. Brain Resuscitation Clinical Trial II Study Group. A randomized clinical study of a calcium-entry blocker (lidoflazine) in the treatment of comatose survivors of cardiac arrest (see comments). N Engl J Med 1991;324:1225–1231.

61. Jastremski M, Sutton-Tyrrell K, Vaagenes P, Abramson N, Heiselman D, Safar P. Glucocorticoid treatment does not improve neurological recovery following cardiac arrest. Brain Resuscitation Clinical Trial I Study Group. JAMA 1989;262:3427–3430.

62. Roine RO, Kaste M, Kinnunen A, Nikki P, Sarna S, Kajaste S. Nimodipine after resuscitation from out-of-hospital ventricular fibrillation. A placebo-controlled, double-blind, randomized trial. JAMA 1990;264:3171–3177.

63. Thel MC, Armstrong AL, McNulty SE, Califf RM, O'Connor CM. Randomised trial of magnesium in in-hospital cardiac arrest. Duke Internal Medicine House staff. Lancet 1997;350:1272–1276.

64. Longstreth WT Jr, Copass MK, Dennis LK, Rauch-Matthews ME, Stark MS, Cobb LA. Intravenous glucose after out-of-hospital cardiopulmonary arrest: a community-based randomized trial. Neurology 1993;43:2534–2541.

65. Longstreth WT Jr, Fahrenbruch CE, Olsufka M, Walsh TR, Copass MK, Cobb LA. Randomized clinical trial of magnesium, diazepam, or both after out-of-hospital cardiac arrest. Neurology 2002;59:506–514.

66. Bleyaert AL, Nemoto EM, Safar P, Stezoski SM, Mickell JJ, Moossy J, Rao GR. Thiopental amelioration of brain damage after global ischemia in monkeys. Anesthesiology 1978;49:390–398.

67. Mullie A, Lust P, Penninckx J, Vanhove L, Vandevelde K, Vanhoonacker G, Krier M. Monitoring of cerebrospinal fluid enzyme levels in postischemic encephalopathy after cardiac arrest. Crit Care Med 1981;9:399–400.

68. Roine RO, Kaste M, Kinnunen A, Nikki P. Safety and efficacy of nimodipine in resuscitation of patients outside hospital. Br Med J (Clin Res Ed) 1987;294:20.

69. Allen GS, Ahn HS, Preziosi TJ, Battye R, Boone SC, Chou SN, Kelly DL, Weir BK, Crabbe RA, Lavik PJ, Rosenbloom SB, Dorsey FC, Ingram CR,

Mellits DE, Bertsch LA, Boisvert DP, Hundley MB, Johnson RK, Strom JA, Transou CR. Cerebral arterial spasm—a controlled trial of nimodipine in patients with subarachnoid hemorrhage. N Engl J Med 1983;308:619–624.

70. Longstreth WT Jr, Diehr P, Cobb LA, Hanson RW, Blair AD. Neurologic outcome and blood glucose levels during out-of-hospital cardiopulmonary resuscitation. Neurology 1986;36:1186–1191.

71. Hicks SD, DeFranco DB, Callaway CW. Hypothermia during reperfusion after asphyxial cardiac arrest improves functional recovery and selectively alters stress-induced protein expression. J Cereb Blood Flow Metab 2000;20:520–530.

72. Xiao F, Safar P, Radovsky A. Mild protective and resuscitative hypothermia for asphyxial cardiac arrest in rats. Am J Emerg Med 1998;16:17–25.

73. Sterz F, Safar P, Tisherman S, Radovsky A, Kuboyama K, Oku K. Mild hypothermic cardiopulmonary resuscitation improves outcome after prolonged cardiac arrest in dogs. Crit Care Med 1991;19:379–389.

74. Safar P, Xiao F, Radovsky A, Tanigawa K, Ebmeyer U, Bircher N, Alexander H, Stezoski SW. Improved cerebral resuscitation from cardiac arrest in dogs with mild hypothermia plus blood flow promotion. Stroke 1996;27:105–113.

75. Zeiner A, Holzer M, Sterz F, Behringer W, Schorkhuber W, Mullner M, Frass M, Siostrzonek P, Ratheiser K, Kaff A, Laggner AN. Mild resuscitative hypothermia to improve neurological outcome after cardiac arrest. A clinical feasibility trial. Hypothermia After Cardiac Arrest (HACA) Study Group. Stroke 2000;31:86–94.

76. Bernard SA, Jones BM, Horne MK. Clinical trial of induced hypothermia in comatose survivors of out-of-hospital cardiac arrest. Ann Emerg Med 1997;30:146–153.

77. Ginsberg M, Belayev L. The effects of hypothermia and hyperthermia in global cerebral ischemia. In: Maier C, Steinberg G, eds. Hypothermia and Cerebral Ischemia. Totowa, NJ: Humana Press, 2004.

78. Kramer RS, Sanders AP, Lesage AM, Woodhall B, Sealy WC. The effect profound hypothermia on preservation of cerebral ATP content during circulatory arrest. J Thorac Cardiovasc Surg 1968;56:699–709.

79. Welsh FA, Sims RE, Harris VA. Mild hypothermia prevents ischemic injury in gerbil hippocampus. J Cereb Blood Flow Metab 1990;10:557–563.

80. Busto R, Globus MY, Dietrich WD, Martinez E, Valdes I, Ginsberg MD. Effect of mild hypothermia on ischemia-induced release of neurotransmitters and free fatty acids in rat brain. Stroke 1989;20:904–910.

81. Cardell M, Boris-Moller F, Wieloch T. Hypothermia prevents the ischemia-induced translocation

and inhibition of protein kinase C in the rat striatum. J Neurochem 1991;57:1814–1817.

82. Dempsey RJ, Combs DJ, Maley ME, Cowen DE, Roy MW, Donaldson DL. Moderate hypothermia reduces postischemic edema development and leukotriene production. Neurosurgery 1987;21:177–181.

83. Toyoda T, Suzuki S, Kassell NF, Lee KS. Intraischemic hypothermia attenuates neutrophil infiltration in the rat neocortex after focal ischemia-reperfusion injury. Neurosurgery 1996;39: 1200–1205.

84. Kumar K, Wu X, Evans AT, Marcoux F. The effect of hypothermia on induction of heat shock protein (HSP)-72 in ischemic brain. Metab Brain Dis 1995;10:283–291.

85. Kumar K, Wu X, Evans AT. Expression of c-fos and fos-B proteins following transient forebrain ischemia: effect of hypothermia. Brain Res Mol Brain Res 1996;42:337–343.

86. Cummins RO, Chamberlain DA, Abramson NS, Allen M, Baskett P, Becker L, Bossaert L, Delooz H, Dick W, Eisenberg M, et al. Recommended guidelines for uniform reporting of data from out-of-hospital cardiac arrest: the Utstein Style. Task Force of the American Heart Association, the European Resuscitation Council, the Heart and Stroke of Canada, and the Australian Resuscitation Council. Ann Emerg Med 1991;20:861–874.

87. Cummins RO, Chamberlain D, Hazinski MF, Nadkarni V, Kloeck W, Kramer E, Becker L, Robertson C, Koster R, Zaritsky A, Ornato JP, Callanan V, Allen M, Steen P, Connolly B, Sanders A, Idris A, Cobbe S. Recommended guidelines for reviewing, reporting, and conducting research on in-hospital resuscitation: the in-hospital "Utstein style." American Heart Association. Ann Emerg Med 1997;29:650–679.

88. Cone DC, Jaslow DS, Brabson TA. Now that we have the Utstein style, are we using it? Acad Emerg Med 1999;6:923–928.

89. Cummins RO. Why are researchers and emergency medical services managers not using the Utstein guidelines? Acad Emerg Med 1999; 6:871–875.

90. Morimoto Y, Kemmotsu O, Kitami K, Matsubara I, Tedo I. Acute brain swelling after out of hospital cardiac arrest: pathogenesis and outcome. Crit Care Med 1993;21:104–110.

91. Torbey MT, Selim M, Knorr J, Bigelow C, Recht L. Quantitative analysis of the loss of distinction between gray and white matter in comatose patients after cardiac arrest. Stroke 2000; 31:2163–2167.

92. Iida K, Satoh H, Arita K, Nakahara T, Kurisu K, Ohtani M. Delayed hyperemia causing intracranial hypertension after cardiopulmonary resuscitation. Crit Care Med 1997;25:971–976.

93. Mullner M, Sterz F, Binder M, Hellwagner K, Meron G, Herkner H, Laggner AN. Arterial blood pressure after human cardiac arrest and neurological recovery. Stroke 1996;27:59–62.

94. Longstreth WT, Inui TS, Cobb LA, Compass MK. Neurologic recovery after out-of-hospital cardiac arrest. Ann Intern Med 1983;98:121–132.

95. Mullner M, Sterz F, Binder M, Schreiber W, Deimel A, Laggner AN. Blood glucose concentration after cardiopulmonary resuscitation influences functional neurological recovery in human cardiac arrest survivors. J Cereb Blood Flow Metab 1997;17:430–436.

96. Kagansky N, Levy S, Knobler H. The role of hyperglycemia in acute stroke. Arch Neurol 2001;58:1209–1212.

97. van den Berghe G, Wouters P, Weekers F, Verwaest C, Bruyninckx F, Schetz M, Vlasselaers D, Ferdinande P, Lauwers P, Bouillon R. Intensive insulin therapy in the critically ill patients. N Engl J Med 2001;345:1359–1367.

98. Zeiner A, Holzer M, Sterz F, Schorkhuber W, Eisenburger P, Havel C, Kliegel A, Laggner AN. Hyperthermia after cardiac arrest is associated with an unfavorable neurologic outcome. Arch Intern Med 2001;161:2007–2012.

99. Takasu A, Saitoh D, Kaneko N, Sakamoto T, Okada Y. Hyperthermia: is it an ominous sign after cardiac arrest? Resuscitation 2001;49: 273–277.

100. Sterz F, Safar P, Diven W, Leonov Y, Radovsky A, Oku K. Detoxification with hemabsorption after cardiac arrest does not improve neurologic recovery. Review and outcome study in dogs. Resuscitation 1993;25:137–160.

101. Lawes EG, Baskett PJ. Pulmonary aspiration during unsuccessful cardiopulmonary resuscitation. Intensive Care Med 1987;13:379–382.

102. Lee-Chiong TL Jr, Stitt JT. Disorders of temperature regulation. Compr Ther 1995;21:697–704.

103. Powers JH, Scheld WM. Fever in neurologic diseases. Infect Dis Clin North Am 1996; 10:45–66.

104. Madl JE, Allen DL. Hyperthermia depletes adenosine triphosphate and decreases glutamate uptake in rat hippocampal slices. Neuroscience 1995;69:395–405.

105. Globus MY, Busto R, Lin B, Schnippering H, Ginsberg MD. Detection of free radical activity during transient global ischemia and recirculation: effects of intraischemic brain temperature modulation. J Neurochem 1995;65:1250–1256.

106. Ginsberg MD, Sternau LL, Globus MY, Dietrich WD, Busto R. Therapeutic modulation of brain temperature: relevance to ischemic brain injury. Cerebrovasc Brain Metab Rev 1992;4:189–225.

107. Krumholz A, Stern BJ, Weiss HD. Outcome from coma after cardiopulmonary resuscitation:

relation to seizures and myoclonus. Neurology 1988;38:401–405.

108. Wijdicks E, Parisi J, Sharbrough F. Prognostic value of myoclonus status in comatose survivors of cardiac arrest. Ann Neurol 1994;35:239–243.

109. Werhahn KJ, Brown P, Thompson PD, Marsden CD. The clinical features and prognosis of chronic posthypoxic myoclonus. Mov Disord 1997;12:216–220.

110. Voipio V, Kuisma M, Alaspaa A, Manttari M, Rosenberg P. Thrombolytic treatment of acute myocardial infarction after out-of-hospital cardiac arrest. Resuscitation 2001;49:251–258.

111. Ruiz-Bailen M, Aguayo de Hoyos E, Serrano-Corcoles MC, Diaz-Castellanos MA, Ramos-Cuadra JA, Reina-Toral A. Efficacy of thrombolysis in patients with acute myocardial infarction requiring cardiopulmonary resuscitation. Intensive Care Med 2001;27:1050–1057.

112. Safar P, Behringer W, Bottiger BW, Sterz F. Cerebral resuscitation potentials for cardiac arrest. Crit Care Med 2002;30:S140–S144.

9

Efficient Dose-Response Finding Strategies for Acute Neuroemergency Treatments

Tom Parke, Michael Krams, Peter Mueller, and Don Berry

The objective of this chapter is to make a case for simulation-guided clinical trial design. We develop the concept of adaptive dose-response finding studies with real-time learning about the research question, continuously updating the choice of treatment, and assessing whether to continue or terminate the trial. We present a software package that allows both the simulation and the conduct of a real-time learning trial. Finally, we use the example of developing a neuroprotective treatment for acute ischemic stroke to develop our ideas.

BACKGROUND: THE ASTIN STUDY

In 2001–2002 we ran a clinical trial exploring the benefit of neutrophil inhibitory factor to acute ischemic stroke patients (1). In designing the study we attempted to solve a number of the problems inherent in stroke treatment clinical trials, in particular the following:

- In a novel treatment with little prior art, how does one select the correct dose for confirmatory phase III trials?

- Very early treatment (within less than 6 or preferably 3 hours) is important, that is, the decision as to which treatment to administer must not prolong the onset-to-treatment time.
- A low incidence of suitable candidate patients requires the use of a large number of centers worldwide.
- The subjects' responses need to be measured in a way that is sufficiently sensitive to discriminate between responses at different doses, is clinically meaningful, and yields the maximum information from each subject.

We developed a Bayesian adaptive dose allocation scheme. It allowed the use of a large number of doses ($n = 16$). Rather than allocating subjects equally to each dose arm, the system learned continuously about the dose response and allocated subjects either to placebo (>15%) or adaptively to the dose that would yield most information about the ED_{95} (minimal dose providing maximal treatment effect) of the dose-response curve. The system continuously assessed the observed effect size and the variability associated with it and was built to terminate the trial at the earliest possible time point, adapting sample size

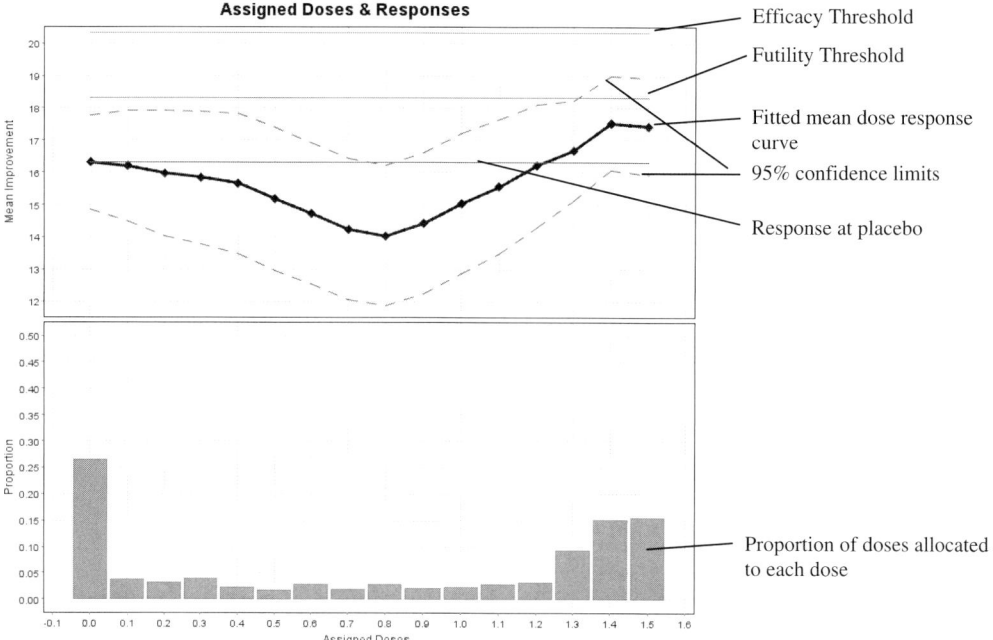

FIGURE 9.1   Final dose response and allocation subjects.

to the data observed in the trial. There was an upper cap for sample size (1300).

To assess initial stroke severity and recovery over time, we chose to use the Scandinavian Stroke Scale (SSS), which consists of nine subitems. We regarded the total score as a continuous variable and used the Copenhagen Stroke Study database (2) to learn about the properties of this scale in describing recovery in an acute stroke population.

Figure 9.1 shows the final position of the estimated dose-response at the end of the trial. Rather than looking at the information at each dose as an independent data point, we modeled all data to estimate the dose-response, using the Normal Dynamic Linear Model (Fig. 9.2).

In acute stroke trials the recruitment frequency per center is usually low (one to five subjects per month). We used close to 100 centers worldwide and implemented a central electronic system that performed the randomization in real time and communicated the results to the center within minutes. Communication was via fax, and

the centers were provided with preformatted fax forms for the various messages they might need to send to the center, options and values being specified by selecting multiple-choice options. These were then read and responded to automatically by the central system.

"Response" was measured by assessing change from baseline to day 90 as measured on the SSS, which was measured at baseline, week 1, week 4, and week 13 (Fig. 9.3).

As well as adapting treatment allocation to maximize the learning about the ED_{95}, the system also continuously assessed whether to stop or to continue the trial (3). This was by evaluating the posterior probability of whether the mean response at the ED_{95} was above an efficacy threshold or below a futility threshold with a certain probability (Fig. 9.4). This termination rule uses the mean response at ED_{95} and the standard deviation (SD) of that response as key defining features of how the trial is progressing. Note that the larger the SD of the mean, the larger the interval below

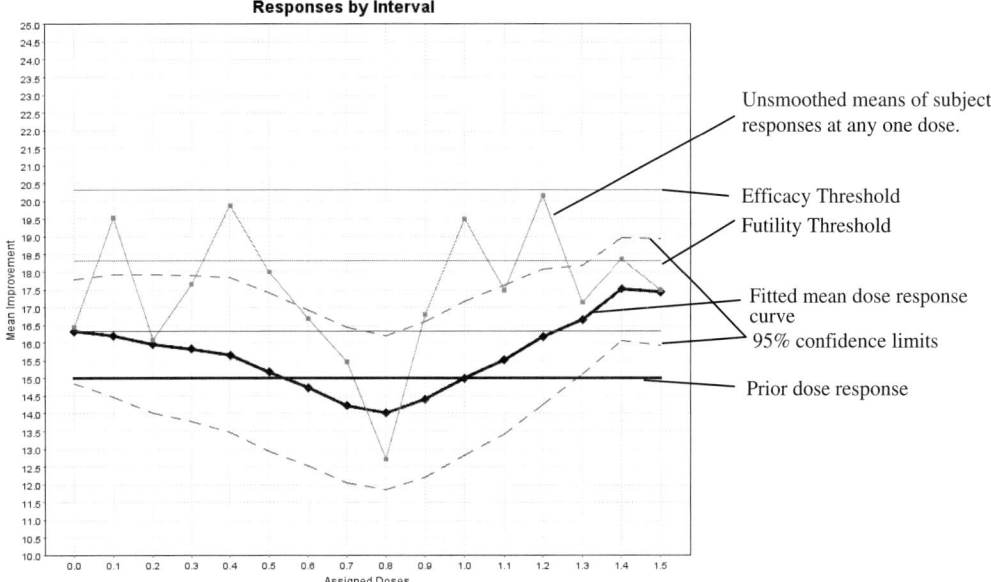

Unsmoothed means of subject responses at any one dose.

Efficacy Threshold

Futility Threshold

Fitted mean dose response curve

95% confidence limits

Prior dose response

FIGURE 9.2 Raw and smoothed final dose response.

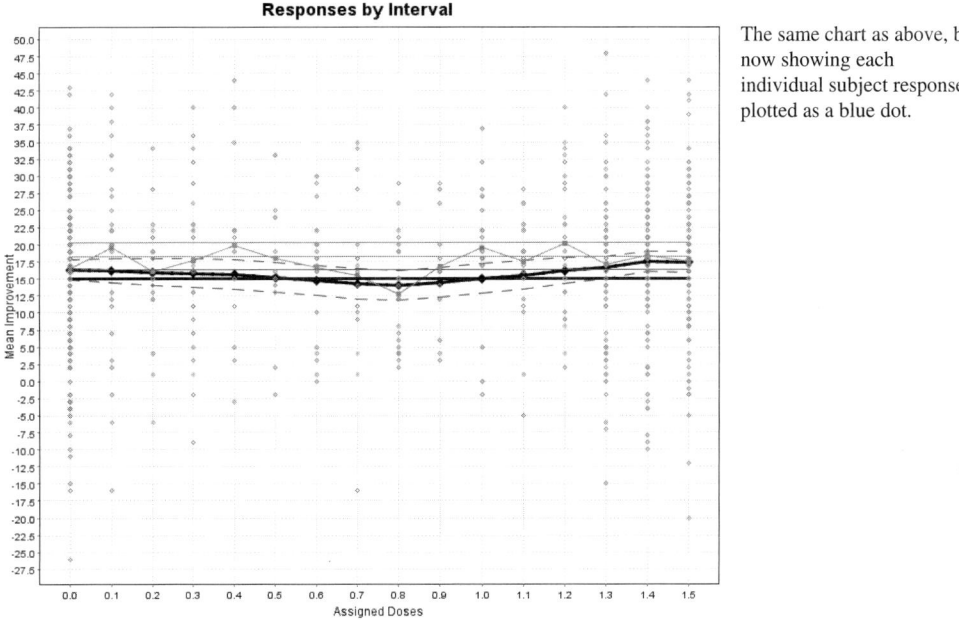

The same chart as above, but now showing each individual subject response, plotted as a blue dot.

FIGURE 9.3 Final dose response showing individual outcomes.

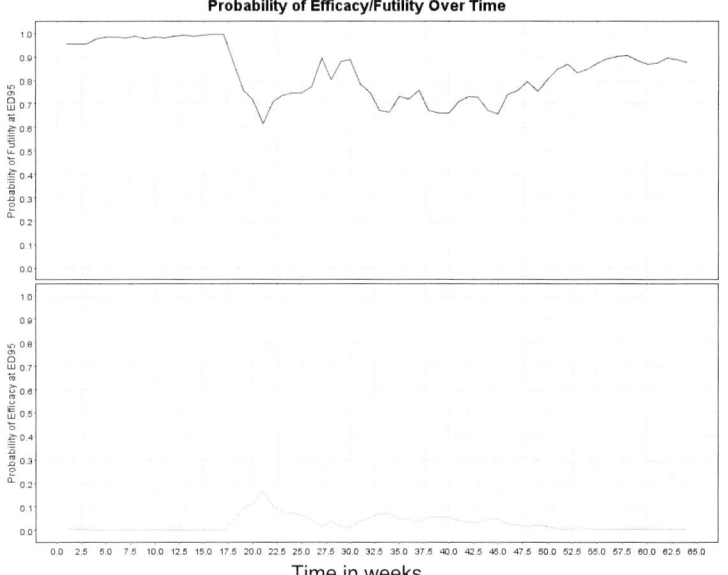

This chart of the termination probabilities over time clearly shows the study likely to fail, and that finally towards the end, the probability of futility exceeded 90%.

FIGURE 9.4   Estimated probability of efficacy and futility during the trial.

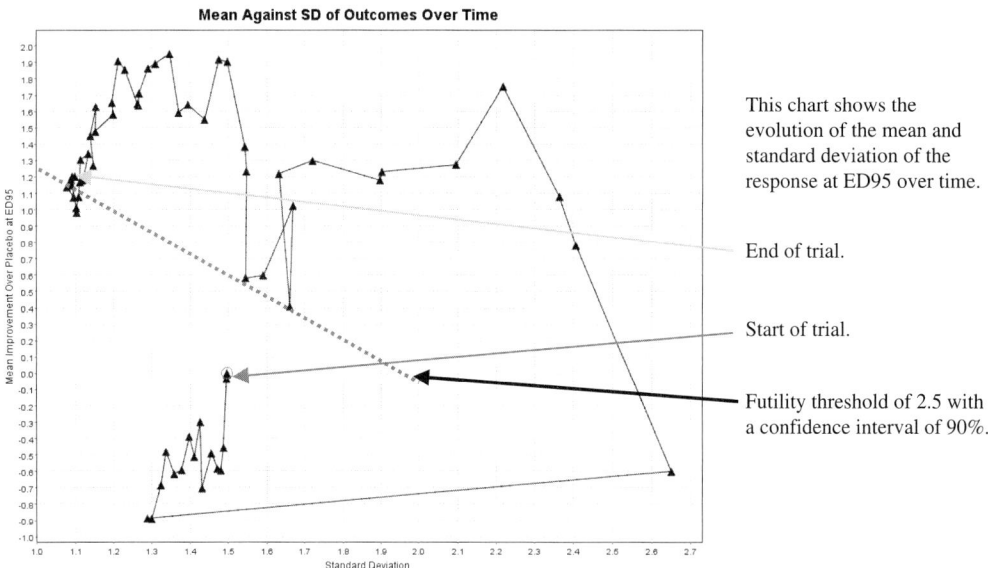

This chart shows the evolution of the mean and standard deviation of the response at ED95 over time.

End of trial.

Start of trial.

Futility threshold of 2.5 with a confidence interval of 90%.

FIGURE 9.5   Mean and standard deviation of response during the trial at the selected dose (ED95).

the futility threshold required by the 90% probability requirement (Fig. 9.5).

The independent data monitoring committee is a key requirement in running

trials with novel designs to overlook the behavior of the computer-guided treatment allocation and termination recommendation. In ASTIN, the independent data monitoring committee was composed of clinical experts who were intimately familiar with the novel aspects of the design and a statistician with expertise in Bayesian approaches, who had not been involved in the development of the system, to enable an independent oversight of the performance of the model.

GENERALIZED ADAPTIVE DOSE ALLOCATION TOOL

We have taken the software developed for the ASTIN study and generalized it by removing specific constraints or assumptions of that study. However, it still retains the essential model used on the ASTIN study. The system is applicable to studies where the doses can be treated as equally spaced, there is a linear scale for measuring response, and the goal is to learn about improvement from baseline of the various doses compared with placebo.

The system is built for dose-response finding trials in indications with a continuous endpoint measure. We illustrate the operating characteristics using a pharmacological intervention in acute stroke as an example.

The design process involves three steps:

1. Exploration: simulating large number of trials to explore how the system performs under a number of different conditions (different dose-response curves, different parameter settings);
2. Setting the parameters and running large-scale simulations to determine the type I and II error rates;
3. Conducting the real trial using the parameters established in step 2.

A major area where we have extended the tool is in providing support for simulations, and nearly all the illustrations in this chapter are taken directly from the tool.

When we simulate trials we are usually trying to answer one of these fundamental questions:

- How likely is this clinical trial to succeed?
- Which settings will minimize type I and II errors for most scenarios?

We can also ask the more general question:

- What are the operating characteristics of my study?

EFFICACY AND FUTILITY

One effect of studying simulations is that it forces one to think clearly at the outset of the trial about what the success criterion is. Conventionally, clinicians have an intuition that a such-and-such improvement over placebo would be sufficient to be clinically meaningful. What is often overlooked is that for a phase II trial, success is being able to undertake a phase III trial, and to undertake a phase III trial we need sufficient confidence at the end of the phase II trials that at the end of the phase III trial we will have a clinically meaningful drug effect.

The apparent effect size alone is not enough to decide whether we can successfully run a phase III trial; rather, the combination of the effect size and our uncertainty about it determines the likelihood that the phase III trial will succeed and the number of subjects the phase III trial requires. Practical and financial considerations then dictate whether a phase III trial can be undertaken. Conventionally, the sample size calculation for phase III is based on the effect size observed in phase II, assuming that it exactly predicts what will be found in phase III. The sample size will be determined to provide a required level of type I and II errors (e.g., alpha = 0.05 and beta = 0.9). To control

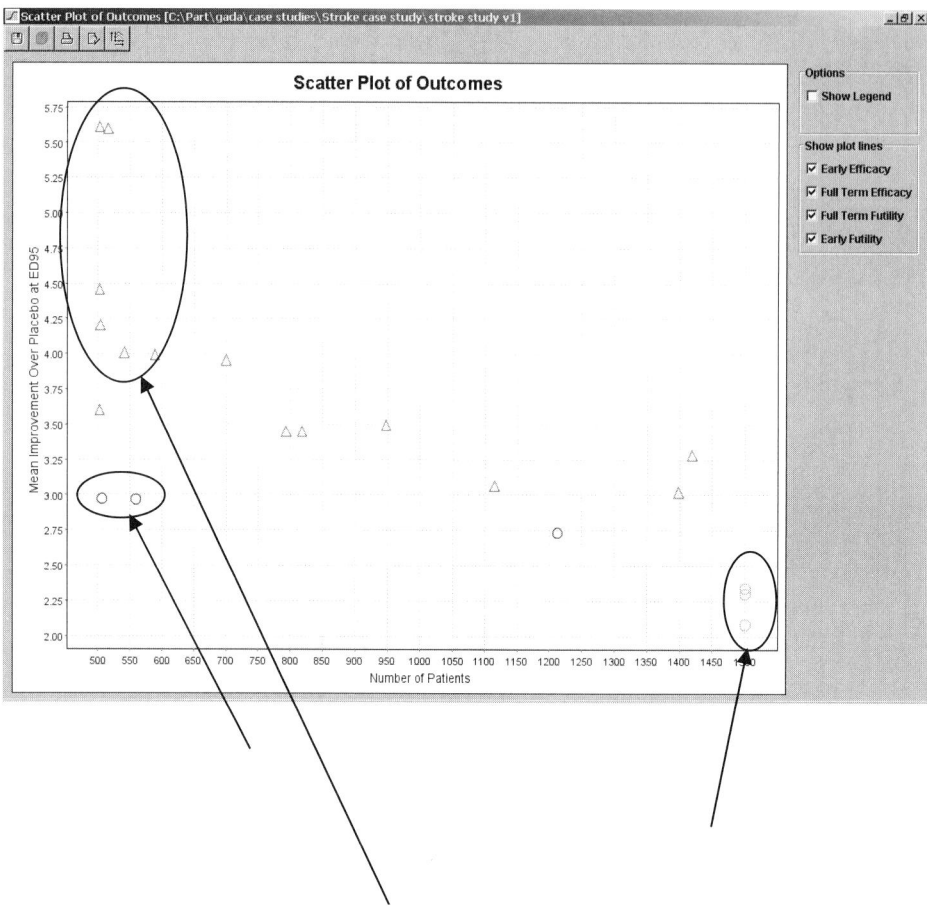

FIGURE 9.6   The statistical significance boundary to control "alpha".

type I errors, statisticians use a frequentist analysis that considers the null hypothesis: that the compound has no advantage over placebo. The probability of such a compound successfully passing the phase III trial (false positive) must be less than some limit "alpha," usually 0.05.

This means that we need fix a "statistical significance boundary" where the phase III trial will only be a success if the mean result is above this boundary. Note this boundary is the level of response necessary for statistical significance—not clinical significance, which may be different (Fig. 9.6).

The level required for statistical significance depends on the SD of the mean response of the phase III trial, s_3. This in turn depends on the SD of the subject responses, σ, and the size of the phase III trial, n_3:

$$s_3 = \sqrt{(2 \cdot \sigma^2 / n_3)}$$

To control the type II errors, we look at the alternative hypothesis: We assume that the compound has a mean improvement over placebo of m as determined by the phase II trial. We require that the probability of such a compound failing the phase III trial (false positive) as being less than some limit β, usually 0.1. This means that we need "statistical significance boundary" $<m$, and the phase III trial will be failure if the mean result is below this boundary (Fig. 9.7).

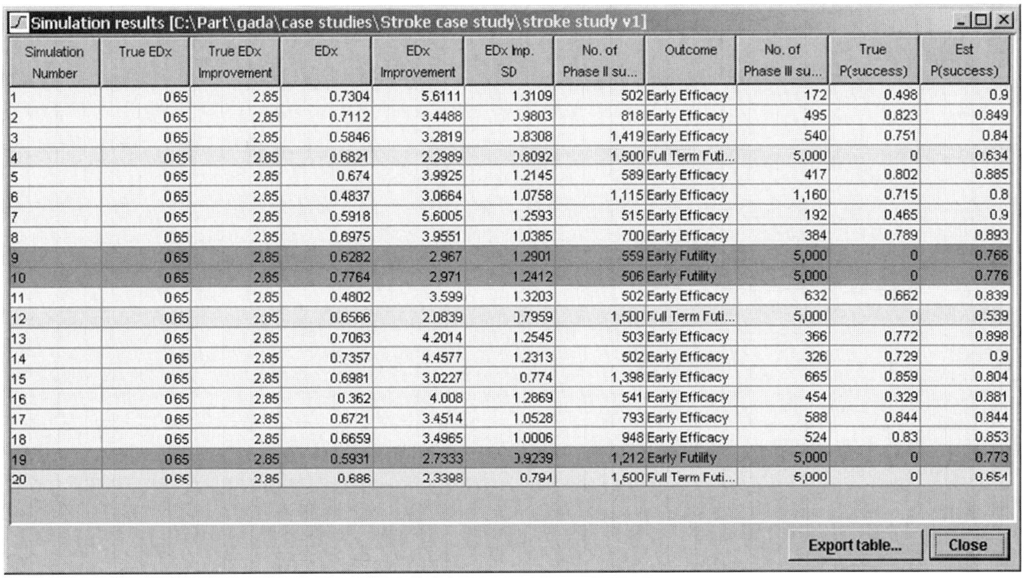

_ □ x|

Simulation Number	True EDx	True EDx Improvement	EDx	EDx Improvement	EDx Imp. SD	No. of Phase II su...	Outcome	No. of Phase III su...	True P(success)	Est P(success)
1	0.65	2.85	0.7304	5.6111	1.3109	502	Early Efficacy	172	0.498	0.9
2	0.65	2.85	0.7112	3.4488	0.9803	818	Early Efficacy	495	0.823	0.849
3	0.65	2.85	0.5846	3.2819	0.8308	1,419	Early Efficacy	540	0.751	0.84
4	0.65	2.85	0.6821	2.2989	0.8092	1,500	Full Term Futi...	5,000	0	0.634
5	0.65	2.85	0.674	3.9925	1.2145	589	Early Efficacy	417	0.802	0.885
6	0.65	2.85	0.4837	3.0664	1.0758	1,115	Early Efficacy	1,160	0.715	0.8
7	0.65	2.85	0.5918	5.6005	1.2593	515	Early Efficacy	192	0.465	0.9
8	0.65	2.85	0.6975	3.9551	1.0385	700	Early Efficacy	384	0.789	0.893
9	0.65	2.85	0.6282	2.967	1.2901	559	Early Futility	5,000	0	0.766
10	0.65	2.85	0.7764	2.971	1.2412	506	Early Futility	5,000	0	0.776
11	0.65	2.85	0.4802	3.599	1.3203	502	Early Efficacy	632	0.662	0.839
12	0.65	2.85	0.6566	2.0839	0.7959	1,500	Full Term Futi...	5,000	0	0.539
13	0.65	2.85	0.7063	4.2014	1.2545	503	Early Efficacy	366	0.772	0.898
14	0.65	2.85	0.7357	4.4577	1.2313	502	Early Efficacy	326	0.729	0.9
15	0.65	2.85	0.6981	3.0227	0.774	1,398	Early Efficacy	665	0.859	0.804
16	0.65	2.85	0.362	4.008	1.2869	541	Early Efficacy	454	0.329	0.881
17	0.65	2.85	0.6721	3.4541	1.0528	793	Early Efficacy	588	0.844	0.844
18	0.65	2.85	0.6659	3.4965	1.0006	948	Early Efficacy	524	0.83	0.853
19	0.65	2.85	0.5931	2.7333	0.9239	1,212	Early Futility	5,000	0	0.773
20	0.65	2.85	0.686	2.3398	0.794	1,500	Full Term Futi...	5,000	0	0.654

Simulation results [C:\Part\gada\case studies\Stroke case study\stroke study v1]

Export table... Close

FIGURE 9.7 The statistical significance boundary to control "beta".

These two constraints allow us to calculate the required SD of the observed mean at the end of the phase III trial, which gives the minimum phase III trial size. *So we have efficacy at the end of the phase II trial if we can size a phase III trial within some given upper bound, otherwise we have futility. We also allow the user to specify a required likelihood that the mean response at the end of the phase III trial is above a specified clinically significant threshold, given the m determined in phase II.

SIMULATING A STROKE TRIAL

To explore the characteristics of any one particular design setup, we start by

---

*We believe that this conventional way of calculating the minimum phase III overlooks two factors and is not conservative enough. First, the response m at the end of phase II is only our measure of the response and not the actual response to the compound, about which we are uncertain (measured by the SD of the mean at the end of the phase II trial). Second, the phase III trial is often larger than the phase II trial, and we can expect to see an SD in the subject responses greater than in the phase II trial.

running a small number of simulated trials, say 20. One of the most useful starting points to analyze the behavior of a set of simulations is the scatter plot of outcomes (Fig. 9.8). This shows one trial simulation that has terminated early and failed but at a fairly high response (*square*); a small group of simulations that have run to the end, failing to show sufficient response (*circles*); and a number of simulations that have terminated early, overestimating the true response. These could result in under-powering the phase III trial, reducing our chance of success in phase III (*triangles*).

This gives us a number of problems we can try to address by tuning the protocol's parameters (we define "protocol" as the set of parameters modulating performance of the simulated trial). Table 9.1 shows the results of the individual simulations. Looking at the results overall, the average dose response is close to the simulated curve and dose allocation is predominately where we want it to be: at placebo and the higher doses (Fig. 9.9). The average curve can be misleading, however; it is salutary to see the dose-response curves of the individual simulations (Fig. 9.10).

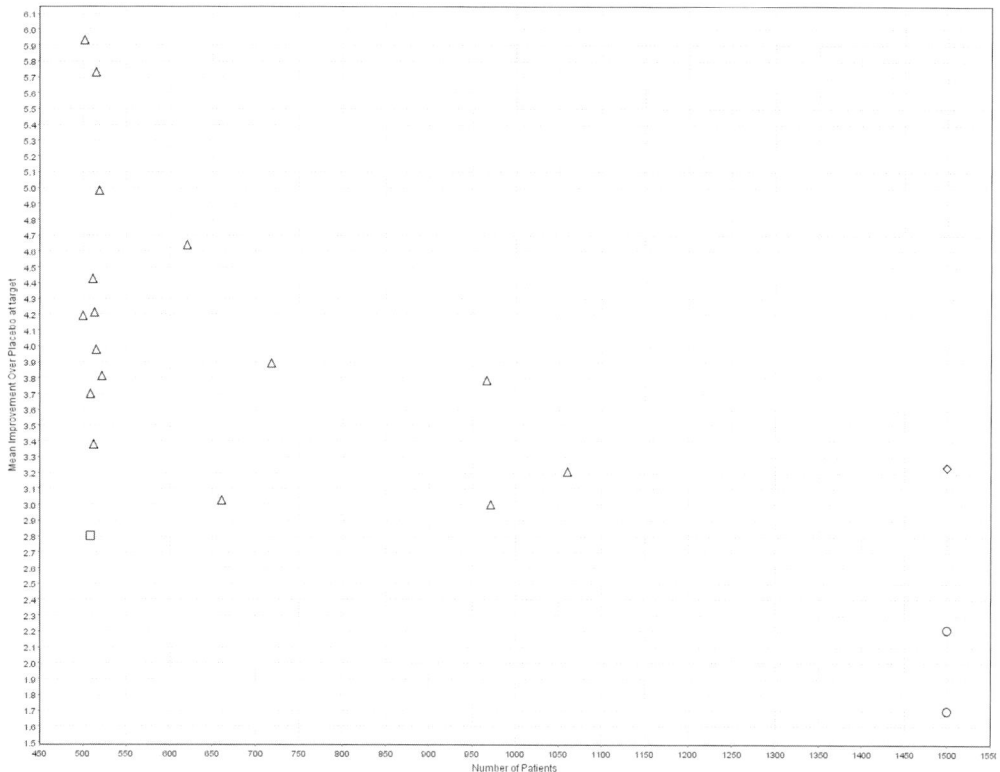

FIGURE 9.8   Scatter plot of simulation outcomes.

Early Unsuccessful Terminations

Looking at the early unsuccessful termination (Table 9.1, simulation 12) and its probability of efficacy/futility over time (Fig. 9.11), we can see that in this simulation, the decision to stop was taken in week 24, that the decision was to stop for efficacy, but that as the true final results came in for subjects already recruited, the probability of efficacy fell until it was below the level required for a phase III trial.

Looking at the dose-response data for this simulation (Fig. 9.12), we can see how noisy the real data is (the unsmoothed data is the feint line joining the gray squares) and what a good job the curve smoothing does of extracting something close to the true curve from this.

Looking at how the mean and SD evolved, we can see how the mean fell after the decision to terminate. Note the

problem is not the mean per se but the mean and SD, which taken together do not give us the required level of confidence in a successful phase III (Fig. 9.13).

Four solutions suggest themselves:

1. Raise the efficacy early termination threshold.
2. Raise the efficacy early termination confidence level.
3. Raise the minimum number of subjects before early termination is allowed.
4. Reduce the noise in the subject responses.

Incorrect Dose Determination

Simulations 3 (Fig. 9.14) and 5 (Fig. 9.15) both identified too low a dose as the dose for ED_{95}. Particularly in the latter case, one can clearly see how the patient responses

TABLE 9.1

Simulation number	True target	True target improvement	Target	Target SD	Target improvement	Target improvement SD	No. of phase II subjects	Outcome	Subject noise	Baseline regression coefficient
1	0.65	2.85	0.76	0.07	5.93	1.33	502	Early efficacy	10.95	0.22
2	0.65	2.85	0.77	0.09	2.21	0.77	1500	Full term futility	11.76	0.27
3	0.65	2.85	0.54	0.12	3.24	0.86	1500	Full term efficacy	11.18	0.30
4	0.65	2.85	0.64	0.10	3.81	1.29	522	Early efficacy	11.45	0.37
5	0.65	2.85	0.39	0.15	3.78	1.02	966	Early efficacy	11.29	0.31
6	0.65	2.85	0.58	0.11	3.38	1.28	512	Early efficacy	11.49	0.26
7	0.65	2.85	0.58	0.12	3.21	0.93	1061	Early efficacy	11.56	0.30
8	0.65	2.85	0.72	0.10	4.64	1.30	621	Early efficacy	10.97	0.37
9	0.65	2.85	0.75	0.11	3.70	1.27	509	Early efficacy	11.32	0.09
10	0.65	2.85	0.78	0.16	4.98	1.29	519	Early efficacy	11.21	0.22
11	0.65	2.85	0.78	0.12	3.00	0.88	971	Early efficacy	11.24	0.21
12	0.65	2.85	0.78	0.16	2.81	1.24	509	Early futility	11.65	0.30
13	0.65	2.85	0.77	0.17	3.98	1.33	515	Early efficacy	11.19	0.37
14	0.65	2.85	0.67	0.17	3.03	1.14	661	Early efficacy	11.31	0.41
15	0.65	2.85	0.72	0.17	1.70	0.81	1500	Full term futility	11.31	0.29
16	0.65	2.85	0.63	0.11	4.19	1.30	500	Early efficacy	11.5	0.30
17	0.65	2.85	0.77	0.07	4.21	1.28	513	Early efficacy	11.77	0.23
18	0.65	2.85	0.72	0.11	5.73	1.21	515	Early efficacy	10.93	0.25
19	0.65	2.85	0.78	0.12	3.89	1.41	718	Early efficacy	11.41	0.34
20	0.65	2.85	0.54	0.12	4.43	1.34	511	Early efficacy	11.08	0.19

Simulation 12 is the one that terminated early but failed. Simulations 1 and 18 significantly overestimate the response and so under-power their subsequent phase III trials. Simulations 3, 5, 6, 7, and 16 select a lower dose with a lower underlying true response and hence lower chance of success in their subsequent phase III trials. Simulations 2 and 15 selected the correct dose but did not detect a high enough response to justify continuing to phase III.

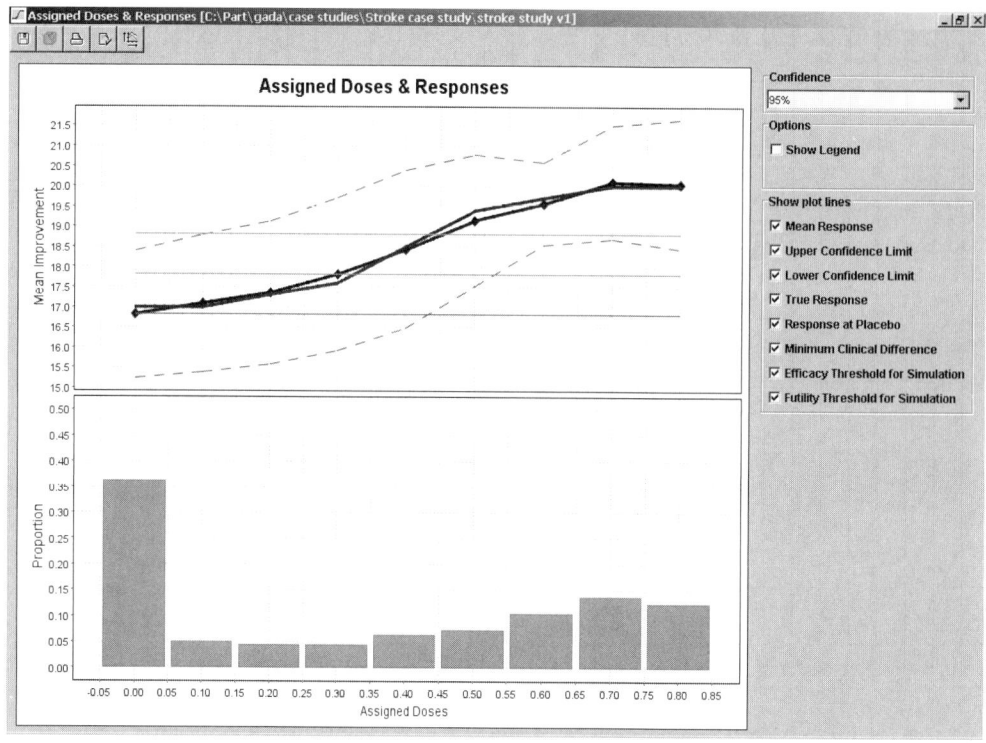

FIGURE 9.9   Assigned doses and responses.

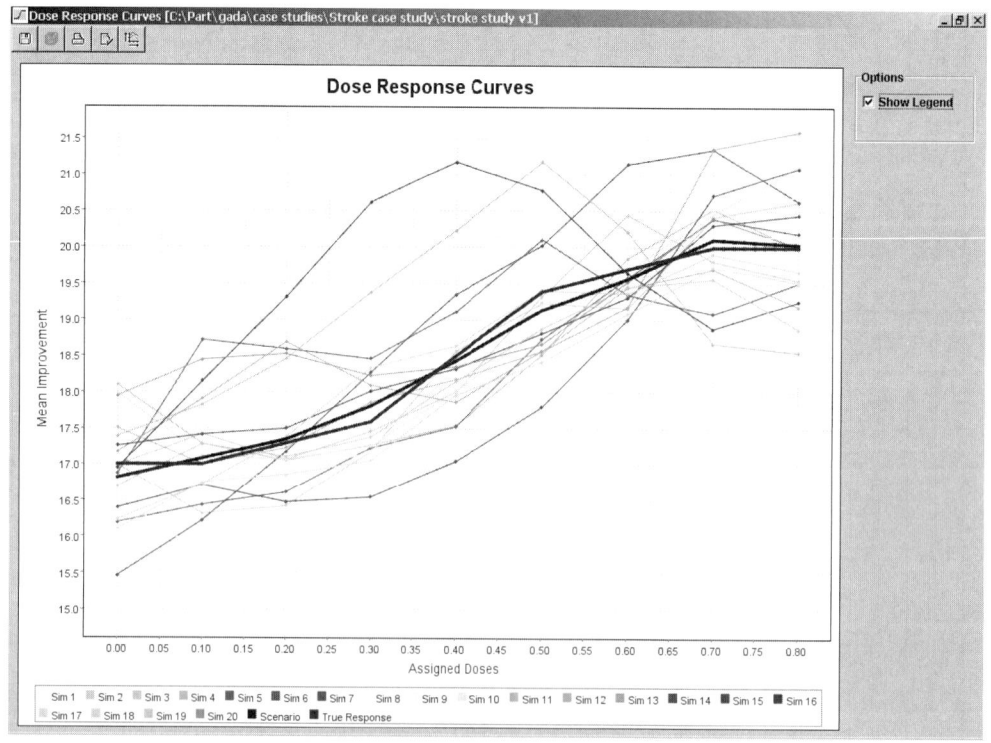

FIGURE 9.10   Dose response curves.

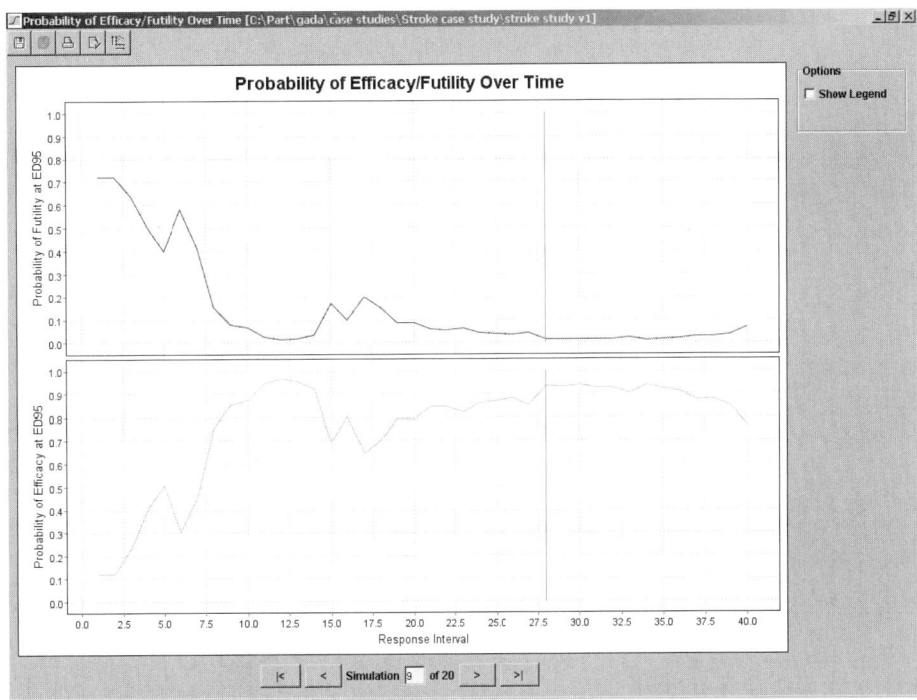

FIGURE 9.11   Probability of efficacy/futility over time.

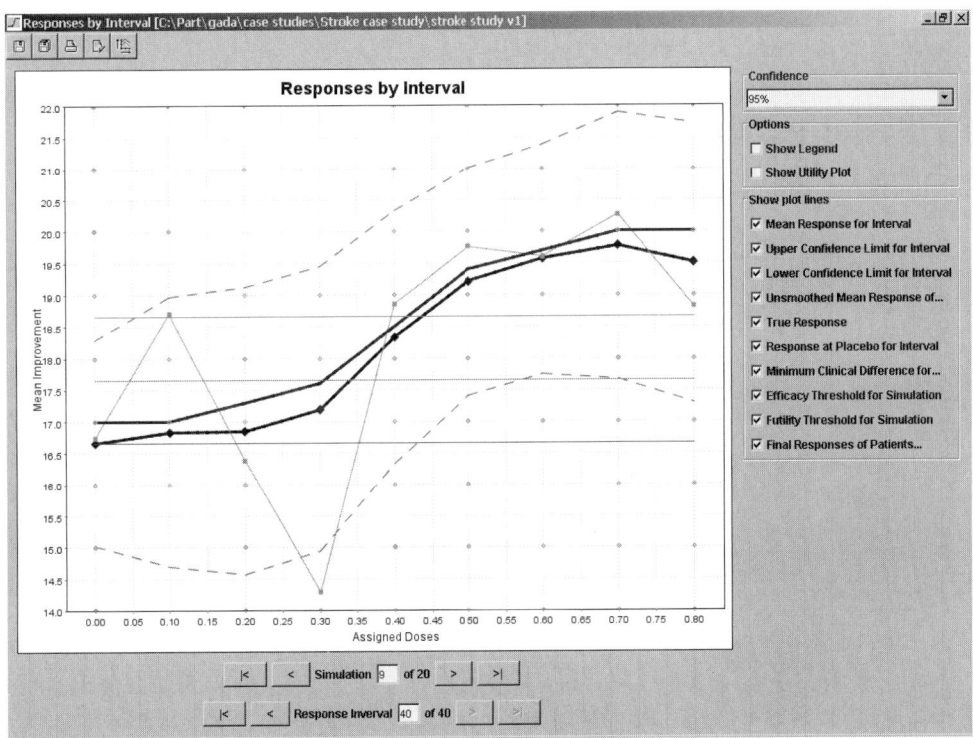

FIGURE 9.12   Responses by interval.

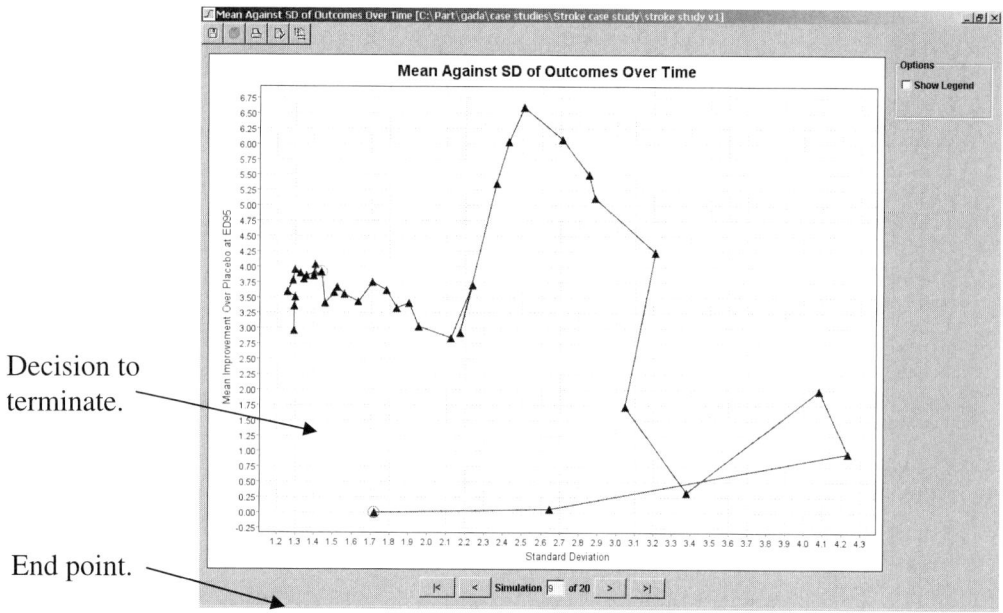

Decision to terminate.

End point.

FIGURE 9.13   Mean against SD of outcomes over time.

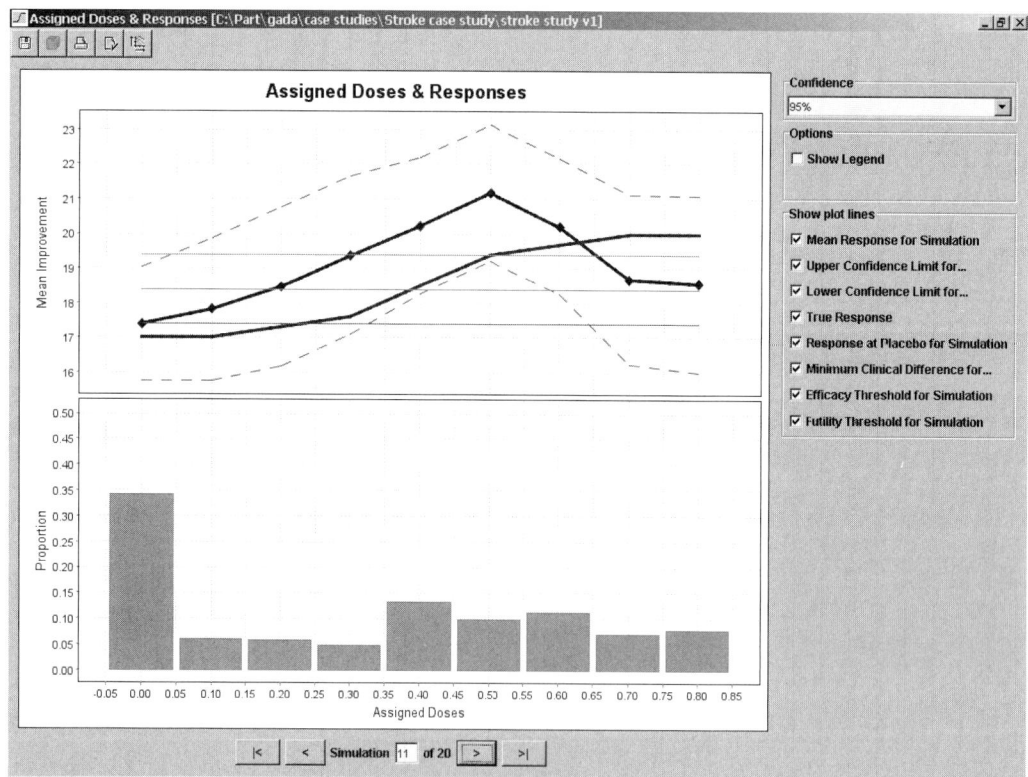

FIGURE 9.14   Assigned doses and responses.

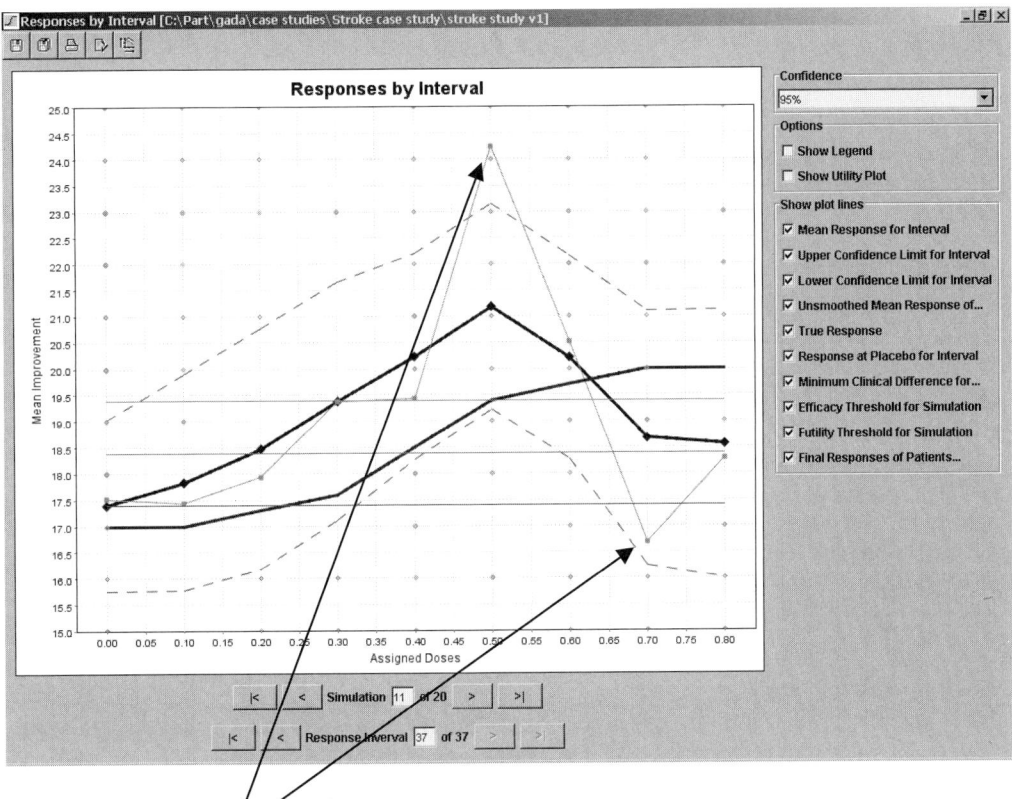

FIGURE 9.15 Responses by interval.

that have been simulated have been unusually distributed and how this has misled the system. These results away from the average underlying curve are all too possible when the SD of the variation in the subjects' responses is as high as four times the signal we are looking for. The solutions to this problem could be as follows:

1. Raise the required minimum number of subjects before early termination and hope for "regression to the mean."
2. Reduce the percentage of doses allocated to placebo so that the response at the doses is better characterized (see later text).
3. Reduce the noise in the subject responses.

Early Over-Optimistic Termination

Simulations 1 and 18 both stopped early for efficacy with an estimated response at ED_{95}, considerably higher than the true underlying response, risking underpowering the phase III trial (Fig. 9.16). Notice how the problem is a poor placebo response, in simulation 1, two points below the curve being simulated from (Fig. 9.17). If we look at how likely or unlikely that is,

SD of the mean of the placebo arm (m_0)

$$= \sqrt{(\sigma^2/N)}$$

where σ is the SD of the distribution of patient responses around the underlying curve (12) and N is the size of the placebo arm.

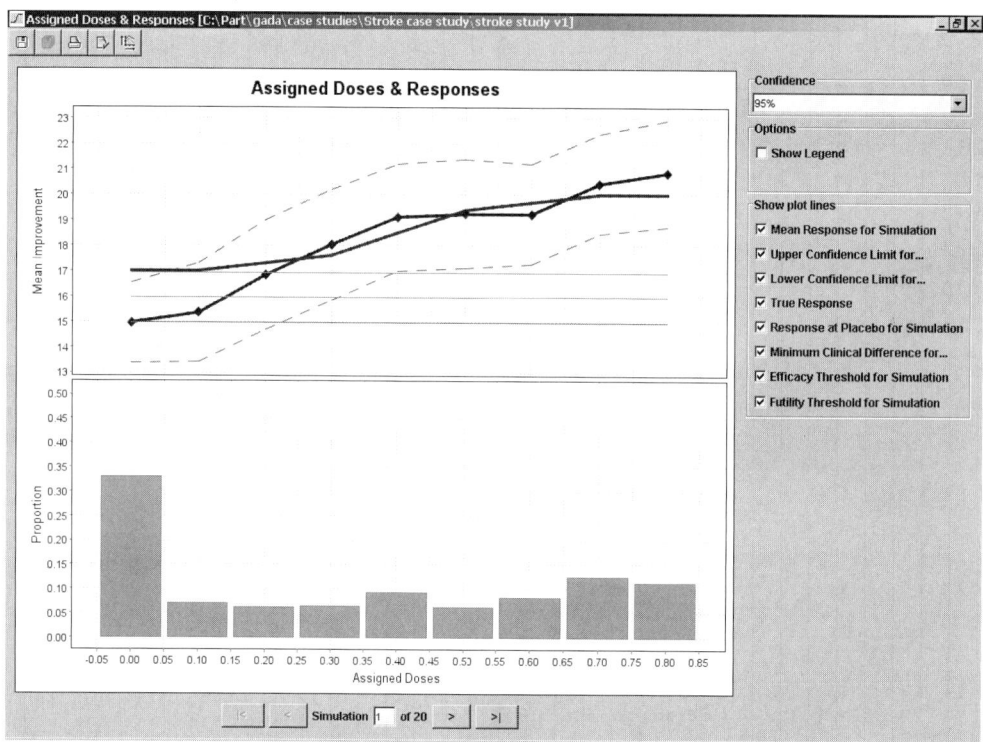

FIGURE 9.16   Assigned doses and responses.

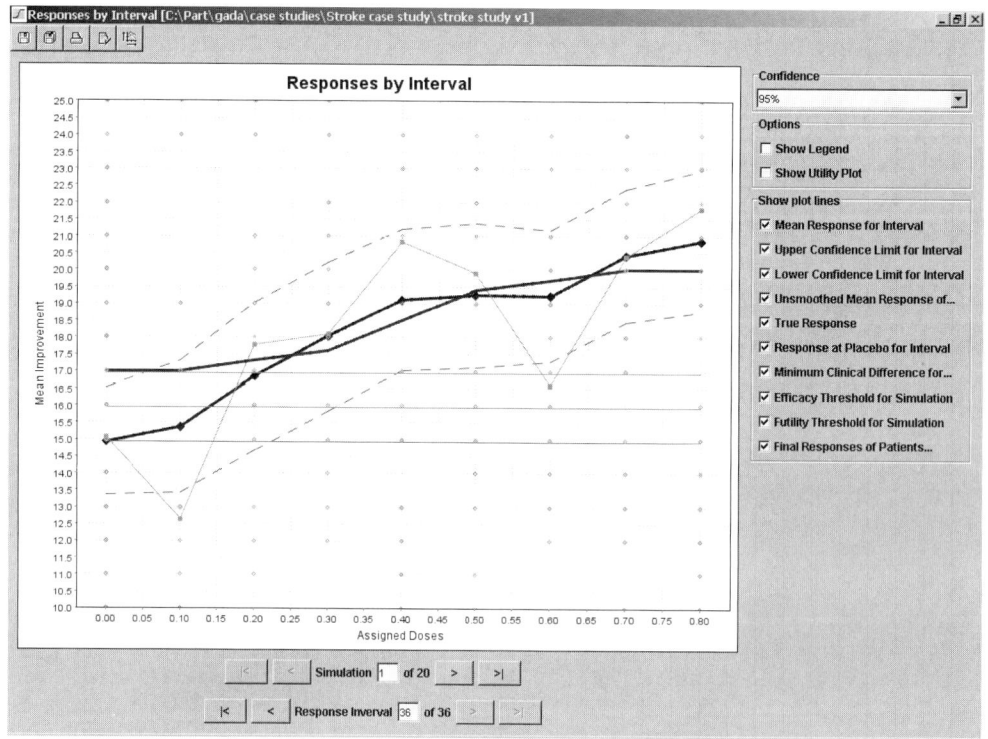

FIGURE 9.17   Responses by interval.

TABLE 9.2

No. of phase II subjects	Size of placebo arm	SD of m_0	Probability that observed response is > or < true by	
			1 point	2 points
500	175	0.91	0.27	0.027
750	263	0.74	0.17	0.007
1000	350	0.64	0.12	0.002
1500	525	0.52	0.06	0.000

TABLE 9.3

SD of distribution of subject response	Size of placebo arm	SD of m_0	Probability that observed response is > or < true by	
			1 point	2 points
12	175	0.91	0.27	0.027
11	175	0.83	0.23	0.016
10	175	0.76	0.19	0.008
9	175	0.68	0.14	0.003
8	175	0.60	0.10	0.001

In these simulations the system is typically allocating ~35% to placebo, so where trials stop with 500 or so subjects, there is a ~3% chance that the placebo response will be wrong by 2 points or more and a 27% chance that the placebo response will be wrong by 1 point or more (Table 9.2). This suggests that a minimum trial size of 500 is too low and that 750 would be more prudent.

An alternative solution is to accept the risk of overestimation of response and mitigate it by fixing a conservative minimum phase III trial size. A further alternative is to improve the study design to reduce the variability of the subject response. If the minimum trial size is kept at 500 but the variability of the subject response reduced from 12 to 10, this is equivalent to a minimum trial size of 750, and if it is reduced from 12 to 8, this is better than a minimum trial size of 1000 (Table 9.3).

WHAT WE NEED TO BE ABLE TO SIMULATE

To set up and run simulations of adaptive clinical trials we need to define a number of characteristics.

Doses

1. The number of doses of the compound. The dose strengths should be equally spaced on a linear or logarithmic scale. Administering equidistant doses is easily accomplished when the treatment is an infusion for which any dose strength can be prepared. For oral treatments we can use the following approach: We ask for four types of tablet strengths: 0 (placebo), 1×, 3×, and 4×. Combining two of each allows us to administer dose strengths of 0–8×. We could even cover a dose range of 0–243× equidistant on a half-logarithmic scale by asking for tablet strengths of 0, 1×, 9×, 81× and combining three tablets into any one dose.
2. Whether using placebo, an active comparator, or both. A minimal proportion to allocate to placebo/comparator can be specified.

Response Scale

1. The minimum and maximum values of the scale.
2. Whether the minimum value is a good or bad outcome for the subject.
3. The minimum and maximum value at baseline for a subject to be entered into the trial.

For instance, in stroke a subject with a baseline SSS score of over 40 has such a good chance of fully recovering (achieving a score of 58 by day 90) that it is difficult to observe any effect due to the drug. Conversely, a stroke subject with a score of 10 or less has such poor chances of survival that again it is difficult to observe any

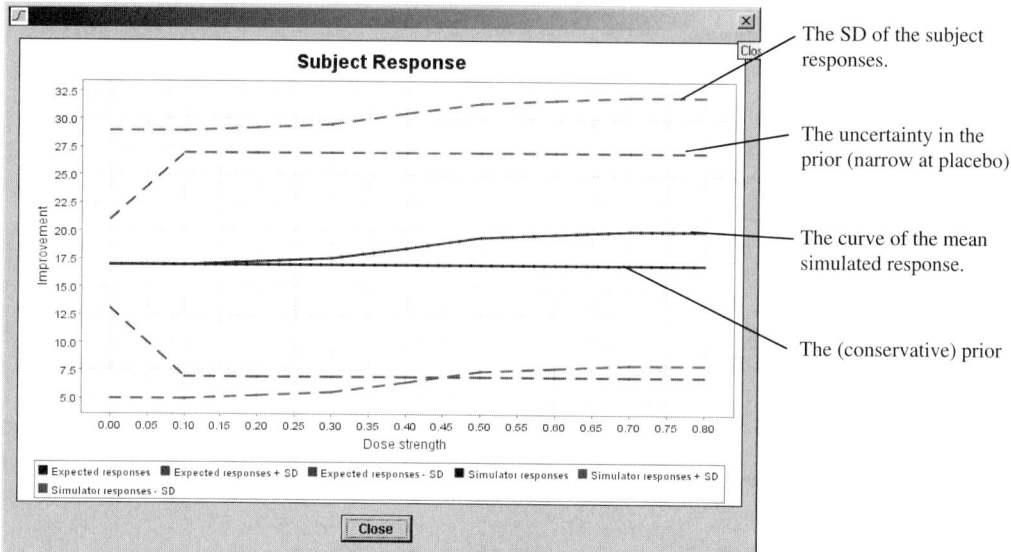

FIGURE 9.18   Subject response.

effect due to the drug. The distribution of the subject baseline scores is also specified.

"Prior"† *Response*

1. What we expect the mean level of response to be for placebo and the different doses of drug. As the final result has to convince regulators, the prior is typically conservative. However, as the regulators become more accustomed to this sort of trial, it may be possible to use priors informed by phase I trial results and animal studies, as long as they are not given too much weight.
2. The certainty of that prior, expressed as an SD of a distribution about that mean. An SD the same as the expected SD

of the subject responses means the prior will have the same weight as a single subject response. As we reduce the SD of the prior, it has greater weight, is equivalent to a larger number of subject responses, according to the following formula:

Equivalent number of responses =

$$(\text{SD of subject responses/SD of prior})^2$$

3. We specify a separate certainty of the prior response for placebo or active comparator as there is likely to be a significant body of evidence for these, and it could be argued that the response is already well understood (Figure 9.18). A word of caution, however; in the ASTIN study we had access to the database of hundreds of subjects in the Copenhagen Stroke Study, but whereas all these, effectively placebo, subjects on average recovered by 10 points on the SSS, in ASTIN the average placebo response was 17 points! With hindsight we could see all sorts of reasons why this difference could arise, but the lesson

---

†The prior is a term from Bayesian statistics and reflects our beliefs and expectations about the probabilities of the outcomes before gathering experimental data, as a result of which our "prior" becomes updated to a "posterior." A prior can be conservative or informed; it can be weakly held or strongly held. It is important that the prior reflects not the beliefs and understandings of those running the experiment but of those that the experiment must convince.

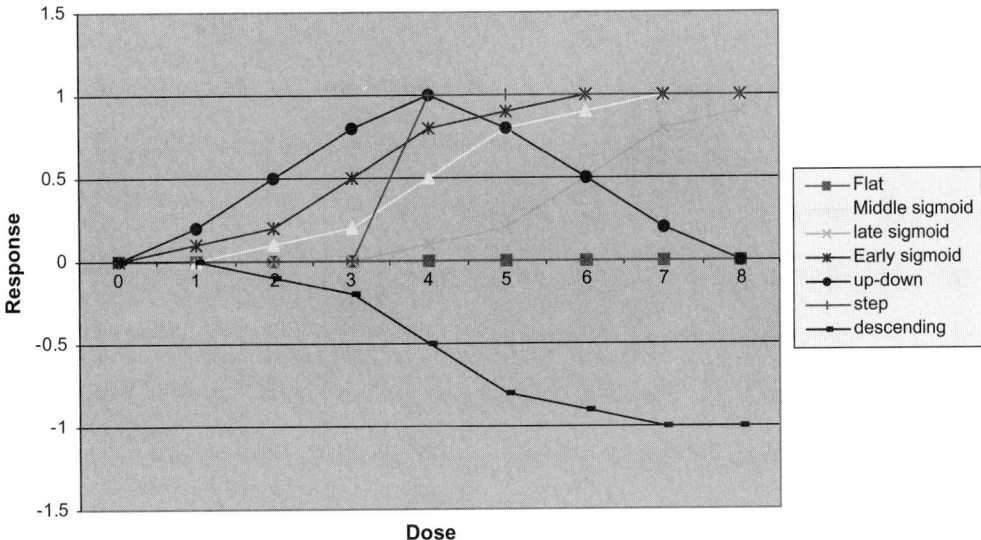

FIGURE 9.19   Example dose response curves.

is do not over-weight prior data no matter how solid it looks—the study it came from will be different from yours.

4. The expected level of improvement in the response due to the drug. This is used to set the smoothing of the curve fitted to the actual data. The higher the expected response (relative to placebo), the less smoothing is applied, and the lower the expected difference from placebo, the more smoothing is applied.

Likely Responses

1. The mean response to simulate at placebo and each dose. It is important to simulate the protocol parameters over a range of plausible curve shapes and effect sizes. This is where input from pharmacokinetic (PK) modeling is vitally important, defining plausible curve shapes and sizes. For instance, we may use a set of curves comprising a flat curve; sigmoids with early, middle, and late rises; sigmoids with very steep and very shallow rises; and a rising then falling curve. The main effect sizes we

study are "just effective," "just ineffective," and "most likely" (if that isn't one of the other two); these help distinguish between the different parameters settings in ways that "clearly effective," "clearly ineffective," and "borderline" do not.

2. The variability in the subject responses, as the SD of the subject response about the mean response. This is one of the most crucial parameters to the outcome of the study but the hardest to estimate in advance and the hardest to influence.

3. The proportion of the subject population that responds to the drug. Currently, having a drug to which only a proportion of the subject population responds just has the effect of increasing the variability in the subject response curves. But the model could be extended to include it and hence improve the fit of the response curve to subjects that do respond.

Trial Size

1. The maximum permitted size of the phase II trial in terms of the maximum

number of subjects. Finances and the amount of drug manufactured require the study to have an upper limit, a point at which it must be stopped and a decision made.

2. The earliest the phase II trial can be stopped, in terms of the minimum number of subjects. With very small numbers of subjects it is possible to get wildly misleading results; to stop this leading to an erroneous early termination, a minimum recruitment threshold is set.

3. The number of subjects to allocate randomly before allocating adaptively. With very small numbers of patients the early dose-response curve can be very misleading; if we then adaptively allocate to the wrong doses, it could prolong the period for which the system is mislead. Particularly for situations where the subjects' responses are very variable, an initial period of random dose allocation is recommended.

4. The likely recruitment rate, the number of subjects per day (this can be <1). This affects how well the system can learn from past responses; the slower the subjects are recruited, the better from this point of view. However, clearly there are limits on how long the trial can take overall. There maybe a trade-off to evaluate between having a study with slower recruitment, fewer centers, and less variability in subjects' responses versus a larger trial with more subjects, more centers, faster recruitment, but more variability in patient responses.

In our adaptive allocation model, it is important to have response data as soon as possible. If we wait until the final response of a subject before using their data in the response model, we are likely to have to perform a lot of dose allocation based on no response data at all and likely to never use a proportion of the response data in allocating. In many treatment areas a subject's early response (after 1–2 weeks) provides a good indication of their final response, and the system's longitudinal model is a mechanism for exploiting this.

Longitudinal Model Parameters

1. Whether to use the longitudinal model or simply last observation carried forward.

2. If using the longitudinal model, we need to enter a prior expectation and a model to be used in the simulation. Currently, our preferred approach is to derive a longitudinal model from historic data from a past trial or study and use this to simulate from, while using a simple approximation to it as a prior.

Termination Conditions

1. The minimum clinically significant improvement over placebo. This is the level of improvement over placebo that the phase III trial will have to demonstrate. Depending on the required probability of success in phase III, the phase II result may have to be higher than this to proceed.

2. The required probability of success in the phase III trial. This is a clinical and managerial judgment and varies with clinical area. For instance, where there are already successful treatments we may need to have a high confidence of success before proceeding. Conversely, in clinical areas where there are no effective treatments, a phase III trial may be undertaken where the probability of success is lower.

3. The efficacy threshold is a response and confidence level, which if exceeded during the trial (after the minimum number of subjects has been recruited) will cause the trial to be stopped early for efficacy. The futility threshold is a response and confidence level, which if the observed response falls below it (after the minimum number of subjects has been recruited) will cause the trial to be stopped early for futility.

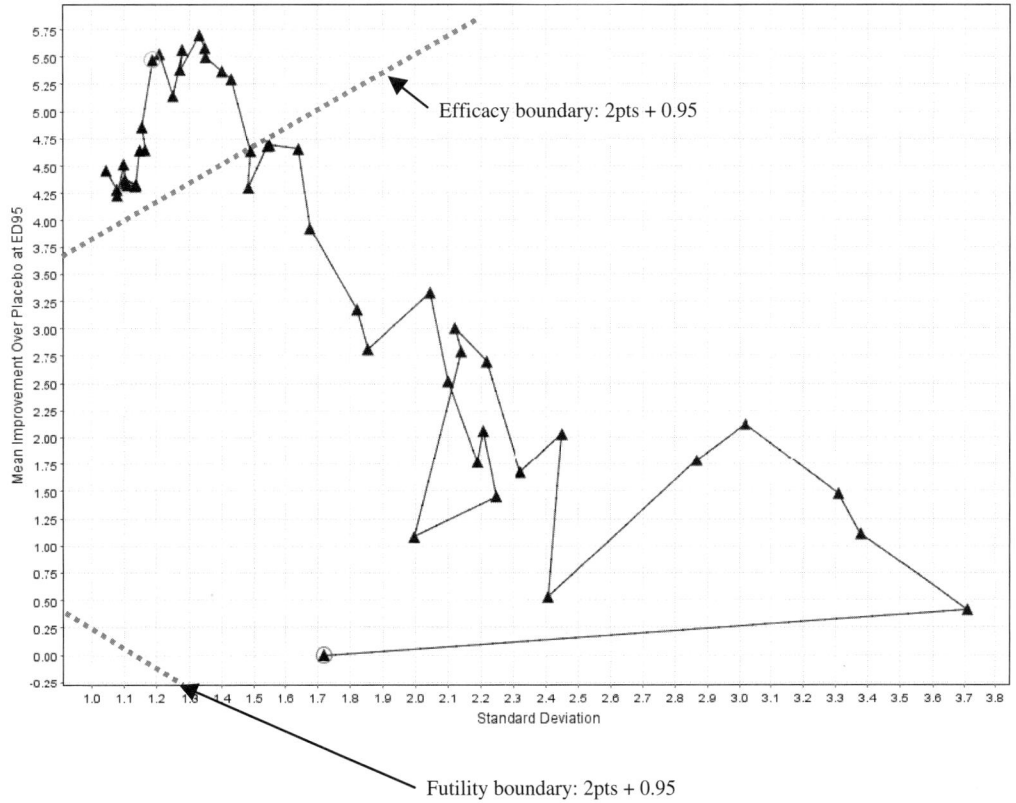

FIGURE 9.20   Mean against SD of outcomes over time.

In our experience, a sensible first guess for the early termination boundaries is the 0.95 confidence of being above (efficacy) or below (futility) the minimum clinically significant improvement (Fig. 9.20).

Optional Features

1. Covariates: factors that can also influence the subject's response. Our system currently fits a simple linear regression of the response against the covariate value. The dose response is then determined from the responses with the covariate effects removed.
2. Extra response variables: additional outcomes can be modeled. It is possible to replace the response value with a combined score that uses both the response (principal endpoint) and the additional outcomes. We currently support a simple linear function to combine these.
3. Death: if death or drop-out is not included in the main response point, there a are number of approaches possible to treating subjects that die or are lost to follow-up. They can be ignored by the dose-response model or given a score such as worst possible score, worst observed score, or last observed score or death can be modeled and then included in the gain function.

WHAT WE LEARN FROM SIMULATING

The most important result from clinical trial simulation is to give a realistic expectation of the likelihood of success of the trial. This can be a salutary experience. Its is easy to have a number of difficulties in a clinical trial—the marginal effect of the drug, the noise of the subject response, the un-informative response scale being used—but not realize how in combination these render a successful outcome not just difficult but impossible.

This should result in redoubled efforts to reduce the noise in the subjects' responses, improve the endpoint being measured, and characterize the expected effect of the drug. But if in the end it is all for naught and the trial is canceled, you always have Peter Thall's words of comfort "I have always considered it more desirable to kill computer generated patients than real ones" (4). Look at the history of failed undertakings and you will see that "cutting your losses" is one of the hardest decisions to take, and simulations can help you take it.

So typically the first thing we learn from simulating our intended clinical trial is that we have a problem! The next thing we learn is which parameters (varied within the bounds of what we think is possible) yield most benefit, or how much things have to be improved to have some hope.

It is also important to look beyond the averages of hundreds of simulations and check some of the individual runs; you will only get one trial and this may be how it looks. For instance, it was looking at some overall successful results that we spotted one simulation that had it wrong. The system had seen a fluke high response at a low dose early on in the trial, then never allocated at a high dose, and so missed the true response. As a result, we made a number of changes, including adding the ability to specify a period of completely random allocation to better characterize the whole curve before adaptively allocating.

THE PROBLEMS OF SIMULATING

For simulations to be convincing and carry weight, they need to be based on good prior data. In particular, it is important to know the following:

1. The likely level of placebo response;
2. The likely variability (noise) in the subject response;
3. The likely profile of subject recovery from baseline to endpoint.

It is important to not only derive the best estimate of these, but also to understand the uncertainty in the estimate and include in the simulation scenarios alternative estimates. It is not necessary to categorize the system across a range of these estimates but to check whether the performance is significantly worse if any of these parameters turn out to have values at the extremes of your confidence limits.

It is also useful (but of secondary importance) to know the following:

1. The likely distribution of the subject's baseline scores;
2. The likely response due to the various covariates.

It is normally not necessary to simulate variations in these parameters.

From animal studies, PK models, and the safety study it is important to establish a realistic range of possible effect sizes for the drug and realistic dose-response curve profiles. Normally, we produce a set of curve shapes each scaled by a range of response values. For instance, we might have a range of sigmoid curves varying in the ED_{50} dose, the slope, and the maximum effect. The curves will fall into two groups: those where the trial should succeed and those where it should fail, normally dependent on the maximum effect size.

It is necessary to decide on a number of protocol-related issues or at least initial values and limits for them. Some of these you will vary across the range of simulations:

1. The minimal clinical effect size;
2. If using a gain function, the relative weights to give to the different outcome measures (response and the extra response variables);
3. The number of doses;
4. The maximum and minimum size of the phase II and phase III trials;
5. The recruitment rate;
6. The trial early termination conditions.

Next, it is necessary to determine how you are going to evaluate the outcomes of the following simulations. A number of measures are possible:

1. Minimize the type I and II errors in decisions to go to phase III.
2. Maximize the likelihood of a successful phase III trial.
3. Maximize the accuracy with which the correct dose is found.
4. Maximize the accuracy with which the correct response is found.
5. Minimize the number of phase II subjects used.
6. Minimize the time the phase II study takes.
7. Minimize the number of phase III subjects used.

Possibly, you will wish to use some weighted combination of these.

The problem that remains is to optimize the selected outcome across huge combination of possible parameter values. We normally start by setting up our initial guess at the protocol and the example curves to simulate. We then run 10 simulations of this initial protocol for each example curve and assess the results.

- Is the model behaving sensibly for this set of parameters?
- Given all the assumptions, does the outcome of the trial look sufficiently hopeful?
- Which of the example curves are most informative? Are there any that can be

dropped? (For instance, curves where the maximum effect is well below the clinical minimum for success tend to give the same results regardless of the curve shape).
- Is our means of evaluation satisfactory? Are we including everything that is in fact important to us, and weighing things correctly?
- Given that we will aggregate the evaluation over all the curves, how many simulations should be run for each scenario?

We should check the model parameters, the degree of smoothing used in the curve fitting and options on the adaptive allocation and randomization that best determine the dose and response.

We next need to choose which protocol properties to optimize:

1. Number of doses, percentage to allocate to placebo.
2. Shape and strength of prior.
3. Minimum and maximum trial size, number of subjects to recruit before adaptively allocating, recruitment rate. These may be interrelated, for instance, the larger the study, the faster the required recruitment rate. They may also affect the SD of subject score; a larger recruitment rate may imply the use of more centers or more investigators and hence greater variability in subject score.
4. The length of monitoring (a shorter length of monitoring may result in greater variability of in subject score) and number of intermediate responses measured.
5. Minimum and maximum phase III trial size and required probability of success for proceeding to phase III.

Simulating every possible combination of parameters over every curve to be simulated (a fully factorial approach) may require a prohibitive amount of simulation.

A modern PC typically takes 30 minutes to 2 hours to run 100 simulations using our model. Varying eight parameters through three possible values each—over 10 curves—could require 30,000–120,000 hours of simulation.

Two approaches allow us to reduce the number of simulations:

1. Optimize each parameter in turn and use the "best" value found in all subsequent simulations.
2. Use a fractional factorial design to arrange the simulations, so that each value of a parameter is simulated with each value of every other parameter, but not every combination.

As long as the effects of the various parameters are independent, either of these allows us to analyze the results for each parameter in turn, averaging over the others. For instance, with four parameters (I–IV), each taking three values (a, b, c), we can arrange a full set of fractional factorial simulations with 9 runs rather than 81 (Table 9.4). It should be noted that this is fully factorial for any two parameters; the minimum number of runs required will be the product of the two parameters that take the largest number of values; and if there are a number of parameters, which take the same number of values, the required number of runs will be larger than this.

TABLE 9.4

Run	Parameter I	Parameter II	Parameter III	Parameter IV
1	a	A	a	a
2	a	B	b	b
3	a	C	c	c
4	b	A	c	b
5	b	B	a	c
6	b	C	b	a
7	c	A	b	c
8	c	B	c	a
9	c	C	a	b

A detailed exploration of fractional factorial design is well beyond the scope of this chapter, but it is a well-researched field with many texts available (5). However, for our purposes we are not attempting to fully understand the effects of the parameters but simply trying to optimize our choice. Furthermore, some of the parameters have costs associated with them (number of subjects, length of trial), and we may be more concerned with finding minimum values for which the results are acceptable than fully optimizing them.

ADAPTIVE ALLOCATION

We believe that our simulations show that using an adaptive allocation scheme allows phase II trials to be run with more doses than a conventional study without requiring any more subjects. For example, we ran 100 simulations of two similar scenarios. In one we allowed the adaptive dose allocation scheme to evaluate placebo and a range of eight doses and to terminate early if it had 95% confidence that it was above or below the required efficacy threshold. In the other subjects were randomly allocated to placebo or one of two doses: the middle and highest doses of the range in the other trial. Both studies were capped at 1500 subjects, but the adaptive study was allowed to terminate early after 500 subjects.

Common to both scenarios was the use of the SSS (0–58) with an endpoint of change from baseline after 12 weeks. The placebo response was 17 points and the response at the highest dose 20 points. The minimum response improvement over placebo for clinical significance was two points. The required confidence of success before proceeding to phase III was 80% (Fig. 9.21, Table 9.5).

The true ED_{95} is at a dose of 0.65, the average adaptive result was 0.67, and the average conventional result was 0.73. The true improvement over placebo

Average over 100 Adaptive Simulations

Average over 100 Conventional
Simulations

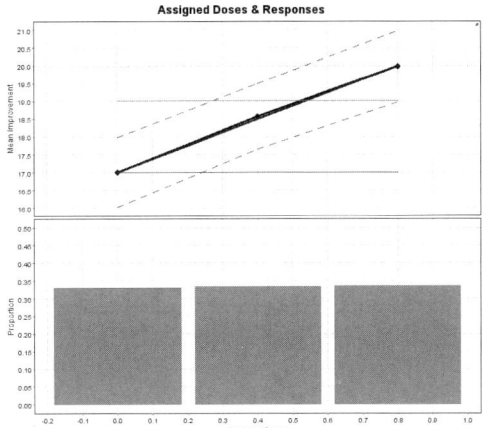

FIGURE 9.21   Simulation averages.

TABLE 9.5

	Found dose	Improvement over placebo at found dose	SD of improvement at found dose	No. of phase II subjects	% of phase III studies	Avg. no. phase III subjects	Avg. phase III success	% phase III success
Conventional	0.73	2.84	0.72	1500	56	1026	0.90	50.2
Adaptive	0.67	3.17	0.96	1152	61	1012	0.87	53.3

at ED$_{95}$ was 2.85, the average adaptive improvement found was 3.17 with an SD of 0.96, and the average conventional improvement found was 2.84 with an SD of 0.72. The average adaptive phase II trial size was 1152 subjects, and the conventional studies were all 1500 subjects.

The adaptive phase II studies were recommended to proceed to a phase III study 61% of the time, and the average actual likelihood that the phase III studies would succeed was 0.87, whereas the conventional phase II studies were recommended to proceed to a phase III study 56% of the time and the average actual likelihood that the phase III studies would succeed was 0.90.

The rate of successfully proceeding to phase III may seem low, because the phase II study is adequately sized to find

a greater than two-point improvement. However, we have stipulated that we require 80% confidence that the phase III study will be successful before it will be undertaken, and this in turn this requires an improvement of 2.7–2.8 to found in the phase II trial. As the simulated response gives an improvement very close to this, the average success rate is only around 50%.

The adaptive studies used fewer subjects and more accurately found the desired dose at the expense of the accuracy and certainty in the determination of the response. Not shown by this simple comparison of two scenarios is that the adaptive study would continue to find the right dose across a range of dose response curves, whereas in this example the conventional study has been simulated

TABLE 9.6   The same simulations with the noise reduced from 12 to 8 points

	Found dose	Improvement over placebo at found dose	SD of improvement at found dose	No. of phase II subjects	% of phase III studies	Avg. no. phase III subjects	Avg. phase III success	% phase III success
Conventional	0.75	2.87	0.51	1500	78	1020	0.97	75.37
Adaptive	0.68	3.23	0.79	897	82	1014	0.95	77.93

Found dose response curves with 12-point noise

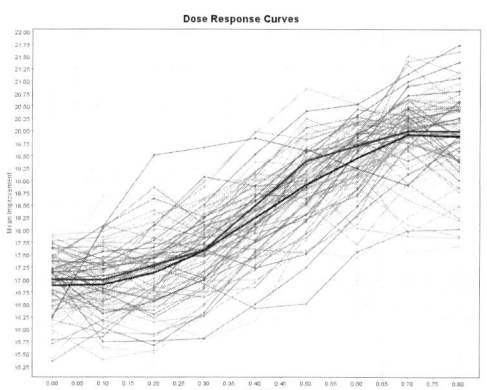

Found dose response curves with 8-point noise

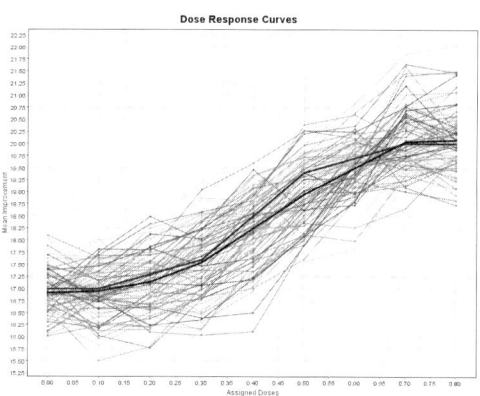

FIGURE 9.22   Dose response curves.

with one of its doses close to the ED$_{95}$. With other choices of the doses to use, the conventional study cannot do any better than the results shown here and can do a lot worse.

Clinical trials often have a problem because of the variability in the subject response (or the coarseness of the subject response) relative to the improvement due to the drug. Adaptive allocation is additionally sensitive to this because the noise in the response can cause the allocation to adapt to the wrong dose. We can see the remarkable impact of reducing variability on efficiency, if we repeat the above simulations but assume that work is done to reduce the variability in the subject scores from 12 points to 8. This could be by using fewer centers, better investigator training, requiring that subjects are always assessed by the same investigator,

or videotaping the assessments and rerating them centrally (Table 9.6).

Figure 9.22 shows the reduction in outliers among the simulations with the reduced noise. Though outliers are not eliminated, eight points is still a large variation compared with the three-point effect.

WHAT NEXT?

We are researching a number of new areas to include in the model.

1. Combinatorial dose response, running phase II trials to investigate the combined effect of two drugs, in particular a new drug and an existing drug, poses new challenges in trial design. We believe that an adaptive dose allocation scheme will be an efficient way to

evaluate a large number of possible combinations and find the most promising combination without unfeasibly large trial sizes. Comprehensive simulation of proposed trials will be essential to establish the best trial design and demonstrate the practicability of the approach.

2. To incorporate pharmacokinetic models, currently we use pharmacokinetic data "at one remove" to set up priors and curves to simulate from. There are two ways we would like to better exploit pharmacokinetic data:

   a. Use the subject's exposure to the compound or the measured concentration of the compound within the area of interest (e.g., the brain, for central nervous system indications) in the subject rather than simply the strength of dose administered.

   b. Use subject responses predicted by the pharmacokinetic model directly in the trial simulations.

3. Adapting to a biomarker (nonclinical endpoint), a major problem in adaptive dosing is the variability in subject responses. Clinically significant endpoints can be too subjective (e.g., quality of life questionnaires) or too coarse grained (e.g., the modified Rankin scale) to measure relative dose response. We are exploring whether we can use a highly sensitive biomarker or a gain function including biomarkers (for instance, in stroke imaging information pertinent to the development of infarct size, S100, etc.) to guide the adaptive treatment allocation, while continuing to use a clinical endpoint to guide the assessment whether to terminate the trial and deciding whether to proceed to a phase III study.

4. Modeling and adapting to side effects as well as response. Our adaptive dosing model could obviously be improved by modeling the occurrence of side effects and incorporating them in dosing. Currently, we rely on manual intervention during a trial to disable dose levels that need to be disallowed. Side effects could be allowed to rule out doses, or they could be combined with the dose score to allow benefit and risk to be jointly evaluated.

5. Implementing a curve-fitting algorithm that does not require equally spaced doses.

6. Implementing a fully decision-theoretic approach to terminating the trial. We have done initial work looking at identifying the true utility of running the trial and using it to guide the decision to terminate the trial. Defining the true utility is a challenge, because it is highly dependent on one's perspective: There may be differences in the utility of the patient, the treating physician, society, the drug company, and regulatory agencies. The beauty of the approach is that it forces us to think about these issues up front and integrate them all into one overriding concept. Ideally, even the termination rule can be allowed to learn over time, and it would be fascinating to develop a truly dynamic terminator that is not blind to new developments. The most exciting aspect of working on this approach has been the need to gather together all parties relevant to the drug development process from the earliest design discussions onward.

ACKNOWLEDGMENT

We have had many fruitful discussions with a large number of experts from different fields whom we thank for their input and support.

References

1. Krams M, Lees KR, Hacke W, Grieve AP, Orgogozo JM, Ford GA. ASTIN: an adaptive dose-response study of UK-279,276 in acute ischemic stroke. Stroke 2003;34:2543–2548.

2. Jorgensen HS, Nakayama H, Raaschou HO, Vive-Larsen J, Stoier M, Olsen TS. Outcome and time course of recovery in stroke. The Copenhagen Stroke Study. Arch Phys Med Rehabil 1995;76: 399–412.

3. Berry DA, Mueller P, Grieve AP, Smith MK, Parke T, Krams M. Bayesian designs for dose ranging drug trials. In: Gatsonis C, Kass RE, Carlin B, Carriquiry A, Gelman A, Verdinelli I, West M, eds. Case Studies in Bayesian Statistics, Vol 5. New York, NY: Springer-Verlag, 2002:99–181.

4. Thall PF. Bayesian Clinical Trial Design in a Cancer Center, Vol. 14, No. 3, 2001.

5. Montgomery DC. Design and Analysis of Experiments, 3rd ed. New York: John Wiley & Sons, 1991.

10

Biostatistical Issues in Neuroemergency Clinical Trials

Wayne M. Alves

A persistent theme of "failed" clinical trials runs through the recent clinical drug development experience with neuroemergency populations. The weight of the evidence provides little support for the failure of therapeutic hypotheses as the underlying cause of these failed trials (1–6). Alternative explanations have centered around biopharmacology considerations (e.g., drug not delivered to target site, possibly due to failure to cross the blood–brain barrier; therapeutic windows not adequately defined; or dose or dose regimen not adequately defined), disease complexity (e.g., multiplicity of damage and outcomes, the clinical populations included in phase III studies), and biostatistical considerations (e.g., methodological flaws in phase I or II trials, suboptimal study design or execution, insufficient statistical power, premature study termination that does not stand up to final scrutiny, and even inappropriate statistical analysis plans). Although not validated, many trials design and outcomes assessment strategies have been offered to end the series of failed studies, including

- Censoring excessively good or poor prognoses, either through inclusion and

exclusion criteria or statistical modeling procedures (4);
- Treatment group shift analyses over a range of ordinal outcome categories;
- Expanding outcome categories of standard clinical rating scales and controlling interrater reliability through structured interviews;
- Use of *early* "surrogate" endpoints for later clinical outcomes (e.g., intracranial pressure changes or clinical neurological worsening);
- Focus on concentration-outcome (drug-clinical endpoint) relationships, although sometimes this approach has been reparative rather than prospective;
- "Improved" preclinical drug development programs.

All these strategies have merit but are not yet validated in the context of clinical trials. The purpose of this chapter is to raise many of the unresolved biostatistical issues in the neuroemergency clinical trials arena. There are no easy solutions or answers. Many of the tools we use come from a long and rich history evident in the evolution of statistics relating to clinical trials methodology. Table 10.1 displays some of the early milestones in the history

TABLE 10.1   Early milestones in clinical trials methodology

Book of Daniel, Old Testament 1:11–16	Early documented comparative study
1747 James Lind, HMS Salisbury	Experiment with lime for scurvy
1789 Edward Jenner	Smallpox vaccination
18th Century	Growth of hospitals in United States allows larger case series
1770s LaPlace	Development of Probability Theory
Early 19th century Pierre-Charles-Alexander Louis	La Charite Hospital, use of statistical comparisons
1835 Pierre-Charles-Alexander Louis	Comparative study of bloodletting
1840s Ignaz Semmelweis	Puerperal ("childbed") fever: washing hands decreased rate
19th Century Claude Bernard	Experimental method (counter to Koch's enumeration method)
1917 Blinding (Torald Sellman)	"The crucial test of therapeutic evidence"
1920 Sir RA Fisher	Chance assignment and use of "control" ("placebo pill")
1932 Use double blind	Placebo control with fixed dose regimens (Harry Gold *et al.* Cornell Medical College
1935 Sir RA Fisher	*The Design of Experiments* first published
1936 "Student"	Matching or balanced assignment methods
1944 CH Hinshaw and WH Feldman	Apply probability to assign treatment
1946 Austin B Hill	Randomization and double-blind procedures

of statistics relevant to the evolution of modern clinical trials. The art of biostatistics, rather than its science, may be the more important consideration. The future success of clinical development programs for neuroemergencies may depend on whether we take to heart many of these considerations and seek novel solutions that can be validated and added to the neuroemergency clinical trials toolkit.

INHERENT HETEROGENEITY OF NEUROEMERGENCY POPULATIONS

Despite their significance as a public health issue, neuroemergencies are relatively rare events when compared with other areas of medicine. As mentioned frequently throughout this volume, neuroemergency patients are initially evaluated in the emergency department and then typically managed in intensive care settings that involve numerous supportive therapeutic interventions and extensive examination points. We have also seen that in

most neuroemergency areas, "gold standard" assessment scales are typically nominal or ordinal categorical scales (7). In addition, the complexity of neuroemergencies requires multiple measures to completely appreciate the clinical outcomes in a given patient, and multiple assessments (e.g., extended Glasgow Outcome Scale, modified Rankin Scale [mRS], National Institutes of Health Stroke Scale) may be needed to provide a comprehensive characterization of the patient. This results in extensive data collection processes and a significant number of case report forms. The available clinical assessment tools are also limited in their sensitivity and specificity to fully reflect clinically relevant endpoints for study outcomes.

These aspects, along with other features, of neuroemergency populations contribute to an inherent variability in patient inputs and outcomes observed in modern clinical trials and a resultant need for large sample sizes and many investigative sites. The emergence of well-defined standard treatment protocols have emerged, for example, the treatment guidelines for traumatic brain injury (TBI) offered by both American and European neurosurgical associations (e.g., Joint Section on Trauma of the American Association of Neurological Surgeons and the Congress of Neurological Surgeons). These guidelines offer us some ability to standardize background patient management and thus to reduce or control the extraneous sources of variation due to idiosyncratic treatment decisions.

CLINICAL TESTING OF NOVEL DRUGS AND TREATMENTS

Neuroemergency clinical trials are complex activities, and enrollment and study operations can be difficult to manage. The diseases and disorders involved may require that significant time elapses before having reliable estimates of clinical outcomes. This has significant impact on the planning for such studies where follow-up periods may require 6 to 12 months. Novel therapies are always exciting, and the urgency surrounding the discovery of new therapies for long unmet medical need can lead to attempts to cut short or speed up the laborious process of adequately testing a new drug for marketing approval. Table 10.2 outlines the traditional clinical development paradigm.

A problem with recent failed trials in neuroemergencies is at least partially a function of the limited supportive preclinical data available to support the clinical development program. The limitations of these data can range from adequacy of drug levels in the target organ, assumptions about the biological activity, to the magnitude of the expected effects on the efficacy endpoints of choice. In part, some portion of the problem of failed trials in neuroemergencies lies in limited supportive preclinical data that a particular drug has the assumed activity and efficacy implied by those who advocate its use. This motivated, among others, the STAIR

group to issue optimal preclinical guidelines to support clinical drug development in stroke (8). Because competition for scarce patients exists, we can expect to see even greater pressure to poorly perform or even to skip over critical steps in the drug development process.

Careful consideration, and control to the extent possible, of the background therapy underlying a neuroemergency clinical trial is essential. This has *not* been an explicit feature of many of neuroemergency clinical trials, although in TBI it is recognized that the lack of a standardized management regimen is an important factor contributing to the limitations of TBI clinical trials (1–5). The trade-off is simplicity of study design versus adequate and standardized management, reflecting the standard of care currently available for the patient. It should also be noted that continued improvements in neurointensive care provide for continual changes in the "standard of care." In addition, issues such as the duration of dosing, adequacy of dosing, and concomitant medications allowed in the trial are all important.

A major issue in neuroemergency clinical trials is that current outcome measures are imprecise and may not be sufficiently sensitive to assess treatment effects of novel drugs (7,9). Clinical trials to determine the effectiveness of a drug therefore require larger numbers of patients, and the outcome determination is typically delayed long after treatment (e.g., at 3 or 6 months). In theory, it would be possible to decrease the numbers of patients needed if outcome measures were improved or if the effects of the drug could be assessed closer to its time of action. For example, it may be possible in some circumstances to assess the physiological or metabolic response to a drug, and further research and regulatory acceptance of this approach are needed. Thus far, proven surrogate endpoints that meet fairly stringent requirements are not widely available for neuroemergencies,

TABLE 10.2   Traditional phases of clinical drug development

Phase	Typical size*	Synonyms/study types	Focus	Key study objectives
I	20–80	Clinical pharmacology Dose tolerability Pharmacokinetics (PK) Pharmacodynamics (PD)	First introduction or use in humans Short-term clinical safety Normal volunteers, sometimes patients Drug metabolism and/or structure activity relationships	Determine metabolism and pharmacological actions, especially PK/PD Identify side effects associated with increasing doses Establish mechanism of action Define drug–drug interactions Explore biological or disease processes
II	100–200	Dose finding Dose ranging Maximum tolerated dose	Maximum tolerated dose Short-term safety and efficacy Patients with target disease or disorder	Minimum effective dose Sufficiently effective dose without undue toxicity Degree of drug tolerance Dose response for tolerability PK/PD in patient population PK/PD variability Target organ (brain) penetration Dose-related biological effects
III	300–1000s	Definitive efficacy Pivotal trials	Efficacy and safety in a well-defined controlled patient population	Establish efficacy in intended disease Identify potential adverse drug reactions Evaluate benefit/risk Provide package insert data Appropriate dose for more diverse disease population
IV	Very small to 1000s	Postmarketing Pharmacovigilance Pharmacoepidemiology	Ongoing safety surveillance	Continued assessment of safety and efficacy Possible new indications
V	1000s	Effectiveness studies	Effectiveness when used in general treatment population	Support of indication Possible new indications

*Sample sizes indicated are largely for illustrative purposes only. Considerable variation in sample sizes exists, in part depending on study objectives, expected effect sizes, and other factors such as limitations of subject availability.

Adapted in part from Mathieu M. New Drug Development: A Regulatory Overview, 3rd ed. Waltham, MA: PAREXEL International Corporation, 1994.

although there have been notable efforts such as the documentation of neurological worsening in TBI (see Chapter 4).

The typically narrow therapeutic window is a challenge in assessing the efficacy or effectiveness and safety of a novel drug. Therapeutic doses often produce side effects. Because the brain is the target organ, these side effects are typically central nervous system events or impairments. One strategy for dealing with the narrow therapeutic window dilemma is to start with a recommended or higher therapeutic dose and allow up- or down-titration of study dose by the investigator. Because the brain is the target organ, it is typical to start low and slowly titrate the dose ("start low, go slow"). It has been the author's experience that many early dose-ranging studies are often crude in design and underpowered to detect meaningful dose-response effects.

DIAGNOSIS AND SUBPOPULATIONS

Achieving a clinically adequate scientifically valid description of disease populations, especially key subpopulations for which treatment effects may be demonstrated, is essential for clinical drug development. The inclusion of patients with excessively good prognosis or excessively poor prognosis will add little to our ability to demonstrate a treatment effect. These outcomes occur at the tails of the outcome distributions and constrict the size of the intermediate outcomes categories where we might expect to see the greatest impact of the therapeutic intervention. The net result is that there will be fewer patients who could move along the outcome continuum toward improved recovery, thus blunting the effect of a potentially efficacious drug.

The use of appropriate inclusion and exclusion criteria that "trims" the extreme prognoses at both ends is a desired solution. It must be kept in mind that the definition of the treatment population must remain targeted to the desired indication.

For example, excluding TBI patients with an isolated epidural hematoma not overly alters the "moderate-to-severe TBI" population, it also removes patients who would have a "good recovery" outcome, independent of the therapeutic intervention, after surgery (10,11). Including such subpopulations is relevant from an effectiveness point of view but could serve to blunt measured treatment effects.

Acute neuroscience populations are heterogeneous and complex. As such, the design of neuroemergency clinical trials is complicated by interactions of pathological mechanisms that are difficult to define and overlap and, in some cases, the significance of which is not fully understood (Fig. 10.1) (1). Consider, for example, the evolution of our understanding of the relevance of traumatic subarachnoid blood in TBI clinical trials (12). Primary and secondary brain injuries, complicating factors (e.g., hemorrhage or extracranial injuries), and competing risk factors (comorbidities) all influence clinical course and outcomes. Careful identification of relevant prognostic factors (i.e., input or baseline variables that may be used as covariates in the analyses plan) for defining appropriate inclusion and exclusion criteria is also a challenge. Among the sometimes extensive set of potential prognostic factors are

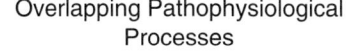

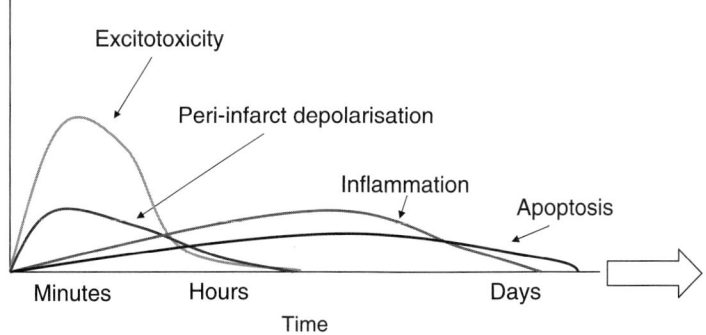

Source: Dirnagel U et al. Pathobiology of ischaemic stroke: an integrative view. *Trends Neurosci* 1999; 22: 391-397.

FIGURE 10.1 Overlapping pathophysiological processes characteristic of neuroemergencies.

those that in essence "swamp" the effects of others.

Neuroemergency patients often require multiple clinical endpoints to adequately describe outcomes. Complicating this is that patients who apparently have the *same* brain injuries can have *different* outcomes, whereas patients with apparently *different* injuries can have the *same* outcomes. Our choice of endpoints may then need to depend on the "expected" outcomes associated with the disease along with intermediate endpoints that might help with interpretation of findings of primary outcomes. It is difficult to balance the effect size of a primary endpoint required for approval or success with the effect that is considered clinically acceptable to be the minimum needed for a treatment to become part of routine neurological or neurosurgical treatment. It would be helpful if some consensus could be reached to define the important clinical outcomes relevant for clinical trials. It would also help if the clinical community could prioritize outcomes in terms of their relevance for clinical trials, for example, what is the current view regarding measures of neurological worsening? Another example is the relevance of the diagnosis of clinical vasospasm as an important endpoint for subarachnoid hemorrhage clinical trials. Angiographic vasospasm is not a reliable endpoint in humans, and rescue or "salvage" therapies are efficacious and thus can blunt treatment effects reflected in clinical endpoints.

DOSING AND TIMING CONSIDERATIONS

How dosing regimen decisions were arrived at is seldom described in neuroemergency clinical trials reports. Investigators sometimes acknowledge considerations such as the timing of study drug administration ("early is best"), which is typically measured in hours but

generally in vague terms surrounding optimal timing after the initial neuroemergency event (1). Dosing regimens may be critical in neuroemergency situations and may depend on numerous pharmacological factors such as the drug's half-life, bioavailability, and its subsequent pharmacokinetics and pharmacodynamics (1). To a lesser extent, the administration route may be relevant, although, for example, venous irritation at an intravenous infusion site may be less of an issue in some neuroemergency populations because of the common use of a central line.

The duration of study drug administration may also be critical. The time from symptom onset to the end of treatment during which we need to ensure a therapeutic level of the "protective agent" depends on the natural history of the disease. These periods may have been accepted with limited knowledge of current therapeutic interventions and thus may be inappropriate. Overlapping pathologies vary in their time course, and although we may successfully target these pathological entities, we may treat too short to blunt or ameliorate others (Fig. 10.1). In general, drugs evaluated in neuroemergency populations seldom have had a rich pharmacology (multiple mechanisms of action).

Time to diagnosis and definitive treatment is not controversial in neuroemergencies. Initiation of treatment simply must be achieved as quickly as possible (13). There are serious limitations to our current knowledge as to the absolute optimal time to treat for which a reasonable expectation exists for a therapeutic impact. Where we are seriously limited is our knowledge of the optimal time to treatment in which we can reasonably expect a drug or treatment intervention to be effective. In neuroemergencies we must presume the earlier the better. Dosing windows of 3 to 6 hours after onset are common. In practical or operational terms, the average time to study drug administration is often seen to "clump" at the upper end of the protocol dosing window. This, combined with the common imprecision of the exact time of symptom onset, makes it questionable whether drugs can be delivered as quickly as necessary to have a chance to work. What is achievable is another matter. Efforts to define appropriate inclusion and exclusion criteria for neuroemergency populations will help, especially to allow timely first dose of study drug while definitive evaluation of inclusion and exclusion criteria take place. Ischemic stroke is a disease where this has proved quite beneficial for the conduct of clinical trials. The treatment of "suspected stroke" often defines the dosing regimen until definitive diagnosis is possible. The implication for hypothesis testing is that the eligible study population is likely to be a smaller subset of the "suspected" population and thus requires multiple analysis populations.

In the setting of neuroemergency trials, one mechanism to aid the rapid initiation of therapy is the use of waived consent, where the ability to secure "informed" consent in the neurological disease/disorder is compromised and the imposition of extended time for attempting to obtain consent works against the critical time window for dosing. Careful consideration of the current regulations should be given to allow for the rapid assessment and treatment of such neuroemergency patients using waiver of consent provisions.

Allowing adequate time to achieve positive response to treatment is a necessary ingredient for successful studies. In time, a treatment group may "catch up" to other treatment groups, and the time course of effect or treatment response, long-term impact of continued dosing, and whether the early effects of treatment are predictive of long-term clinical changes remain to be explored. Typically, longer term outcome studies focus on prevention of relapses, but even here there is no clear body of evidence regarding the most effective drugs. There are inadequate data on long-term quality of life outcomes or other clinically relevant functional outcomes. Global and social functioning endpoints require a relatively longer time to detect improvement unambiguously.

KEY DESIGN FEATURES OF NEUROEMERGENCY TRIALS

The gold standard for providing proof of the therapeutic efficacy of new drugs and treatments is the randomized controlled clinical trial. It is not likely that we would ever be able to subject every drug with putative benefit in neuroemergency populations because of the complexity of modern clinical trials. The number of subjects required for the trials places practical limits on how many drugs can be tested in phase III trials. Before the 1990s, controlled clinical studies to evaluate neuroemergency drugs seemed to make allowances for smaller numbers of patients than is typically desired or required for definitive proof of efficacy (1). Although these earlier trials were notable attempts to resolve a specific question of clinical utility, investigators seldom explicitly took into consideration the heterogeneity and complexity of neurological damage and likely clinical outcomes, which drives statistical power and sample size considerations in modern clinical trials. In the evolution of a therapeutic hypothesis, study designs in which small numbers of patients receive a new treatment are of some initial value. In our experience, however, in smaller studies the treatment success-to-failure ratio is near to 50:50. We can imagine the extreme case in which no treated subjects improve and either we have a potential safety issue or we are unable to decide whether we have an ineffective treatment. This may be true for potentially efficacious drugs or treatments as well as those that do not have any likelihood of clinical value.

Good clinical study design practices for neuroemergency clinical trials require the careful planning of the methodology to compare interventions that involve design features to increase sensitivity (i.e., statistical power) of the study within the practical constraints imposed by realistic sample size considerations and the naturally occurring heterogeneity of neuroemergency populations. Throughout this volume, it is evident that heterogeneity is observed in severity, pathology (i.e., damage), and the assessment outcomes in such neuroemergencies. Key study design issues should include consideration of prestratification and other randomization strategies, testing of multiple treatments, and using between- and within-subject design features to improve inferences in the presence of the *expected* heterogeneity of neuroemergency populations.

As often mentioned in this volume, achieving an adequate scientifically valid description of neuroemergency populations is especially difficult (14,15). This has importance for providing clear definitions of relevant prognostic factors (i.e., input or baseline variables and their potential use as appropriate covariates). The importance of these definitions for inclusion and exclusion criteria and in data analysis plans cannot be underestimated. Further, subsetting neuroemergency populations, whether for planned or post-hoc comparisons, remains difficult, and current approaches are likely to be controversial.

Neuroemergency populations require multiple clinical endpoints to adequately describe the complete story of damage and outcome. However, from a statistical perspective, the use of multiple endpoints can create almost insurmountable hurdles given the required adjustments to statistical significance and the need to account for the intermeasure correlations.

The use of surrogate outcome measures, for example, brain imaging endpoints (e.g., diffusion-weighted imaging (DWI)/perfusion-weighted imaging (PWI) with

magnetic resonance imaging), although of significant use in designing of larger phase III trials may have significant regulatory limitations and issues of generalizability. Similarly, the ever-present possibility of treatment-induced pathological outcomes creates further complexity in properly assessing the benefit-to-risk ratios. So far, these "surrogate" endpoints have not been successfully used. In part this is due to the stringent statistical requirements required of a surrogate endpoint (16,17). In addition, there can be considerable noise in these measures, and it is not clear how to handle patients that do not display the necessary inclusion features on the surrogate endpoint, for example, ischemic stroke patients who fail to show a mismatch between diffusion and perfusion brain imaging studies. In addition, regulatory authorities have been reticent, in recent years, to consider approval based on such measures.

Simple standards exist for data analysis in neuroemergency clinical trials. A relevant patient population for testing a new therapy can typically be fairly easily identified. In spite of this, there are a number of reasons noise is introduced into the data set. Contributing factors are practicality of study design and conduct, measurement error, patient selection bias, and so on. The standard data analysis population is the *intent-to-treat* analysis, that is, *all* patients randomized to the arms of the trial whether or not there are errors of classification, randomization, or treatment. There are also compelling reasons to require a "no missing data" efficacy analysis (i.e., we must impute outcomes when they are missing) (18). The imputation rules used are likely to be scientifically controversial, and they need to be spelled out clearly and uniformly followed. Because complete follow-up is difficult to achieve in some neuroemergency populations, imputation rules must exist, for example, the "last observation carried forward" procedure. There are limited solutions, and in the case of neuroemergencies, subjects with extreme outcomes (functional independence or death) are the most likely to be lost to follow-up.

In summary, considerable benefit to the design of future neuroemergency clinical trials could be achieved from clinical research aimed at

- Improving "clinical phenotyping" of treatment populations;
- Improving assessment of intermediate drug effects (mechanistic endpoints) that can serve as supportive evidence;
- Focusing on *clinical benefit* and crisper more reliable endpoint assessment;
- Applying statistical design analysis strategies more in tune with the inherent complexity and heterogeneity of neuroemergency populations;
- Considering "novel" approaches to neuroemergency trials design (e.g., adaptive designs) and randomization/stratification strategies.

HANDLING MULTIPLICITY OF ENDPOINTS

The multifactorial nature of neuroemergencies and the need for multiple outcome measures lead to the statistical problem of multiple comparisons. This is especially true when we do not clearly specify all *a priori* hypotheses we wish to test. From a statistical viewpoint, corrections must also be made for the planned multiple comparisons and for the many post-hoc comparisons that are inevitable as data are analyzed. In fact, these post-hoc corrections may be just as strict as those applied to preplanned multiple comparisons. There are several practical options for evaluating multiple endpoints. Although we cannot discuss them in detail, readers are urged to consult their biostatistics colleagues. The most frequent adjustment used is a Bonferroni procedure, although under certain circumstances compelling arguments can be made for using Hotelling's

T^2 procedure, the sum of normalized scores, stepwise procedures, or some form of combined test (19). Statistical methodology for coprimary endpoints (e.g., Hochberg's procedure) is another option, and one that could be quite useful in neuroemergencies where "win–no win rules" are simultaneously applied (10). We should indicate that with few exceptions, past and present (completed and published) neuroemergency trial data were analyzed without consideration of this issue and use of these procedures (1).

PRECISION OF OUTCOME MEASURES

An important issue facing neuroemergency clinical trials is that current outcome measures are fairly imprecise and are not designed to assess either physiological responses or functional changes due to drug treatment effects. It should be kept in mind, however, that *clinical* benefit is more important, certainly to patients and also to regulatory agencies (20). Consequently, in determining the effectiveness of a drug, the outcome evaluation may be delayed long after the treatment, which often times requires significantly greater sample sizes. It may be possible to decrease the numbers of patients needed if primary outcome measures were improved (e.g., attempts by the American or European Brain Injury Consortia to improve the reliability of the original and the extended Glasgow Outcome Scale) or if the drug could be assessed closer to its time of action.

Solely measuring functional changes, however, does not alleviate the difficulty because of the multifactorial functional impairments that follow most neurological emergencies (e.g., TBI). The issue is whether determining a more direct effect of a drug will truly allow a conclusion about the drug's effectiveness using fewer patients. For example, measurements of infarct volume in stroke patients may not correlate with final clinical outcome, even if the usual noise in brain imaging data is overcome. If a measurable physiological effect of the drug being tested is not present, will a new drug be likely to prove efficacious? The key to this reasoning lies in the ability to choose an appropriate measure of the drug's physiological effect, which is not straightforward in neuroemergency populations. On the other hand, we may still demonstrate long-term clinical improvement in the drug treatment group without such a measure. The "cost," however, may be an increase in sample size requirements.

The preceding discussion points to a critical research issue in neuroemergency trials in general. If an incorrect or less than optimal measure is chosen for evaluating a drug, then by that measure the drug may prove ineffective when in actuality it has a benefit when measured against a more appropriate endpoint. This reasoning contributes to the use of numerous secondary and exploratory endpoints. It may be more reasonable to determine whether sufficient evidence based on the measures expected to be used in later studies can be obtained earlier in clinical development, indicating that it is worth proceeding with increasingly more powerful and more expensive testing of the drug.

VARIATION IN PROGNOSTIC FACTORS

Variation in prognostic factors is one major reason for inconclusive findings when one tries to draw conclusions from a set of reported clinical trials. Interpretation of the results from a given "sample" of neuroemergency patients can be ever so subtly influenced by shifts in these prognostic factors (14,15). Major prognostic factors in neuroemergency populations are age and gender; injury severity, for example, as defined by the Glasgow Coma Scale and selected features of the neurologic examination; hypotension; hypoxia; incidence of extracranial injuries in the case of TBI; and the nature of the injuries themselves (i.e., the underlying pathology). For example, there is a well-established synergy between intracranial and extracranial injuries in TBI patients with multiple systems trauma (10,11). The need to categorize and possibly prestratify subjects according to underlying pathology has been recognized by several neuroemergency investigators. All appreciate that randomization is no guarantee in diseases as complex and multifactorial as neuroemergencies where even subtle imbalances can have a major impact on whether a clinical trial succeeds or fails.

Two main options available from a design perspective are to prestratify neuroemergency patients before randomization, generally according to injury severity, and/or to perform a covariate adjustment in the analysis of the study data (21,22). These considerations should be developed in the study protocol and elaborated in the formal statistical plan of the trial *before* the blind being broken for analysis. How prognostic factors contribute to the observed findings of a specific neuroemergency clinical trial will be uncertain but nonetheless necessary to consider and evaluate. For example, intuitively we know that small changes in the average age or the proportion of males in a TBI population can alter the observed mortality rate. Similar arguments can be made regarding the distribution and balance of pathology among treatment groups. What we need to understand is how subtle multivariable imbalances impact on the findings of neuroemergency clinical trials. This is in part the rationale for the randomization of larger numbers of subjects. What to do about the issue in each study is not so clear. Although a variety of statistical adjustment procedures exist, they typically have not been used extensively in neuroemergency trials. In part, this is because it is often difficult to draw clear clinical information relevant to drug safety and efficacy from the results of "adjusted" models, but also because many studies were probably inadequately powered to allow such analysis (i.e., too few subjects were enrolled in the trials). In some cases, we simply do not have sufficient prior clinical trial experience.

WIN–NO WIN RULES

Evaluating new drugs in neuroemergency populations is complicated by the potential for pathological outcomes. For example, a drug that improves mortality or "good recovery" only to simultaneously increase the proportion of vegetative or severe disability outcomes is undesirable. The general statistical solution to this problem is to test the hypothesis that the intervention or drug improves favorable outcomes while simultaneously testing that there is not a concomitant rise in the pathological outcomes in favor of the untreated or placebo group. The practical consequence of imposing win–no win rules for present neuroemergency clinical trials is the difficulty in successfully demonstrating efficacy and no unfavorable trend in the same patient population (i.e., we are testing hypotheses at both ends of the outcome distribution). In general, demonstrating strict equivalence when showing there is no difference in pathological outcomes between the treatment groups is not required. Rather, we only need to show that we do not exceed a clinically acceptable but *predefined* threshold for the observed difference (e.g., no more than 10%). Unfortunately, this method only tests a subset of the studied population at either end of the scale, thus reducing the utility of the full sample and the associated power. Some studies have begun to analyze clinical outcomes using the cumulative odds ratio approaches that can first provide for some prespecified combinations of ordinal levels of the selected scale

(e.g., in the case of the mRS, the collapsing of death and severe disability). Such an approach assesses the full spectrum of outcomes on the selected outcome scales, providing for a more sensitive and powerful analytical approach.

As an example, although several categories are available in traditional outcome scales (e.g., the Glasgow Outcome Scale [23], Extended Glasgow Outcome Scale, or mRS [24]), formal statistical analysis has historically grouped these outcome scales into two categories by selecting a score that delineates a favorable outcome from an unfavorable outcome. There are two major disadvantages of using this approach of dichotomizing an ordered ordinal outcome scale.

First, dichotomizing a multiple category endpoint is not consistent with current clinical practice. For example, an outcome after a major stroke is not a binary outcome but fundamentally exists in a continuum, with shifts on this continuum being quite relevant to an individual patient, even if the shift is not such that it crosses the arbitrary boundary separating a purportedly bad outcome from a purportedly good outcome (25). A discussion of this concept is discussed in the *Points to Consider* document on acute stroke produced by the Committee for Proprietary Medicinal Products (CPMP) (26) which states the following: "Alternatively it may be shown that an agent effectively moves patients from the severe outcome to the moderate disability group and from the moderate disability to the recovered group, i.e., that the drug effect applies across all grades of severity of stroke, moving patients to a higher grade of independence in their activities of daily living. Again, for this kind of analysis, clinically relevant shifts need to be defined and justified in the study protocol. In this case a *categorical analysis* provides more information on the drug effect than a dichotomous analysis." The problem of ignoring the information in multiple category endpoints by making them dichotomous, although identified by CPMP, has plagued acute neuroemergency research for decades (27). With the emergence of modern statistical methods that can appropriately examine this additional information in multiple category endpoints, the continued reliance on such inefficient methods should be reconsidered.

Second, use of an approach that has less (perhaps much less) statistical sensitivity in detecting clinically relevant benefit or harm from an intervention exposes a larger number of patients to the experimental process of a clinical trial without compensating benefit. This limitation in the of use of the information provided by each participating subject goes against the principle of conducting the most efficient trial possible, a requirement by most ethical guidelines for research in human subjects.

The mRS is regarded in studies of acute stroke as the most appropriate measure of clinical outcome. In many ongoing acute stroke trials, a shift in the proportions of residual disability, across the whole range of mRS outcomes, has been planned to provide a more robust and clinically appropriate analysis than a simple dichotomized approach of good versus poor outcomes. This analysis using the proportional odds approach therefore reflects not only the probability of good recovery, but can also reflect another important treatment effect, namely a shift to an increased likelihood of moderate versus severe residual disability. This is particularly relevant for those studies where predicted outcome is poor and where an important treatment benefit may not be detected by the simple dichotomy approach.

As a specific example, the mRS has seven categories (see below) whose purpose is to provide a means of evaluating both the individual patient's level of disability and, in a clinical study context, the proportion of patients who shift categories in any study of a therapeutic intervention. The mRS is an important and universally used clinical outcome measurement in most stroke trials. The mRS is a combined outcome scale that measures neurological function and disability, differentiating seven categories, including death as an mRS of 6 (28,29):

mRS	Neurological symptoms	Disability
0	None	None
1	Yes	None
2	Yes	Mild
3	Yes	Moderate
4	Yes	Moderate severe
5	Yes	Severe, bedridden
6	—	Death

The proportional odds model with seven ordered response categories can be used in an analysis of the primary endpoint. The advantages of using this approach (7), in comparison with the binary outcome ("bad" or "not bad" outcome) model are the following: (a) the proportional odds model with seven ordinal categories uses much more information from the mRS score than a binary model; (b) this approach can make it less likely that superiority is claimed due to a positive effect on the bad end of the mRS scale while at the same time there is a negative effect on the good end; and (c) the proportional odds analysis will be more sensitive to changes on any part of the mRS scale than a binary model. Therefore, the proportional odds method requires a smaller sample size than is needed for a simple chi-squared test. This ensures that the general principle of conducting an efficient trial is performed, thereby adhering to the ethical guidelines for research in human subjects.

In summary, for clinical and statistical reasons, the planned analysis of the primary endpoint of scales such as the mRS should be performed, preserving the ordered categorical data (category by category only potentially combining severe disability and death) that will provide a full picture of potential effectiveness of any therapeutic intervention in the neuroemergency clinical setting.

STATISTICAL PLANNING CONSIDERATIONS

Statistical Power

Several statistical design issues must be weighed when planning neuroemergency clinical trials (1). A sufficient number of patients are necessary to detect whether a potentially efficacious drug has in fact the desired effect in the treatment population. Whether we should prestratify patients to improve sensitivity is a difficult decision. For example, we may minimally want to stratify randomization to ensure we achieve balance in important patient subsets (e.g., injury severity defined as "moderate" vs. "severe" using the Glasgow Coma Scale). On the other hand, there are compelling reasons to suggest that factors defining subpopulations of patients reflect possible sources of significant risk for mortality and morbidity. In many cases, it has been very difficult to determine whether the neuroemergency clinical trials performed to date were adequately powered for their primary hypothesis in light of competing risk factors. The current standard in neuroemergency clinical trials must be for statistical power considerations to be given greater weight in the planning and the early phases of the clinical trials.

Control of Statistical Errors

In general, one of the most serious problems in clinical trials is true control of the type I error rate (which is the probability of rejecting the null hypothesis when in fact the null hypothesis is true). This issue can arise from multiple outcome measures, repeated measures at multiple time points, use of alternative analysis techniques for the same data, and many other designed and sometimes "underdesigned" facets of a clinical trial.

Although there are ways to address multiple comparisons, none of these works very well in neuroemergency clinical trials. We can designate a particular variable and a particular analysis at a particular time point as a primary outcome and allow the study to live or die based on that choice. If we choose the wrong variable/time/analysis, we lose the opportunity to analyze that except in an exploratory manner. We can use multivariate techniques to test hypotheses simultaneously regarding a number of variables, but the power of such tests is usually low (30,31). We can form composite measures and use them as first steps in the analysis (19). If we fail to show an effect on the composite measure, we then lose our opportunity to test hypotheses about the variables that comprise it. We can adjust the significance level for the individual tests, so that the overall error rate equals the nominal error rate, but this usually results in a serious degradation of power, and most clinical trials choose the set on which to base the adjustment unrealistically, not reflecting the true set of hypothesis tests which might have been performed.

Of the various methods that may be used, the Bonferroni procedure is most often used, probably because it is the simplest yet most conservative. It is almost always the case that there are better and more modern methods we could use (e.g., Hochberg's procedure) (30).

Type II error is the probability of failing to reject a null hypothesis when it should be rejected; that is, failure to declare an effect to exist when in fact it does. Statistical power is achieved by minimizing type II error (i.e., power = 1 − type II error), and we use the trial design to provide a trial with high power. Power is, however, partly a function of outcome measures chosen, the size of the effect that we can achieve, the variance (or noise) in the outcome measures chosen, the sample size, the analysis method chosen, and the significance level (type I error). We can design trials to have any statistical power desired, but it is often the case that the sample size required would be unacceptably large. Unrealistic assumptions are often made about variance or the ability to recruit subjects that render trials that appear to be designed to have high power to in fact possess much lower power. Better outcome measures, with higher validity and reliability, which would possess lower variance (random noise), would contribute to higher power and, by extension, smaller sample sizes.

Type I Error and the Multiplicity Problem

Many of the chapters in this volume indicate that multiple primary variables, multiple secondary endpoints, and multiple analyses are primary sources of multiplicity in neuroemergency trials. It is typical to select limited (single) primary endpoints and treat the remaining tests as secondary or even exploratory. For example, one strategy is to use "domains" of outcomes and to control type I error within each domain. However, this approach does not guarantee study-wise error protection because defining "independent" domains may be difficult. For example, it is hard to discriminate between generalized clinical changes and the effects of treatment on secondary measures such as cognitive improvement on neuropsychological tests. It is understood that cognitive deficits are not independent of other aspects of the clinical picture.

As stated before, the multiple comparisons issue is inherent in modern neuroemergency clinical trials. Multiple analyses at multiple time points pose a serious inference problem. Multiple treatment comparisons also arise due to a multiplicity of treatment arms and analysis populations. Modern clinical trials generally require that "intent-to-treat" (ITT) analyses be primary, generally involving "last observation carried forward" (LOCF) methods to deal with subject dropout or other missing data. Modified ITT populations often exclude patients for which there are no post-dose observations. Results based on observed cases (i.e., number of patients at each time point) may be used to complement the ITT analysis. Very few neuroemergency clinical trials, however, also report findings for completers or "per protocol" subjects, largely due to small subsample sizes or very high dropout rates in larger studies.

Demonstrating Superiority or Equivalence

The absence of proven treatments for neuroemergencies relegates the evaluation of novel treatments to a test of superiority over placebo treatment. Initially, convincing evidence of efficacy requires comparison with an untreated (i.e., placebo) group. For that matter, even in chronic neuroscience diseases, comparative studies of newer medications seldom focus on "superiority" of one treatment over another, except in the case of evaluating safety or side-effect profiles. In terms of efficacy, in most cases "equivalence" or "noninferiority" claims are the focus (32,33). Further, "active control" equivalence trials require that both the upper and lower effect margins be defined, whereas "noninferiority" studies require only definition of a lower margin. As such, well-designed equivalence or noninferiority trials can require much larger sample sizes that superiority trials.

For those neuroemergency indications where approved drugs exist, it should be pointed out that statistical sensitivity is also required to distinguish an effective treatment from less effective or ineffective treatments. In the case of a test for superiority, there is *de facto* sensitivity if the trial is positive. Studies that focus on noninferiority claims can mistakenly lead to a drug being deemed "noninferior" depending on the efficacy of the comparator in the context of the trial. The recognized solution to this has been the three-arm trial with both active and placebo controls. Also, there might be a need for multiple dose levels of the comparator drug if a superiority claim is to be adequately supported. Unfortunately, this problem is not an issue for many neuroemergency indications where few if any drugs are approved.

An active comparator can serve as a control arm. However, known disadvantages of a comparator can be accentuated by the dose selection for the active comparator. For example, unnecessarily high doses of a comparator would increase the likelihood of side effects. Too low a dose would blunt the differences between the comparator and the test drug of interest. In principle, one could use a dose-response trial design to confirm efficacy, but to date there are very few neuroemergency trials that seek to define the shape and location of the dose-response curve, estimate appropriate starting doses, or identify optimal strategies for individual dose adjustments.

The initial efficacy trials for a novel drug compares the new treatment with placebo and seeks to demonstrate via hypothesis testing that treatment is superior to placebo. However, demonstrating that one treatment has superior efficacy to another is difficult, because the difference between them on efficacy measures is often small. Further, if one treatment is not better than another, it may still be useful if it is superior to placebo, for many reasons including different side-effect profiles. Hypothesis testing evolved from a desire to demonstrate the superiority of one treatment over another. It is in fact impossible to demonstrate that two treatments are the same using traditional hypothesis testing. Thus, clinical trials have borrowed

methodology from bioequivalence studies, in which two treatments are said to be bioequivalent if the difference between them falls within some small predetermined interval. In clinical trials, such methodology is called equivalence or noninferiority methodology. Successful use of such methodology depends critically on designation of a sufficiently small difference between the treatments (the larger the difference, the easier it is to declare two treatments equivalent), with a large enough sample size, and a well-designed and well-conducted trial. One problem is that in superiority analyses, all the mistakes one might make (too much variation, poor measurement, protocol violations, too few subjects) lead to a tendency to fail to reject the null hypothesis, whereas in equivalence or noninferiority trials, care has to be taken to ensure that such errors do not lead mistakenly to a conclusion that the treatments are equivalent. The methodology used is to do two one-sided tests or to fit a confidence interval (sometimes two one-sided confidence intervals) and concluding that the drugs are equivalent only when this interval is sufficiently small. There is no guarantee of observing consistent "noninferiority" between treatments over several trials, and for this reason and others, regulatory agencies are somewhat more reluctant to accept equivalence or noninferiority trials as a basis for approval.

Even if two drugs differ in their efficacy in a substantial and important way, it is always possible that a clinical trial, via random variation alone, will fail to find a significant effect. Even when comparing a new treatment with placebo, it is often helpful to have an additional arm with a known effective treatment. This way, if the new drug fails to demonstrate superiority to placebo but the established drug did, this demonstrates that the trial was capable of demonstrating a drug–placebo difference, and we may attach more importance to the failure of the new treatment to separate from placebo. If the established known effective drug fails to separate from placebo, the study is said not to possess "assay sensitivity" and is often termed a "failed study" rather than a "negative study." However, although placebo control is often desirable, it is sometimes unethical. In that case, in trials comparing active drugs, it is logically impossible to demonstrate that either is better than placebo, whether there is a difference between them or not. On rare occasions, regulatory agencies may require that a new drug demonstrate superiority to an established drug.

SELECTION BIAS

Selection biases that can cloud analysis in neuroemergency trials arise from several sources. Recruitment of patients is largely dependent on postinjury time of arrival in emergency departments and the mechanism of informed consent. If the treatment window is narrow, say 4 hours, there may be no time to find the appropriate representative to provide informed consent. Ideally, we would want every eligible patient at the site enrolled in the trial. If the site is not properly organized, many eligible patients will be excluded. It will then be difficult to know whether the patients included at the site are representative of that site or the neuroemergency population in general. Patient characteristics and treatment and care of patients are other potential sources of selection bias. Of importance is whether multivariable selection effects occur that may be subtle and not apparent in the overall analysis population. Present neuroemergency trials must address this question in preplanned analyses of data. At minimum, the practical regulatory requirement of two independent positive trials to demonstrate efficacy provides investigators with a test data set as long as the blind on the second trial remains intact.

MINIMIZING POTENTIAL SOURCES OF BIAS

The principles of modern clinical trials design are well established. The gold standard is a well-controlled and closely monitored study that involves concurrent control subjects, adequate blinding of treatment, and the random assignment of patients to treatment groups (1). In modern neuroemergency clinical trials, concurrent controls may consist of subjects who receive a placebo or vehicle treatment or standard care or management, or both. Preserving the study blind in a clinical trial is essential given the attendant problems of bias, admittedly at times theoretical and subtle, that can influence patient selection for the trial and the evaluation of clinical outcomes. The current standard is to use double-blind procedures to ensure that at least both the subject and the investigator are unaware of what treatment was received. Assessment of medical management or device trials (e.g., intentional hypothermia therapy) may not permit the use of double-blind procedures. In this case, one can and should use *blinded evaluators* for the purpose of outcomes assessment.

Bias can enter into the results and interpretation of clinical trials in numerous ways: study designs, selection of endpoints, trial conduct (protocol violations, dropouts or discontinuations), and definition of analysis groups (e.g., exclusion of subjects based on outcomes). Randomization and treatment blinding are the two key tools for minimizing bias. If successful, randomization and treatment blinding minimize differential bias, but they do not eliminate other biases that threaten the internal validity of studies. Randomization addresses bias in the long run, whereas treatment blinding has to be effective to address bias.

Randomization procedures help to preserve the study blind in a well-designed and executed clinical trial. The current standard of randomizing patients into the respective arms of the clinical trial is specifically intended to avoid or minimize bias by increasing the likelihood that prognostic factors and unknown or extraneous sources of variation are controlled or balanced across the different treatment groups. It is important to remember that balance of prognostic factors through randomization procedures is not guaranteed. Random assignment to treatments also allows us to validly use statistical tests or procedures for treatment group comparisons. Equal randomization is the most theoretically attractive procedure for random assignment of subjects to a trial because this approach often maximizes statistical power (1). The simplicity of equal randomization for achieving desired statistical balance and for subsequent data analysis is worth noting. Experience with neuroemergency clinical trials indicates that the desired statistical balance between treatment groups may be difficult to achieve with neuroemergency populations. Relatively large sample sizes (e.g., 1200 subjects or more) have apparently not solved the problem. Consistent with this approach of randomization is the opportunity to evaluate relevant baseline variables (e.g., initial lesion size) as covariates in to the planned analytic procedures.

APPROACHES TO RANDOMIZATION

Poorly applied diagnostic criteria, biased endpoint assessment (leading to misclassification), and the inherent variability of time at onset and course of disease may also introduce bias. Age, gender, time to hospitalization, and time to treatment are important factors that may have unexpected effects on clinical outcomes. This creates a censoring problem (left-censored data) that often is not addressed in practice except through reliance on randomization.

Bias can also be introduced by failing to properly exclude subjects. Balancing these factors across treatment groups as best possible is desired.

Simple random assignment is accomplished by use of random numbers to generate a list of assignments for subjects as they enter the trial. However, for multicenter trials, it is typical to do randomization separately within each center to avoid confounding center with treatment assignment. This is really a special case of stratification, in which randomization is done in such a way as to ensure that the groups are equal "on average" with respect to one or more important characteristics, such as gender or duration of illness. Stratified random sampling involves randomizing patients within two or more predefined strata based on important prognostic consideration. This approach has been seldom applied in neuroemergency trials except perhaps to balance severity of injury at the time of randomization (e.g., moderate vs. severe TBI).

There is no guarantee in neuroemergency clinical trials that randomization will be successful. Numerous TBI trials, for example, are thought to have failed because of subtle imbalances in key pathological prognostic factors.

Global competitive randomization strategies using central interactive voice response systems are also possible but have seldom been used in neuroemergency trials, often times because of the additional delays imposed by such systems. In the past, adaptive randomization methods have been proposed in which the probability of assignment to treatments depends on (or is informed by) the outcome in earlier subjects. Adaptive randomization has been used only rarely, both because of ethical concerns and because outcomes may not be evident until many or most subjects have already been randomized.

Subjects drop out or discontinue study medications. Discontinuation can be due to lack of clinical efficacy or to intolerable side effects. In neuroemergency populations it is more typical for safety considerations or mortality to lead to early study discontinuation. Other reasons may exist; for example, the treatment regimen or study procedures could be too burdensome to subjects or their families. The issue is whether the study discontinuation is non-random or "informative" (34). Informative missing data are a serious problem for all statistical procedures handling missing data when the "missingness" itself carries information about efficacy or safety. The closer to randomly missing the missing data are, the better missing data techniques are able to cope with them.

HANDLING INCOMPLETE DATA

The analysis of clinical trial data with missing values remains perhaps the thorniest problem yet to be adequately addressed in modern clinical trials. Neuroemergency clinical trials are longitudinal, with measures assessed at baseline, before randomization and treatment, and then at intervals after treatment until either the trial ends or the subject drops out. Established and accepted methods (e.g., repeated measures analysis of variance, mixed models analysis, random regression models) for analyzing longitudinal data are straightforward when there are no missing data and all subjects complete the trial. Unfortunately, clinical trials have missing data, typically because the patient died or was unable to complete assessments due to the severity of postinjury impairments.

Statistical methods robust to the presence of missing data include repeated measures analysis, list-wise and pair-wise deletion of observations, mixed models, generalized estimating equation models, LOCF analyses, observed cases or completer analyses, per protocol analyses, and imputation procedures such as substituting in means or worst-case values. Each of these

models may be useful in clinical trials depending on the type and amount of missing data (18,21,35).

The most frequent approach is carrying nonmissing values forward (i.e., LOCF). Regulatory agencies prefer LOCF because it tends to be conservative. One could also try bracketing results using LOCF procedures and completer analyses, but this has seldom been found in neuroemergency study reports.

SEQUENTIAL ANALYSIS AND STOPPING RULES

Stopping a clinical trial can occur for safety reasons, early detection of efficacy, or the futility of continuing the trial due to the low likelihood of a positive trial if completed as planned (i.e., the primary endpoint cannot possibly pass statistical criteria for a positive study). Methods and procedures are available for stopping a trial at an interim analysis point for any of these reasons, and selection among them depends on the primary objective of the interim analysis. Because this is a form of multiple comparisons requiring statistical adjustments, careful consideration of the amount of significance level one is prepared to "spend" in performing the interim analysis is a critical requirement for proper use of these procedures so that a sufficient significance level is left for the final hypothesis test should the clinical trial continue to completion.

DATA ANALYSIS CONSIDERATIONS

Current statistical issues in neuroemergency clinical trials relating to data analysis include the precision of outcome measures, multiplicity of endpoints, selection biases, and the effects of adjustments to data and subsequent interpretations. Data analysis planning for neuroemergency clinical trials must weigh which analysis populations are relevant. In defining appropriate data analysis populations, both scientific and regulatory concerns are at issue.

The *ITT* population is the *randomized* population, that is, subjects are analyzed in the groups into which they were randomized. In present neuroemergency trials, this typically serves as the primary analysis population even if there are errors in screening, randomization, or study drug dosing. Screening occurs, a judgment is made to enroll, and the study drug is prepared and dosing begins. Generally, we use the time of study drug preparation as the time of randomization, although this can pose practical problems (e.g., a pharmacy mixing error based on being alerted that a *possible* subject is being screened). Sometimes a modified ITT population is defined as "randomized and receives one dose of study drug" (the *dosed* population) or "randomized plus one dose of study drug plus one outcome evaluation" (the *assessable* or modified ITT population).

The data analysis window is also a very important consideration in analysis of neuroemergency trials. A "full window" is all data received regardless of whether it falls within the desired time window for assessment (or "medical window," e.g., 3 months plus or minus 2 weeks). A full window would include an outcome received at 4 months postinjury, whereas the medical window analysis would only include cases within 2 weeks of the 3-month follow-up date.

Regulatory submissions generally require a "no missing data" analysis that typically involves imputing a score for patients with missing observations (e.g., the LOCF procedure) (18,36). In present neuroemergency trials it may be more sensible to use later observations for earlier missing observations. For example, the 12-month Glasgow Outcome Scale may be a better surrogate for a missing 6-month score than the 3-month Glasgow Outcome Scale. Regulatory guidelines also require an

analysis of center (or investigator) effects, either qualitative or a formal quantitative test for homogeneity of treatment effects. For example, we might want to examine treatment outcomes stratified by admission severity classes defined as "severe" and "moderate" across centers using Mantel-Haenszel or other procedures (1).

UNRESOLVED BIOSTATISTICAL ISSUES

Scientific interest and the daunting but exciting challenges of demonstrating efficacy and safety of novel therapeutic compounds create considerable interest in participation in neuroemergency clinical trials. As investigators we are often desirous of participating in each of the trials currently underway. However, the requirements of multiple-project organization, logistical problems such as overlapping eligibility and standard clinical care management protocols, potential conflicts of interest, and possible ethical dilemmas indicate we should exercise prudence in allowing concurrent enrollment in multiple trials at the same investigator site (1).

There are several reasons for defining subsets for subsequent analyses once the primary efficacy analysis has been performed. First, modest but important treatment effects in clinically relevant subpopulations may be hidden. Generally, we consider this problem under the heading of covariate adjustment or stratification (22,37). In the case of neuroemergency, subclasses may not have been well defined or may be overidentified (i.e., too many categories). In addition, mechanistic and prognostic classifications may or may not correspond. These subsets must be specified in advance of analysis (i.e., before breaking the study blind). When there is uncertainty about the relevance of some classifications, the availability of a second independent trial would allow the first to be used as a test set for defining potentially relevant subpopulations, allowing these post-hoc defined subpopulations to be optimized in any follow-up trial.

CONCENTRATION-OUTCOME RELATIONSHIPS

Pharmacokinetics–pharmacodynamics modeling potentially can allow us to identify optimal plasma concentration range. Bayer used this methodology to develop a randomized exposure controlled trial study design in which the optimal dose concentration predictive of the primary endpoint was identified. (See "Repinotan—Randomized Exposure controlled Trial [RECT]." Stroke Trails Directory [www.strokecenter.org].) A bedside diagnostic test was developed to achieve the highest possible proportion of patients in the defined optimal concentration range. Although it appears this approach was reparative in the context of earlier failed studies, it served as a novel approach to phase III studies. A virtue of this approach is it can offer the opportunity to define the best risk-to-benefit ratio and the best chance of success. The role of trial simulations in neuroemergency populations is explained in Chapter 9.

Pharmacogenetics may offer a way forward as well (16). The genetic basis of therapeutics stems from discrete potential differences in drug response. A 0.1% difference in genomes predicts 3×10^6 polymorphisms (16). Perhaps the greatest opportunity lies in genetic variations related to enzymes involved in drug metabolism. It is well known that there are extensive (efficient) metabolizers, poor (deficient) metabolizers, and ultrarapid metabolizers (gene overexpression). For example, in the cytochrome P450 family, CYP2D6 is involved in metabolism of perhaps 25% of all drugs. CYP2C19 is known to be deficient in 18–23% of Asians but in only 2–5% of whites. The potential impact of this on relevant drugs is an important consideration in both trial design and planned analyses.

These approaches to clinical trial designs have been scarcely used in neuroemergency populations but offer some potential. We need to alter the traditional drug development paradigm, and an example of how we might do so is suggested in Table 10.3.

SURROGATE ENDPOINTS

There are currently no proven surrogate endpoints for neuroemergencies. In fact, there are few established intermediate or mechanistic endpoints, that is, endpoints we biologically understand and which manifest the severity of the illness. Not surprising is that there are few if any "true predictors" of clinical outcome, and many of the measures used do not yield quantitative measures of clinical benefit that can be directly weighed against side effects. The criteria for a surrogate marker are stringent such as biological plausibility, prognostic value for clinical outcome, and evidence from trials that the treatment effects on the surrogate correspond to effects on clinical outcome are difficult to achieve (17,38). There are areas in which there is a potential to exploit surrogate measures.

ADAPTIVE TRIAL DESIGNS

The use of adaptive designs in which treatment allocation or assignment depends on accumulating information in the trial has seldom been used in neuroemergency trials. There are several criticisms to these designs: possibility of assignment bias, potential for imbalance in important covariates, inability to use classical statistical tests, unaddressed delays in treatment response, and not achieving informed consent. It may be the case in neuroemergency trials that endpoints occur too far

TABLE 10.3 Possible strategy for clinical development*

Phase I: Tolerance
 Tolerance marker
 Maximum tolerated dose
 Multiple dose tolerance
 Pharmacokinetics (PK)

Phase IIA: Exploratory
 Dose response
 Concentration response
 Risk factor interactions
 PK-pharmacodynamics
 Preliminary efficacy

Phase IIB: Proof of concept
 Risk factor controlled
 "Best" subset
 Two to three arms, adequate and well controlled
 Primary
 Secondary

Phase III: Confirmatory (pivotal) study
 Primary endpoints
 Secondary endpoints

*This classification has benefited from numerous discussions with John R. Schultz, PhD. Any errors are solely the author's.

out in time for these designs to be practical or efficient.

ARE LARGE PRAGMATIC TRIALS NECESSARY?

High internal validity is only one aspect of clinical trial design. High external validity or "generalizability" of findings is also clinically important. Rationales for alternative research strategies have included the possibility of large simple designs, the underlying logic being that more heterogeneous populations may help to minimize protocol-induced selection bias and provide greater external validity or generalizability of findings. This approach to trial design requires much larger sample sizes and may be more difficult to interpret. The fact that neuro-emergency patients are relatively rare compared with other diseases may also create resistance to this approach. This study design may be more useful for evaluating effectiveness of medications, which in practice is different from what is observed in randomized controlled efficacy trials.

Large pragmatic trials are not the same as simple megatrials or "large simple trials" (39–41), although they have some similar elements. Several features of such designs are random assignment of medications on a double-blind basis, an algorithmic-like progression across multiple trial stages depending on response in earlier phases, and longer term follow-up. Large pragmatic trials typically include more heterogeneous samples with fewer exclusion criteria, allow concomitant medicines as in usual practice, are longer in duration (up to 2 years and longer), and involve assessment of drug effects on a wide range of primary and secondary endpoints, especially social and vocational functioning, quality of life, and family/caregiver burden. It remains to be seen whether these types of trials will prove useful in neuroemergency populations.

References

1. Alves WA, Eisenberg HM: Head injury trials—past and present. In: Narayan RK, Wilberger JE Jr, Povlishock JT, eds. Neurotrauma. New York: McGraw Hill, 1996:947–967.

2. Maas AIR, Marmarou A, Murray GD, Steyerberg EW TI. Clinical trials in traumatic brain injury: current problems and future solutions. Acta Neurochir 2004;89(suppl):113–118.

3. Maas AI, Steyerberg EW, Murray GD, Bullock R, Baethmann A, Marshall LF, Teasdale GM. Why have recent trials of neuroprotective agents in head injury failed to show convincing efficacy? A pragmatic analysis and theoretical considerations. Neurosurgery 1999;44:1286–1298.

4. Machado SG, Murray GD, Teasdale GM. Evaluation of designs for clinical trials of neuroprotective agents in head injury. European Brain Injury Consortium. J Neurotrauma 1999;16:1131–1138.

5. Murray GD, Teasdale GM. Quality of randomised controlled trials in head injury. Trials in head injury are more complex than review suggests. Br Med J 2000;321:1223.

6. Unterberg A. The European Brain Injury Consortium survey of head injuries. Acta Neurochir (Wien) 1999;141:223–236.

7. Murray GD, Barer D, Choi S, Fernandes H, Gregson B, Lees KR, Maas AI, Marmarou A, Mendelow AD, Steyerberg EW, Taylor GS, Teasdale GM, Weir CJ. Design and analysis of phase III trials with ordered outcome scales: the concept of the sliding dichotomy. J Neurotrauma 2005;22:511–517.

8. Stroke Therapy Academic and Industry Roundtable. Recommendations for standards regarding preclinical neuroprotective and restorative drug development. Stroke 1999;30:2752–2758.

9. Mendelow AD, Teasdale GM, Barer D, Fernandes HM, Murray GD, Gregson BA. Outcome assignment in the International Surgical Trial of Intracerebral. Haemorrhage. Acta Neurochir (Wien) 2003;145:679–681; discussion 681.

10. Gennarelli TA, Champion HR, Sacco WJ, Copes WS, Alves WM. Mortality of patients with head injury and extracranial injury treated in trauma centers. J Trauma 1989;29:1193–1201; discussion 1201–1202.

11. Gennarelli TA, Champion HR, Copes WS, Sacco WJ. Comparison of mortality, morbidity, and severity of 59,713 head injured patients with 114,447 patients with extracranial injuries. J Trauma. 1994;37:962–968.

12. Servadei F, Murray GD, Teasdale GM, Dearden M, Iannotti F, Lapierre F, Maas AJ, Karimi A, Ohman J, Persson L, Stocchetti N, Trojanowski T, Unterberg A. Traumatic subarachnoid hemorrhage: demographic and clinical study of 750 patients from the European brain injury consortium survey of head injuries. Neurosurgery 2002;50:261–267; discussion 267–269.

13. Mendelow AD, Gregson BA, Fernandes HM, Murray GD, Teasdale GM, Hope DT, Karimi A, Shaw MD, Barer DH, and the STICH investigators. Early surgery versus initial conservative treatment in patients with spontaneous supratentorial intracerebral haematomas in the International Surgical Trial in Intracerebral Haemorrhage (STICH): a randomised trial. Lancet 2005;365:387–397.

14. Hukkelhoven CW, Steyerberg EW, Rampen AJ, Farace E, Habbema JD, Marshall LF, Murray GD, Maas AI. Patient age and outcome following severe traumatic brain injury: an analysis of 5600 patients. J Neurosurg 2003;99:666–673.

15. Lewsey JD, Leyland AH, Murray GD, Boddy FA. Using routine data to complement and enhance the results of randomised controlled trials. Health Technol Assess 2000;4:1–55.

16. Bullock PL. Pharmacogenetics and its impact on drug development. Drug Benefit Trends 1999;11:53–54.

17. Prentice RL. Surrogate endpoints in clinical trials: definition and operational criteria. Stat Med. 1989;8:431–440.

18. Little RD, Rubin DB. Statistical Analysis of Missing Data. New York: Wiley, 1987.

19. Pocock SJ, Geller NL, Tsiatis AA. The analysis of multiple endpoints in clinical trials. Biometrics 1987;43:487–498.

20. Paul S. Clinical endpoint. In: Chow S-C, ed. Encyclopedia of Biopharmaceutical Statistics. New York: Marcel Dekker, 2000:110–113.

21. Laird NM. Missing data in longitudinal studies. Stat Methods 1988;7:305–315.

22. Permutt, T. Adjustment for covariates. In: Chow S-C, ed. Encyclopedia of Biopharmaceutical Statistics. New York: Marcel Dekker, 2000:1–4.

23. Dirnagel U, Iadecola C, Moskowitz MA. Pathobiology of ischaemic stroke: an integrative view. Trends Neurosci 1999;22:391–397.

24. Andrews PJ, Harris B, Murray GD. Randomized controlled trial of effects of the airflow through the upper respiratory tract of intubated brain-injured patients on brain temperature and selective brain cooling. Br J Anaesth 2005;94:330–335. Epub 2004 Nov 5.

25. Berge E, Barer D. Could stroke trials be missing important treatment effects? Cerebrovasc Dis 2002;13:73–75.

26. Committee for Proprietary Medicinal Products. Points to consider on clinical investigation of medicinal products for the treatment of acute stroke. CPMP/EWP/560/98. London: The European Agency for the Evaluation of Medicinal Products, 20 September 2001.

27. Gray LJ, Bath PMW, Collier T, et al. Optimizing the statistical analysis of functional outcome in stroke clinical trials. European Stroke Conference. Bologne, Italy 2005, Abstract.

28. Saver JL, Kidwell C, Eckstein M, et al. Prehospital neuroprotective therapy for acute stroke: results of the field administration of stroke therapy—magnesium (FAST-MAG) pilot trial. Stroke 2004;35:106–108.

29. van Swieten JC, Koudstaal PJ, Visser MC, et al. Interobserver agreement for the assessment of handicap in stroke patients. Stroke 1988; 19:604–607.

30. Hochberg Y, Benjamini Y. More powerful procedures for multiple significance testing. Stat Med 1990;9:811–818.

31. O'Brien PC. Procedures for comparing samples with multiple endpoints. Biometrics 1984;40: 1079–1087.

32. Lin J-P. Enrichment designs. In: Chow S-C, ed. Encyclopedia of Biopharmaceutical Statistics. New York: Marcel Dekker, 2000:185–188.

33. Lin J-P. Equivalence trials. In: Chow S-C, ed. Encyclopedia of Biopharmaceutical Statistics. New York: Marcel Dekker, 2000:188–194.

34. Peto R, Collins R, Gray R. Large-scale randomized evidence: large, simple trials and overviews of trials. Ann N Y Acad Sci 1993;703:314–340.

35. Rodenberg C, Cheng S-C. Dropouts. In: Chow S-C, ed. Encyclopedia of Biopharmaceutical Statistics. New York: Marcel Dekker, 2000:170–175.

36. Ting N. Carry-forward analysis. In: Chow S-C, ed. Encyclopedia of Biopharmaceutical Statistics. New York: Marcel Dekker, 2000:103–109.

37. Fisher RA. Statistical Methods for Research Workers, 14th ed. Edinburgh: Oliver and Boyd, 1970:272–286.

38. Fleming TR, DeMets DL. Surrogate end points in clinical trials: are we being misled? Ann Intern Med 1996;125:605–613.

39. Peto R, Baigent C. Trials: the next 50 years. Large scale randomised evidence of moderate benefits. Br Med J 1998;317:1170–1171.

40. Peto R, Collins R, Gray R. Large-scale randomized evidence: large, simple trials and overviews of trials. J Clin Epidemiol 1995;48:23–40.

41. Yusuf S, Collins R, Peto R. Why do we need some large, simple randomized trials? Stat Med 1984;3:409–422.

11

Data Safety and Monitoring Board: Role in Acute Neurological Trials

Brett E. Skolnick

BACKGROUND ON DATA SAFETY MONITORING BOARDS

Data safety monitoring boards (DSMBs), also known by other names (e.g., data monitoring committees [DMCs]), were introduced in the early 1960s as a means of ensuring the safety of subjects in clinical research. At the time it was determined that a mechanism was needed to establish an effective means of interim data monitoring in clinical trials. A fundamental component of this was the solicitation of experts in the field of study who were otherwise independent of the conduct of such a trial to ensure an objective assessment of issues that might be discovered during trial conduct. Over the last few decades, the functions and frequency of establishment of DSMBs has continued to increase for a variety of reasons, including an increased number of trials with a mortality endpoint, more stringent requirements by regulatory and funding agencies, and a greater appreciation for the potential biases, both in efficacy and safety considerations, which may be introduced in clinical trials of therapeutic interventions. These issues have continued to support the need for an independent monitoring board to objectively evaluate the safety of an ongoing clinical trial. Additionally, and perhaps with increasing attention, a DSMB can assist in the timely determination of whether a new intervention is safe and efficacious. The DSMB may also be important from an ethical perspective, as continued patient randomization may become unethical if equipoise is compromised. The use of such a committee may also aid in maintaining the relevance and scientific validity of a clinical trial. It should be noted that the current U.S. Food and Drug Administration (FDA) regulations impose no requirements for the use of DSMBs in trials except for research studies in emergency settings conducted under 21 CFR 50.24(a)(7)(iv), in which the informed consent requirement may be waived.

In the setting of neuroemergency trials, which includes trials in acute stroke, traumatic brain injury, and status epilepticus (see Chapters 1–3, 4, and 5, respectively), unique issues are present that make the role of the DSMB and its interactions with other trial committees (e.g., steering committees) of particular importance. Little published work is available in this area, but a competent review of the role of the sponsor and investigator with attention to these committees can be found in Donnan and colleagues (1).

Historical Foundations

The National Institutes of Health identified the need to perform ongoing study monitoring of treatment outcomes as part of the ethical responsibility of investigators to the study participants (2–4). The purpose of safety monitoring was established to ensure and maintain the scientific integrity of subjects participating in research projects and to protect the safety of such subjects. Meinert (3) defined safety monitoring as a process during a clinical trial that involves the review of accumulated outcome data for groups of patients to determine whether any of the treatment procedures/interventions should be altered or discontinued. The DSMB guidelines (3,5) specify that all clinical trials should have a system in place for appropriate monitoring and oversight to ensure the continued safety of subjects and to ensure the validity of the data.

The type of safety monitoring activities should be commensurate with the nature, size, and complexity of the trial. For a small single-center study, safety monitoring might be performed by an independent safety officer in conjunction with an independent statistician. For a similar single-site high-risk trial, an independent DSMB may be more appropriate. For large trials conducted either at a single institution or at multiple institutions, monitoring is best performed by an independent DSMB with an appropriate constituted membership. Such ongoing review of the data by an independent individual or committee ensures that the trial can continue with ongoing and appropriate assessments of patient safety. In each of these settings it is important that the role and responsibilities as well as reporting responsibilities of the monitoring entity be clearly described in an appropriate document (see Appendix A).

DSMBs IN NEUROEMERGENCIES

Organization

A typical DSMB principally comprises clinicians and scientists knowledgeable of the clinical indication under study as well as a biostatistician and possibly a bioethicist familiar with the clinical area and with an appreciation for the balance between clinical trials and clinical practice. The role of the selected individuals is to ensure the safety of the subjects in the clinical trial, while performing a primary role of analyzing adverse events and performing interim analyses on relevant clinical outcome variables (6).

Responsibilities

The primary responsibility of the DSMB is to protect the safety of study participants. To perform this responsibility, the DSMB must be empowered to request further information from study investigators, require that the investigators change the descriptions in the informed consent document, and/or recommend the discontinuation of the study. The methods for implementing and/or requesting this information can take many forms.

The DSMB should be empowered to impact on the study sites, by means of the study sponsor, the contract research organization (CRO), or the steering committee. They must have a clear line of responsibility and a clear path of communication to effectively execute their principle charge of protecting subject safety. The communication paths should be fully described early in the study initiation so that study sites, sponsor or CRO, and the steering committee are fully aligned and aware of the communication pathways. An expansive literature exists on the role of the DSMB (7–9), but there is a need to continue to reassess the functions, responsibilities, and impact of the DSMB in clinical trials.

A DSMB's responsibilities may begin before trial recruitment, yet this aspect of their role has received little attention in published work. Issues include whether the DSMB will have input into the trial protocol, whether the DSMB should meet before data start to accrue, and whether any specific issues relevant to the trial exist, such as regulatory implications. Potential DSMB members should only agree to membership if the trial protocol is acceptable to them. However, their role during protocol development is less clear. An early meeting of the DSMB members can be useful to allow members to meet and discuss how the committee will meet during the trial (e.g., face to face, teleconference, video conference) to discuss the protocol in detail and to discuss how they might respond to hypothetical situations. Depending on the nature of the protocol and the study population, this may also afford members the opportunity to discuss the role of an ethicist or other specialty members.

Designing a Monitoring Plan

The specific responsibilities and the details of the data to be made available for evaluation can only begin once the study design and population are specified. It is only after this time point that the DSMB can design, with an independent statistician often times in conjunction with the assigned study statistician, the study safety monitoring plan. The monitoring plan should specify the responsibilities of the DSMB, including frequency of data review, triggers for ad-hoc reviews, and the contents and format of the safety reports. In addition, specific instructions as to whom each report will be sent (e.g., DSMB, steering committee, sponsor) and what procedures, if any, the respective recipient should follow (e.g., DSMB will forward the report with patient-specific identifiers removed for the various parties).

No strict recommendations can be provided as to what should be monitored

or how frequently a DSMB should meet. These decisions are usually set out in the protocol and are reviewed by the DSMB, who may develop a set of DSMB-specific bylaws that govern these activities. To assist the clinical study team and the DSMB in formulating the safety monitoring plan, the following considerations should be reviewed. Appendix A shows a sample DSMB charter.

Study Phase

Phase I, II, and III studies generally require different levels of safety monitoring. For many phase I and II trials, independent DSMBs may only be necessary or appropriate when the intervention is of moderate or significant risk. In some settings, the continuous close monitoring by the study investigator in conjunction with an assigned independent safety officer may be an adequate and appropriate format for monitoring, with prompt reporting of toxicity to the institutional review board (IRB) and/or ethics committee and the appropriate regulatory authorities. In situations involving potentially high risks or special populations, investigators must consider additional monitoring safeguards. As studies progress through phases II and III, a DSMB is typically required independent of the assessment of risk. The intensity and frequency of safety monitoring increases as the number of subjects and sites increase, various dosing levels are tested, and subjects are randomized to interventions. The need to document the safety profile of the therapeutic intervention (e.g., drug, biological, device, or surgical intervention) or likely adverse events and to ensure data integrity requires more frequent and more rigorous views of the data.

Regulatory Considerations

There are additional administrative considerations if the clinical trial requires compliance with FDA and International Committee on Harmonization (ICH) regulations. Monitoring should conform to good clinical practice, and sponsors have a responsibility to design trials in line with the principles of good clinical practices, which were developed by ICH (and set forth in document ICH E6) to address the design, conduct, performance, monitoring, auditing, recording, analysis, and reporting of clinical trials (10).

Study phase (I–III) and plans for NDA/BLA submission also influence the frequency and intensity of monitoring studies. Pivotal studies that will influence the outcome of an NDA/BLA are generally subjected to rigorous monitoring. Although it is often argued that the safety profile of a therapeutic intervention is known by the time a phase III is conducted, this is frequently not accurate because the early studies are generally conducted in small samples where adverse events may remain under-reported or simply undetected. Other safety concerns such as futility of outcome, protocol adherence, site performance, and data quality also need careful scrutiny. The first of these typically are the responsibility of the DSMB, whereas the remainder are the typical responsibility of the steering committee and/or the monitoring group (sponsor or CRO). In many cases the DSMB is in the best position, because of its systematic and timed reviews of the data, to comment on all these components.

Trial Design

The design of the trial is, in part, related to the study phase. As studies progress from phase I through phases II and III, larger sample sizes are required, thus resulting in greater variability in both study implementation and subject sample. In addition, adverse events are more likely to emerge as more people are exposed to the intervention. In multicenter, multinational, clinical trials, there is greater need to examine site-specific data collection and outcomes and intersite differences. Later phase II and all phase III studies are generally designed as randomized, controlled, clinical trials. Because the subject and investigator are blinded, it cannot be determined whether the adverse events that occur are related to the therapeutic intervention. Thus, careful review of the data both in aggregate and by treatment group should take place at regularly scheduled intervals. If adverse events occur in different proportions in the study groups and there are concerns regarding the negative effects of the intervention, then the study statistician and/or the DSMB may decide to unblind themselves to afford a better understanding of the nature of the effects and thus be enabled to best protect the safety of the study participants.

Diseases Under Investigation

The nature of the disease being studied may influence the safety monitoring plan. When the natural history of a disease is known, the investigators and the DSMB are more likely to anticipate the nature and frequency of adverse events. However, for diseases involving potentially high risks or special populations, investigators must consider additional monitoring safeguards. For example, for studies involving children, investigators may consider the use of a consent monitor to ensure that informed consent or assent is properly administered. In addition, those trials with high-risk patient samples or high-risk interventions (neurosurgical interventions, acute stroke, generalized seizures) require a DSMB. In many of these disease states, special attention has not been paid to the underlying safety profile for the current standards of care, which makes the responsibilities of the DSMB more difficult.

A monitoring plan should consider the nature of the therapeutic intervention. The level of scrutiny will depend on the severity of the disease and may require frequent patient assessments and regularly scheduled safety reviews. The same approach may be needed if the disease is serious and/or life threatening and endpoints are anticipated to occur frequently and/or early in the study.

Study Sample

The nature of the neurological disease/injury and the trial design will influence the size and characteristics of the subject sample. Phase I and II studies typically have smaller subject samples, and treatment studies for diseases are likely to include subjects of similar demographic and health status. Phase III studies have larger subject samples. Treatment studies for diseases such as ischemic stroke and traumatic brain injury are likely to include subjects with varying etiologies, medical history, and demographics.

The diversity of a study sample can be controlled, to some degree, by the inclusion/exclusion criteria that determine who is eligible to participate in a study. In some studies, eligibility criteria will increase the homogeneity of the patient sample. Increased homogeneity may decrease the number of confounding variables that may be considered during analysis. However, stringent inclusion/exclusion criteria may also reduce subject recruitment to the study as well as restrict the understanding of the therapeutic interventions safety profile. This is a careful balancing act that must be appropriately considered in trial design. It is important, therefore, to balance these competing demands so that subjects can be recruited to a study in a timely and cost-effective manner so that the study yields results that are of high quality and allows the determination of the safety and efficacy of the intervention(s) and its potential clinical generalizability. These considerations protect the subject's safety in that they are not committed to a study that is unduly extended over time, in which clinical practices may continue to evolve or that shows no hope of successfully evaluating the intervention.

The safety plan should specify a review of the rate of subject accrual by site and by the study overall, the sites' adherence to inclusion/exclusion criteria and other protocol requirements, and the expected compliance rate of the subjects. Although the specific assessments and corrections may often be the responsibility of an appropriately organized steering committee, this information may have inherent value to the DSMB as well. Studies in which the study requirements are invasive, the intervention causes many adverse events, or the study sample is elderly or has significant comorbidities may have difficulty accruing and retaining subjects. Careful monitoring of the subject recruitment, enrollment, and retention will help to protect the safety of study subjects, the integrity of the study, and the quality of the data.

If subject accrual is expected to occur quickly, then safety monitoring should take place early and may be related to a percent of the total planned accrual. For example, if 60 subjects are to be recruited in 6 months, safety review can take place after the first month of enrollment or after the first 10% of the subjects are enrolled, whichever comes first. In many high-risk samples, safety reporting should be provided to the DSMB continuously over the study duration with the planned meetings representing the time when appropriate trending could be performed and discussed by the DSMB. One specific value of the DSMB is to evaluate the significant adverse events (SAEs) trending data in addition to the individually reported SAEs with special attention and knowledge of the underlying factors prevalent in the disease under study.

Study Intervention

Therapeutic interventions that have been previously studied are more likely to have a known safety profile and the frequency and type of adverse events are more easily anticipated. However, the safety of such a therapeutic intervention is also related to the subject sample being treated, the indication for its use, dosage level and dosing regime, the presence of comorbid diseases, and the duration of the intervention (e.g., study drug/biological frequency). All these factors need to be considered in deciding on the frequency and intensity of safety monitoring as well as the types of reports (e.g., individual SAEs, safety trends) that are required.

Endpoints/Outcome Variables

Endpoints that are well defined and immediate are easier to monitor. Acute illnesses are more likely to have these types of outcomes. For example, treatment of an acute intracerebral hemorrhage with a single therapeutic intervention purported to reduce hemorrhage expansion is likely to yield clearer results in a relatively short period of time. In contrast, treatment of traumatic brain injury with repeated dosing of a neuroprotective therapy where 6-month functional outcomes are assessed may require both continual assessments of the impact of repeated treatments and an extended follow-up period. Thus, the subject's time on study intervention and the study duration, from baseline through final follow-up, will influence the type and frequency of safety monitoring.

Study Status and the Review Process

The monitoring plan should describe the review processes and the roles of the data management group, the study statistician, and the DSMB in relation to the content, format, and process of the review. Typically, the data management group will produce administrative reports that describe study progress including accrual by site, demographics in aggregate and by site, as well as subjects' status in aggregate

and by site. Reports might describe outstanding issues or protocol violations by site and in aggregate regarding adherence to inclusion/exclusion criteria and the study protocol. These reports should be reviewed by the steering committee for appropriate action but should also be provided to the DSMB. A carefully described action plan should be generated from these reports, where needed, to provide a means of continual improvement at specific sites or within the regions/countries requiring attention. This is of particular importance in global trials, where initial practice standards may differ but not be known at study start and may change at differing rates. There is also an opportunity to consider the use of a committee focused on adjudication of the patient management during study conduct (see later text).

Safety Reports

The safety reports should also include a list of adverse events, serious adverse events, deaths, and disease- or treatment-specific events by site and in aggregate properly prepared for DSMB meetings. In some studies of treatments with an unexpected high toxicity, independent medical monitor may need to be in place to review each adverse event or treatment description to ensure good clinical care and to identify any potential trends. In some studies, the independent statistician may review data and will alert the DSMB if event rates are of statistical concern, occur in a disproportionate number in one of the treatment groups, or fall outside of predetermined boundaries. The statistician may distribute interim reports to the DSMB between meetings to allow the call of special sessions when appropriate. The review plan should specify the process for reporting safety concerns between the medical monitor, the IRB, DSMB, the sponsor, and, where appropriate, the relevant regulatory authorities.

Typically, the DSMB reviews the safety reports in aggregate fashion and blinded to treatment group. If there are a significant number of adverse events, the DSMB may request that the treatment groups be unblinded to ensure there are not untoward treatment effects. The review plan should specify how data are to be presented and what the various triggers would be for unblinding of the data.

Interim Analysis

The data should be prepared in such a manner as to allow the analyses as designated for the interim analyses. The schedule for interim analyses can be a fixed time frame (e.g., every 6 months), after a certain number or percentage of subjects are enrolled (e.g., 25%, 50%, 75%), or in response to a specific number of occurrences of an event (e.g., n deaths). The interim analyses can have a multiplicity of functions. The primary one would be to assess the underlying safety of the therapeutic intervention in the context of the current clinical trial. This is managed by the continual review of reported serious adverse events and the periodic interim analyses on these events. Additional interim analyses can be constructed to evaluate the appropriateness of the assumptions underlying the original power calculations and in prespecified circumstances recommend the repowering of the study where such assumptions have been demonstrated to be incorrect. Finally, and most importantly, the interim analyses can be designed to evaluate interim efficacy. This is of particular importance in neuroemergency trials where the problems of equipoise are most difficult to maintain in the face of anecdotal evidence and where the ethics of continuing the study could be challenged by either investigators or IRBs/ethics boards. The DSMB has the additional responsibility in these situations to ensure that a critical review of risk-to-benefit be performed so

that they can recommend early stopping of the trial where the risks or benefits demonstrate that clinical equipoise is at risk.

Independence of Review

The DSMB should be separate and independent from the clinical trial staff as well as anyone responsible for subject/patient care in the context of the study. Members of the DSMB should not have scientific, financial, or other conflicts of interest related to the trial. Study clinicians should be blinded to the safety monitoring data, because exposure to emerging trends may influence enrollment and care, thus potentially biasing the clinical trial. Appendix B shows a sample conflict of interest statement.

Potential conflicts of interest or simply competing interests of DSMB members should be discussed before involvement and then in a timely and continual manner, either by reassessment or clear contractual restrictions. Any perception of DSMB membership bias may have important impact on credibility of the decisions made as well as on the overall trial integrity. Conflicts could include stock ownership, additional consulting agreements with the sponsors, involvement in operational aspects of the clinical trial, and many other real and perceived potential conflicts. These potential conflicts should all be carefully managed to ensure that the DSMB has unquestioned integrity. For additional information on this topic, refer to Grant and colleagues (11).

Specific Example

It is considered necessary for all phase III multi-institutional clinical trials as well as any high-risk phase I or phase II clinical trials to include in the protocol a description of the DSMB and a description of plans for data review. At a minimum, such a plan would include a method for adverse event grading and a means of providing attribution (such as a scale described later), a plan for how the periodic safety reviews will be performed, and a plan for the reporting of unanticipated adverse event to the sponsor, the investigators, and their respective IRBs or ethics committees. Grading scales (or adaptations of the common terminology criteria for adverse events developed by the National Cancer Institute [12] for the reporting of adverse events/toxicity) are important in areas of clinical neurology so that a consistent means of assessment of adverse events can be performed. One such system could use a simple scale such as the following:

0 = No adverse event or within normal limits
1 = Mild adverse event; did not require treatment
2 = Moderate adverse event; resolved with treatment
3 = Severe adverse event; resulted in inability to carry out normal activities and required medical attention
4 = Life-threatening or disabling adverse event
5 = Fatal adverse event

An idealized grading system should be developed in a coordinated fashion to make it applicable to all neurological trials, as was done by the National Cancer Institute, or at least a reasonable approach, specific to the disease under study, should be made to define and categorize events so that uniformity in the application and interpretation of reported adverse events can be performed. An alternative to such a grading system being performed by individual sites is to put in place an events adjudication committee that is furnished with all relevant information to make an independent assessment of the nature of the reported adverse events and to classify these in a systematic and consistent fashion.

Early Stopping Rules

As important as it is to specify clinical endpoints and to appropriately power the clinical study, it is of equal important to establish appropriate "stopping rules" for the premature termination of a clinical trial. A stopping rule specifies the outcome differences detected between groups during an interim analysis that can stop a clinical trial. The stopping rules should reflect one of the following conditions:

- Clear evidence of harm or harmful side effects of the treatment;
- No likelihood of demonstrating treatment benefit;
- Overwhelming evidence of treatment benefit.

One of the benefits of stopping rules is that they can prevent overreaction to random highs or lows in treatment response rates and adverse events because they generally require very low threshold probability values in interim analyses to indicate significance.

However, stopping rules, also called "discontinuation guidelines," are not the only data source sufficient to justify stopping a trial for several reasons:

- New information: There may be new information available, including the results of other trials, a change in the understanding of the underlying biology or external to the current trial, and evidence of unacceptable adverse effects.
- Limits of assumptions: Assumptions in the trial design regarding sample size and power, subject recruitment, the adverse event profile, and anticipated treatment effect differences may prove to be unwarranted after the trial is underway.
- Limitations of rules: Rules cannot be developed for all potential study scenarios and contingencies.

There are consequences of stopping a trial early even when such a stop is justified. The scientific purpose behind clinical trials is to calculate with some assurance the size of the differences between treatment outcomes. With less than a full complement of subjects/events, the confidence intervals associated with treatment effect estimate are larger. Another consequence of early stopping is to bias the estimates of treatment effect. This bias occurs because random high values in treatment effect may be used to justify early stopping, but rarely would such random low values be so used. Another important potential consequence of early stopping rules is an incomplete sampling of adverse events associated with the therapeutic intervention. Therefore, the delicate balancing of safety and efficacy that must be effectively managed should be appreciated, which is part of the significant responsibilities of the DSMB.

Stopping rules should be defined early in a study and incorporated into the protocol or into the formal statistical analyses plan. In either place stopping rules require realistic estimates of sample size to be effective. Overly optimistic subject accrual projections often mean that the trial is unable to show the test effect with the necessary assurance. Stopping rules are no more reliable than the data on which they are based. Thus, the quality of the data must be ascertained for the interim analyses, making this a part of the DSMB's responsibilities.

Before using stopping rules, there are a host of issues that should be considered, according to Friedman and colleagues (2):

- Group differences: Possible differences in baseline characteristics and/or prognostic factors between the groups should be explored and necessary adjustments made in the analysis. These are often incorporated into final statistical analyses, as covariates, but are often times not appropriately assessed during interim evaluations.

- Response variables: Potential bias in the assessment of response variables must be considered, especially if the trial is not double blind. This can also occur in the setting in which equipoise has shifted.

- Missing data: Possible impact of missing data should be assessed. For example, could the conclusions be reversed if the experience of participants with missing data from one group were different from the experience with missing data from the other group?

- Protocol compliance: Differences in protocol compliance should be evaluated for possible impact.

- Side effects: Potential side effects and outcomes of secondary response variables should be considered in addition to the outcome of the primary response variable.

- Subgroup consistency: Internal consistency across subgroups and various outcome measures should be examined.

- Off-protocol treatment bias: Patients/subjects who are not enrolled into the study but are treated with one arm of the study suggest that nonequipoise was present.

Relevant statistical methods used in safety monitoring include classical or group sequential methods, flexible group sequential procedures, applications of group sequential boundaries, asymmetrical boundaries, and curtailed sampling procedures. These methods are discussed in numerous textbooks on statistical methods for clinical trials, a few of which are included in the reference list (13–20). Clearly, there are significant complexities in the decision to stop a trial, and this is one of the most difficult roles performed by DSMBs, which is uniquely positioned to be able to consider and respond to all the available evidence (9).

Multiple Clinical Trials

The occurrence of two similarly designed, placebo-controlled, clinical trials that are either accruing or fully accrued and continuing to follow-up with patients while awaiting (e.g., 1-year follow-up information) safety and outcome data poses some interesting ethical considerations. What happens if one trial is stopped at a planned interim analysis by its DSMB due to a statistically significant treatment benefit? A method of communicating this decision to the DSMB of the other trial is needed, but what should be the decision process and responsibility for the second DSMB? How should this DSMB deal with the situation if their analyses of the data do not support such a decision? What responsibilities do the sponsor and the individual DSMBs have in conveying the different decisions? These are some of the dilemmas that can face DSMBs during an active development program (7).

Adjudication of Endpoint or Medical Management

A separate group or committee can be convened whose concern is to review critical endpoints reported by trial investigators to determine whether they meet protocol-specified criteria. Alternatively, such a committee can be responsible for ensuring comparability of medical management across the trial sites (see Chapter 13). The nature of the information reviewed on such endpoints may include laboratory, pathology and/or imaging data, autopsy reports, physical descriptions, and any other data deemed relevant. The committee should be masked to the assigned study arm when performing its assessments. Such committees are of particular value with subjective endpoints or when such endpoints require the application of a complex definition and/or when the intervention is not delivered in a blinded

fashion. Although such committees clearly do not share responsibility with DSMBs for evaluating interim analyses, their assessments (if incorporated in a timely manner into the database) will help to ensure that the data reviewed by DSMBs are as accurate and free of bias as possible. Given the complexity of neuroemergency trials where medical management is at times idiosyncratic, there may also be a role for a medical management adjudication committee. This committee's role will be ensure that either protocol guidelines or published medical management guidelines are being adhered to by the participating sites. One good example of such is the role of the American Brain Injury Consortium technical center in ensuring compliance to AANS and American Brain Injury Consortium guidelines for the management of traumatic brain injury patients. This committee, if properly charged, can be responsible for working to reduce heterogeneity regarding patient management that can in neurointensive care settings significantly impact on clinical outcomes.

DSMB CHARTER

A DSMB charter is a prerequisite to properly define the role, responsibilities, and reporting activities of a DSMB. Guidance documents are available (9,21) that describe a regulatory perspective on a DSMB. The DSMB charter should, at a minimum, identify the members of the DSMB, their reporting functions and responsibilities, and a description of the plans for transmitting summary reports to the principal investigator, the sponsor, and, where appropriately, the regulatory authorities.

Clinical/Ethical Considerations

The study sponsor and the associated external clinicians (via the steering committee or others in advisory roles) have implicitly agreed, in the initiation of a randomized, controlled, clinical trial, that an area of uncertainty exists to clinicians regarding a therapeutic intervention. This is often referred to as "collective equipoise." For patients to be entered into a trial, the available evidence should enable an individual clinician to randomize an individual patient into the trial. The next facet of such a randomization indicates that a "promising" new treatment can be studied where a predefined outcome difference could be anticipated. These differences could be expressed as an improved quality of life and/or a prolongation of life with a reduction in toxicity or costs (health care and other associated costs with the disability). Studies must therefore be designed to be of sufficient power to reliably detect these differences and to be generalizable to clinical practice.

Once these considerations are met, it is then of equal ethical importance to ensure that the study is not prolonged when unequivocal evidence of benefit or harm are detected or, in more subtle conditions, where there is little chance of demonstrating the beneficial clinical effect that justifies the "toxicity" costs of the intervention. In the absence of an efficient form of ongoing data monitoring while working to securing the requisite sample size, important and unpredicted results might be missed that could have impact on the collective equipoise. For example, unexpected beneficial effects might be detected, toxicity may be greater than predicted, or the suggested clinical benefit may simply be too small to outweigh an unexpected toxicity. A functioning DSMB ensures that the benefit-to-risk ratio remains acceptable for participating patients. The role of a DSMB in clinical trials conducted in acutely ill patients indicates the importance of an efficient and reactive DSMB to monitor patients' safety, especially during large multicenter trials. It is also the role of the DSMB to monitor the literature and to make certain that new discoveries in the field do not make the trial's therapeutic interventions or manipulation(s) in the clinical setting irrelevant (e.g., the possible impact of stenting and coiling on endarterectomy).

Risks Associated with a DSMB

The DSMB when reviewing the associated interim analyses may compromise the integrity of a clinical trial if the DSMB's responsibilities are not properly established and managed. As such, the use of DSMBs has become a focal point of assessment by regulatory authorities. Three concepts appear central for DSMB functionality: equipoise, multiplicity, and bias.

In a clinical trial, equipoise exists when there is a lack of evidence for choosing one treatment strategy over another. The principle of clinical equipoise can be defined as the state or condition at the start of a trial where clinical equipoise regarding the merits of the particular regimens to be tested exists and the trial is designed in manner to make it reasonable to expect that, if the trial is successfully conducted, clinical equipoise will be disturbed. Multiplicity is the concern that with multiple looks at the safety profile when multiple types of adverse events are likely to be present, bias may occur during the interim review of adverse events observed in each study arm. A careful assessment of underlying risk associated with the disease/injury under study is an important task. There are events that may result from the disease being treated or as a result of the intervention itself. Well-defined and regulatory-specified definitions should be adhered to and in special circumstances events of special interest identified. For example, individuals with cerebrovascular disease are at an elevated risk of other vascular events (e.g., ischemic stroke, myocardial infarction). Thus, a specific case of myocardial infarction in a participant in a trial of a new neuroprotectant may not be readily attributed to the new therapy. A DSMB, however, will regularly review the number of myocardial infarctions observed in each study arm. If an imbalance between groups emerges, concerns will arise that some of the myocardial infarctions may be due to the therapeutic intervention rather than to the disease itself. Because a potentially large number of adverse event categories may be observed and compared between the study arms, the interpretation of safety findings by the DSMB must be sensitive to the issues of multiplicity.

The DSMB also needs to be concerned that adequate adverse event reporting occurs and that investigators do not under-report such results due to their viewing these as being related to the underlying disease. In this way the DSMB has to be assured that adequate site monitoring and reporting is occurring in an accurate and timely manner. Knowledge of unblinded interim comparisons from a clinical trial is not necessary for those conducting or those sponsoring the trial; further, such knowledge can bias the outcome of the study by inappropriately influencing its continuing conduct or the plan of analyses. In the context of interim data evaluation and the results of interim analyses, these should generally not be accessible by anyone other than DSMB members. Sponsors should establish procedures to ensure the confidentiality of the interim data to reduce any potential bias. These three factors, equipoise, multiplicity, and bias, must be carefully considered and weighed in both formulating the DSMB charter and while executing the responsibilities of the DSMB.

INTERACTIONS

Reporting

Interim reports, prepared to the DSMB specifications, by an independent statistician should include comparative effectiveness and safety data presented by study group, whether coded or uncoded, which should generally be available only to DSMB members during the course of the trial, including any follow-up period, formal close of the study, and database release. These reports are generally not shared with the sponsor, because it may make it difficult for the sponsor to change the trial design or statistical analysis plan in an unbiased manner.

In many cases, the DSMB will summarize its deliberations in two parts: an "open" section, which begins with a single page summarizing the recommendations of the DSMB and subsequent pages that present data in aggregate and primarily focuses on trial conduct issues such as accrual and dropout rates, timeliness of data submission, eligibility rates, and reasons for ineligibility; and a "closed" section, in which the comparative outcome data are presented. The DSMB may share the open section of these reports with sponsors, who may convey any relevant information in these reports to investigators, IRBs, and other interested parties, as the data presented in the open section are not likely to bias the future conduct of the trial and are often important for improving trial management. It is likely in the future that the first page of the open section will provide valuable input to individual IRBs/ ethics boards in managing their continuing responsibility of monitoring the progress of specific clinical trials, particularly those with greater than minimal risk.

Sponsor

Sponsors are required to report any serious events occurring with their drugs/ biologics/devices to regulatory authorities in the various countries. The request by the authorities to be fully informed has led sponsors to allow that a company representative be informed about any adverse events associated with the drug/ biologics/device(s) under study. However, that goes against the general principle that everyone except the DSMB should remain blinded. It would appear to be optimal, in the context of a clinical trial, to report all SAEs that fulfill reporting requirements, including those possibly associated with the treatment comparator or placebo group. It is noted that this reporting is of limited value because it lacks information regarding the actual intervention as well as the denominators. The overall information known to the DSMC is by far more meaningful than that from a single institution. Some of the reported adverse events occur as disease-related outcomes. In many neuroemergency trials there is no way, other than by statistical comparison, to know whether adverse events are a likely result of the study treatment or the disease.

Steering Committee

The general rule should be that the DSMC advises the steering committee (and sponsor) of the trial on continuation, stopping, or extension of the sample size of the trial as well as other actions to ensure the highest quality for the trial. Thus, the ultimate decision rests with the steering committee or sponsor. It has happened that the steering committee does not agree with the recommendations by the DSMC. This can be of particular importance regarding stopping or continuation of a trial. If the DSMC recommends that the trial should be stopped based on safety problems, it is generally expected that the steering committee and/or sponsor also act accordingly. However, this may not always be the case. The trial may still be stopped because the sponsor has access to additional sources of information (other clinical trial information from concurrent trials) and/or because of the sponsors' control regarding the financing of the trial.

The recommendations to the steering committee should include not only the risk-to-benefit assessments for trial continuation and any recommendations

based on prearranged interim analyses for safety and/or efficacy, but may also include recommendations for repowering of the study based on these prespecified analyses. These recommendations are best conveyed in a preestablished manner, if possible, to minimize any introduction of bias or possible misinterpretation by the steering committee or the sponsor.

Sites and IRBs/Ethics Boards

Sites need to provide at least annual information to their respective IRBs/ethics boards to ensure continual review of ongoing clinical trials. Similarly, IRBs/ethics boards are empowered to protect trial participants by reviewing initial research plans and providing continuing review of approved research (1,15). Three to four decades ago, when the structure and function of the IRBs/ethics boards were first being codified, an individual local IRB/ethics board was expected to review the protocol and its informed consent document and to make an initial judgment about the potential risks relative to the potential benefits of the proposed study and the appropriateness of communications about these risks and benefits to potential trial participants (22). It also was expected to review research in progress and determine whether the balance of risks and benefits remained appropriate for the research to continue. These expectations remain today, although they can be impossible to meet for multicenter studies and, increasingly, for complex single-center studies (23,24).

IRBs and ethics boards are not able to perform safety monitoring by review of individual adverse events and are often burdened by duplicative reviews of large multicenter studies. It is the responsibility of the sponsor and/or the DSMB to provide relevant and timely information to these boards to enable them to determine the unique issues related to their site and whether the study should continue at their institution. Special considerations may exist that could result in certain regions/countries discontinuing their participation based on the interim results and/or DSMB safety reviews, which are a unique responsibility of the IRB or ethics board of the respective trial site.

SUMMARY

The role of DSMBs is continuing to expand as both the complexity of acute neuroemergency trials and the need for proper safeguards to ensure protection of patients are being subject to increasing, yet appropriate, scrutiny. A well-defined DSMB charter with careful consideration to the specific events to be monitored and the methods of analyzing such is becoming increasing critical. The use of independent adjudication committees is more common in the neuroemergency settings as the definitions of clinical outcomes and variations in medical management become increasing difficult to discriminate. An excellent functioning DSMB will be ever more important in the coming years to ensure that we do not continue trials with significant risk or fail to recruit a sufficient sample size to adequately demonstrate safety and efficacy.

References

1. Donnan GA et al. Recommendations for the relationship between sponsors and investigators in the design and conduct of clinical stroke trials. Stroke 2003;34:1041–1045.
2. Friedman LM, Furberg CD, Demets DL. Fundamentals of Clinical Trials. Mosby-Year Book, 1996.
3. Meinert CL. Clinical Trials: Design, Conduct, and Analysis. New York: Oxford University Press, 1986.
4. Weiss NS. The Study of Outcome of Illness. New York: Oxford University Press, 1996.
5. National Institute of Health. NIH Policy for Data Safety Monitoring. http://www.nih.gov/grants/guide/notice-files/not98-084.html, 10 June 1998.
6. Ellenberg SS. The use of data monitoring committees in clinical trials. Drug Inform J 1999; 30:553–557.
7. Dixon DO, Lagakos SW. Should data and safety monitoring boards share confidential interim data? Control Clin Trials 2000;21:1–6.
8. Asplund K. The role of the data safety and monitoring committee in stroke trials. Eur Neurol 2003;49:115–119.
9. Baum M, Houghton J, Abrams K. Early stopping rules—clinical perspectives and ethical considerations. Stat Med 1994;13:1459–1469.
10. Guidelines for Good Clinical Practice. http://www.fda.gov/cder/guidance/959fnl.pdf, 1998.
11. Grant AM et al. A proposed charter for clinical trial data monitoring committees: helping them to do their job well. Lancet 2005;365: 711–722.
12. Trotti A et al. CTCAE v3.0: development of a comprehensive grading system for the adverse effects of cancer treatment. Semin Radiat Oncol 2003;13:176–181.
13. Ashby D, Machin D. Stopping rules, interim analyses and data monitoring committees. Br J Cancer 1993;68:1047–1050.
14. Boshuizen F. Comparisons of threshold stopping rule and supreme expectations for independent random vectors. Stoch Anal Appl 1990; 8:389–396.
15. Dorea CCY. Stopping rules for a random optimization method. Siam J Control Optim 1990; 28:841–850.
16. Hughes MD, Pocock SJ. Stopping rules and estimation problems in clinical trials. Stat Med 1988;7:1231–1242.
17. Mukhopadhyay N, Sen PK, Sinha BK. Stopping rules, permutation invariance and sufficiency principle. Ann Inst Stat Math 1989;41:121–138.
18. Pignon JP, Arriagada R. Early stopping rules and long-term follow-up in phase III trials. Lung Cancer 1994;10:S151–S159.
19. Pocock SJ. Size of cancer clinical trials and stopping rules. Br J Cancer 1978;38:757–766.
20. Vanputten W. Stopping rules for small-group sequential trials based on Fisher exact test. Control Clin Trials 1988;9:246.
21. Food and Drug Administration. Guidance for Clinical Trial Sponsors: On the Establishment and Operation of Clinical Trial Data Monitoring Committees: Draft Guidance. http://www.fda.gov/cber/gdlns/clindatmon.htm, 2004.
22. Office of the Inspector General. Department of Health and Human Services, Institutional Review Boards: A Time For Reform. DHHS publication no. OEI-01-97-00193, 1998.
23. Levine RJ. Institutional review boards: a crisis in confidence [see comment]. Ann Intern Med 2001;134:161–163.
24. Morse MA, Califf RM, Sugarman J. Monitoring and ensuring safety during clinical research. JAMA 2001;285:1201–1205.

APPENDIX A. SAMPLE DSMB CHARTER

DSMB Guidelines for Study _____

Background

Brief review of the clinical area of study, including incidence rates, morbidity, and mortality. A review of the preclinical and/or clinical basis for the current study. Status of current medical/surgical management.

Scope of the Data and Safety Monitoring Board (DSMB) as a single or multiple study DSMB. Current status of clinical development of the drug/device/biologic or other intervention. Including a brief review of any available clinical trial information with the study drug/device/biologic or other intervention.

A brief description of the study trial (synopsis). A brief description of the primary focus of the DSMB activities and their ability to request a pause or cessation in enrollment from the sponsor.

A full description of the organization of the ongoing safety surveillance

Data and Safety Monitoring Board

The DSMB has been constituted for the Study XX. The tasks of the DSMB are to
- Independently examine safety data
- Monitor study adherence
- Assessment of (if relevant) clinical efficacy
- Assist the internal [*sponsor's name or internal safety committee*] Safety Committee by providing expert evaluation of all serious adverse events (SAEs) for this study

Composition of the Membership

The DSMB is composed of permanent members who may request assistance from a number of ad-hoc members when needed. The members of the DSMB cover the relevant subspecialities involved in the medical and/or surgical management of the clinical population under study.

Permanent members:

Specialty:
Name:
Academic title:
Address:
Contact information (e-mail, phone):
DSMB role:

Specialty:
Name:
Academic title:
Address:
Contact information (e-mail, phone):
DSMB role:

Specialty:
Name:
Academic title:
Address:
Contact information (e-mail, phone):
DSMB role:

Specialty: Statistician
Name:
Academic title:
Address:
Contact information (e-mail, phone):
DSMB role:

Ad hoc members:

Specialty:
Name:
Academic title:
Address:
Contact information (e-mail, phone):
DSMB role:

Specialty:
Name:
Academic title:
Address:
Contact information (e-mail, phone):
DSMB role:

DSMB Working Procedures

The DSMB will perform ongoing assessments of SAEs as they are reported. For each SAE a case report (CIOMS or similar format) will be provided to each DSMB member for evaluation. The DSMB can at any time request additional information regarding a reported SAE.

At any time the DSMB or an individual DSMB member (via communication from the DSMB chairman to sponsor) can request that the clinical trial is placed on hold while further information about an individual SAE is collected and evaluated.

On an individual basis each DSMB member will evaluate the data correlating any SAE in the treatment group(s) with provide measures or indices of a safety problem.

At telephone conferences held on a regular basis with intervals as decided by the DSMB, the DSMB members will discuss the safety data and reach a conclusion regarding continuation of the trial. In case of disagreement between DSMB members, a simple majority will decide the issue. If no majority can be obtained, the chairman of the board has the final vote.

The Chairman of the DSMB conveys the conclusion to the specify relevant internal unit/division of sponsor/CRO of the sponsor (Attention: Designated Contact Person).

Critical events (a listing of previously determined areas of concern either by prior studies with the drug or with similar drugs in class).

Based on information obtained from the DSMB members and the steering committee, the following events have been defined as critical in the current trial, provided they occur in proximity to study drug (or device or other intervention) dosing:

• Fatalities evaluated as possibly/probably related to study drug
• Specific clinical and/or laboratory events (specify as appropriate)

The background incidence of these events in the general clinical population is expected to be low or medium or high with a cumulated incidence × percent. percent. Additionally, other unforeseen possibly/probably related SAEs should be evaluated as critical events.

Due to the MOA or half-life of the study drug, the period during which an SAE is evaluated as critical has been restricted to a close time proximity of dosing (<7 days; adjusted as appropriate for study drug). A causal relation between an SAE and the study drug after this period may still be possible, although less likely.

Stopping Rules. A sequential safety interim analysis will be performed for the endpoint (define here the relevant endpoint[s] that will be assessed to evaluate safety and/or efficacy). Specify the timing (based on recruited patients, treated patients, or a time interval, whichever is most appropriate). The results of the sequential analyses will be forwarded to the DSMB, by the independent statistician, to evaluate the evidence and if necessary request a delay in further subject recruitment/ randomization.

Communication. Electronic mail will be used to distribute all SAE data between [*specify the group/division responsible for forwarding the SAE reports*] and the DSMB members. If further information is requested by the DSMB (e.g., demographic data, lab data, or other safety related data), it can be provided in appropriate formattedfiles to the DSMB members directly from the [*specify group/devision responsible either sponsor or CRO*].

Electronic mail will be used for communication within the DSMB and between DSMB members and the reporting division. Where necessary, hardcopy will be provided by courier service to the address specified.

Telephone conferences will be used for the DSMB evaluation on a regular basis.

The intervals will be defined by the DSMB. Telephone conferences can be initiated by the Chairman of the DSMB to review any single SAE or group of SAEs that require discussion.

The DSMB is ultimately responsible for the safe conduct of Study XX. The DSMB has reporting responsibilities to *the sponsor's safety division/unit*. This is done to ensure that broader safety issues related to the study drug are also being continually evaluated by an internal safety group within the sponsor or their designee (please refer to later section).

The DSMB may at any time request additional safety information.

The DSMB may change its working procedure in consultation with the sponsor and such procedure changes should be appended, as appropriate to this document.

Internal Safety Surveillance at Sponsor

Division/Unit name responsible at the sponsor

All SAEs occurring during clinical trials are reported by the investigators to the sponsor or its designee within 24 hours (initial information). Data are entered into a global safety database (specify reporting tool) and an appropriately trained Medical Officer in the specify division/unit who is responsible for the specific project will evaluate the event within 72 hours of receipt.

The Sponsors (or designee's) Medical Officer is further responsible for

- Providing data as described above for the external DSMB
- Forwarding expedited SAE reports for submission to authorities according to regulatory timeline requirements
- Study safety surveillance on a day-to-day basis
- Providing data for and chairing the internal safety committee meetings

Internal Safety Committee

The sponsor's (or designees) internal safety committee covers clinical trials exploring study drug/device/biologic. The committee works according to written guidelines and will meet at least monthly to discuss and evaluate the overall safety of study drug/device/biologic under study. A Medical Officer from sponsor's division/unit chairs the committee.

The internal safety committee evaluates each case under blinded conditions. [*Include here a brief description of the processes involved in internal safety review, what events and at what time points reviews occur, the nature of the reviews, and the outcomes from such review. It should also be clear whether the internal safety committee remains blinded to study product and if not what safeguards are in place to avoid potential biasing of clinical development staff. Finally, it should be clear what the internal safety committee is able to do regarding the ongoing trial(s). Are they able to recommend discontinuation of the study trial? Are they able to bring specific information to the attention of the DSMB for further "expert" evaluation? Can they suggest to the steering committee or other governing body a need to amend the protocol? How do their recommendations effect/interact with those of the DSMB?*]

Contact Persons

A complete listing of relevant contact information for the DSMB members, the independent statistician, the sponsor's secretary, data management personnel, and clinical trial personnel.

References

As appropriate but at a minimum to include the protocol, investigator's brochure, and any relevant literature.

APPENDIX B. CONFLICT OF INTEREST STATEMENT FOR DSMB MEMBERS

CONFIDENTIAL

Study Title

Dr. _____
DSMB member
I attest to the following:

- I am not a part-time, full-time, paid, or unpaid employee of any organizations that are: (a) involved in the study under review; (b) whose products will be used or tested in the study under review, or whose products or services would be directly and predictably affected in a major way by the outcome of the study;
- I am not an officer, member, owner, trustee, director, expert advisor, or consultant of such organizations.
- I do not have any financial interests or assets in any organizations meeting the above criteria, not does my spouse, dependent children, nor any organization with which I am connected; and
- I am not a current or past collaborator or associate of the principal investigator.

Having read the above: *(please check the appropriate answer)*

____ I have no relevant interests or activities.

____ I have noted any exceptions in the space below:

I will notify the study sponsor promptly if:

- A change occurs in any of the above during the tenure of my responsibilities, or
- I discover that an organization with which I have a relationship meets the criteria for a conflict of interest.

I am aware of my responsibilities for maintaining the confidentiality of any non-public information that I receive or become aware of through my activities with the Sponsor and the associated Data Safety Monitoring Board and for avoiding using such information for my personal benefit, the benefit of my associates, or the benefit of organizations with which I am connected or with which I have a financial involvement.

_____   _____
Signature   Date

_____
Print name

12

Role of a Project Medical Officer in Acute Neuroemergency Clinical Trials

Joseph A. Kwentus

The conduct of clinical trials in patients with acute neurological injury/disease involves a complex set of clinical conditions, which may require timely and accurate diagnosis, determination of appropriateness for recruitment and randomization, and careful clinical follow-up. This process is labor intensive, fast paced, and clinically demanding. Successful implementation, execution, and finalization of such clinical trials depend on a responsive and knowledgeable clinical team at the research site. The specific nature of neuroemergency clinical trials, however, also demands that the clinical trial sponsor (e.g., pharmaceutical company, medical device company, or other sponsoring party) consists of clinically knowledgeable physicians who can provide appropriate guidance in the design, implementation, and execution of such a trial. In the neuroemergency setting, the unique challenge of recruitment requires that an appropriate and rapid response to issues raised by study personnel be accurately and rapidly managed. It is essential that the project medical officer (PMO) has the appropriate skills and perspectives to perform any number of functions, one of the most critical being their focus on those who require the most timely response to issues raised by the site personnel during active patient screening and recruitment. In any acute neuroemergency trial, such as in acute ischemic stroke, intracerebral hemorrhage, or status epilepticus, one should expect this type of rapid response from the sponsor's medical team and/or its designated contract research organization.

The increasing complexity of clinical trials, particularly in the acute neurology domain, provides unique challenges from the perspective of medical management. Some projects require a larger medical workforce than the sponsor may have available. Neuroemergency trials, in most cases, require 24-hour coverage 365 days of the year during trial execution. Contracting with independent physician consultants or a physician group associated with a contract research organization is one way of addressing this need. Regardless of the source of medical coverage, an appropriate team is needed to manage the medical tasks associated with the project and thus provide the sites and the sponsor with optimal service. Such attention often times results in more appropriate patient selection, higher data quality, and more rapid and appropriate patient enrollment.

The person who assumes this role is often referred to as the project medical officer. The PMO provides an opportunity to supplement existing skills either in the sponsor's organization or within a specific project team. The physician who works as a sponsor's PMO in an acute neurology trial assumes a unique set of responsibilities. Such responsibilities may be performed using a variety of models. The PMO may be defined as the responsible medical authority to make unilateral decisions/recommendations or these responsibilities may be shared with the sponsor's medical team. The responsibilities are described below with the intent of providing an example of a model that could be considered to fulfill these responsibilities.

PMO QUALIFICATIONS

The ideal PMO should have been part of a clinical team that actually performed similar studies and/or managed similar patients in the acute neurology area. If the PMO has not had experience doing clinical research with such patients, then at a minimum relevant clinical experience with the population would be beneficial. Clinical and research experience with the population allows the PMO to play a unique position on the sponsor's team because the PMO may be one of the few team members to have either studied these patients in similar clinical trials or provided direct patient care to such patients. The PMO who brings such relevant experiences to the team may have an especially important role during the development of the protocol. Although the PMO may contribute important suggestions regarding the scientific aspects

of the protocol, the role of the PMO can be quite diverse, potentially affecting many aspects of the study. Thus, the role of the PMO in acute neurology trials encompasses several domains, with the PMO being able to be involved in many if not all of the following components, at the discretion of the sponsor:

- Assessments of study feasibility
- Study design and implementation
- Review of the case report form (CRF) and CRF guidelines
- Providing medical input to regulatory/ ethics and institutional review boards (IRBs)
- Therapeutic training for sponsor's staff
- Site selection
- Medical support at investigator meetings
- Managing emerging site issues during enrollment.
- Managing site initiation process and booster activities
- Guiding serious adverse event (SAE) and CRF completion issues
- Coding of medical events for adverse events (AE) and SAEs
- Writing/contributing to the clinical trial report
- Supplementing sponsor's staff at regulatory meetings.

The following describes the specifics surrounding each of these components of the study for which a PMO may make significant contributions.

ASSESSMENT OF STUDY FEASIBILITY

Drug discovery can often result in a unique compound (biologic or drug) that may have potential application in an area outside the competencies of the sponsor. This may create an opportunity for out-licensing for the sponsor or represent a unique challenge for the sponsor to perform the early steps (*ex vivo*, pre-clinical) or all the steps through regulatory approval. The challenges are to ensure that the compound is studied in appropriate preclinical models as well as evaluated in both the appropriate patient populations and with the proper study design using the optimal clinical outcome tools. A broad perspective on clinical trials in the relevant areas is required to not repeat earlier clinical trial problems. The PMO may provide this expertise, or in some cases the sponsor may create a steering committee to be responsible for evaluating the study feasibility.

STUDY DESIGN AND IMPLEMENTATION

Although many of the acute stroke studies to date share many similarities, critical nuances in protocol design and endpoint selection may make a project more difficult in certain settings and easier in others. Acute stroke studies have extremely tight protocol timelines. The sequences of tests and scheduled events push the coordinator and the investigator to perform study procedures within a few minutes of each other. The sequence of evaluations and procedures in these studies is always extremely rigid. Careful consideration of the timing of such events, the staffing required to provide appropriate adherence to these requirements, and the appropriate training to ensure that the study staff can adhere to the protocol and obtain the necessary trial data are areas where the PMO can provide significant input.

The PMO's guidance during protocol design should directly facilitate the ease of execution of the protocol at the study site level. PMO contributions may include suggestions regarding the practicality of timelines, the consent process, implementation of study instruments, and the logistics of obtaining tests and measures in the acute clinical setting. To optimize this requires knowledge of the types and number of resources at the site level. The PMO must be able to anticipate the issues that the actual procedures of the project will raise and must address whether protocol execution can be appropriately managed at the selected sites. Finally, the PMO should be prepared to answer questions when the sites become concerned with operating the project in their unique local environments. The PMO should have thought through the potential problems and should be available with practical answers that aid the site in the successful execution of the protocol within their hospital environment.

This becomes of particularly importance in the context of global clinical trials. The continued expansion of clinical development programs into global clinical trials provides increasing difficulties in ensuring that any protocol, but particularly acute neuroemergency trials, are managed similarly within a country and, perhaps more importantly, between countries. Particular issues may need careful attention to achieve comparability of the data in these settings. Different clinically accepted medical/surgical management standards might have significant effects on relevant clinical endpoints. As an example, varying management strategies for deep vein thrombosis prophylaxis may introduce bias in both SAE reporting and long-term outcome measures. Differences in the time for approval of recombinant TPA between the United States and Europe may have altered the types of patients entered into studies during this period. Even subtle differences in "standards" of care may also have impact on primary outcome measures. These issues should not be underestimated and need proper attention by using PMOs who have active knowledge of differences in such management strategies and who can play a role in bringing these differences, as they are discovered, to the attention of the sponsor.

The PMO is also in a position of providing suggestions on how to manage such differences in medical management. The provision of endpoint adjudication committees is one method for minimizing definitions of outcomes. Another is the use of medical management adjudication committees to evaluate management strategies at the clinical trial sites to ensure that all sites are in compliance to the dictates of the protocol or to nationally or internationally established recommendations. One example of this is the work performed by the American Brain Injury Consortium in providing quality review of medical and surgical management in the context of traumatic brain injury. Such medical management adjudication can identify critical differences between sites in a region or across sites in a global study that once identified can then be appropriately managed.

REVIEW OF CRF AND CRF GUIDELINES

The CRF provides the primary data collection tool for any clinical trial. The construction of the CRF, although dictated by the data elements in the protocol, may take many different forms and as such may be subject to varying interpretations on how to best design and complete it. The PMO should take part in the design and review of the CRF because the PMO may, as a result of his or her global trial experience, recognize issues that could have an effect on making the data collection more or less difficult when put into practice at the trial sites.

Decisions regarding the design of the CRF should optimize their completion and may take many different paths. The flow of the CRF may be optimized to match the flow of the patient through the protocol but should always take into careful consideration the types of medical/surgical units and their respective personnel

who will be providing data for CRF completion. Otherwise, the CRF may only serve to confuse study personnel and then create issues of accurate and source-verifiable data. A lack of understanding of the nature of the workflow, on the involved medical/surgical units, may allow the development of a CRF and its associated workflow that might increase the number of entry errors and thus add to the number of protocol violations and/or deviations. The PMO along with the responsible trial staff should be able to recognize ambiguous data fields on the CRF and identify data fields out of sequence from the flow of protocol execution. If the CRF or parts of the CRF will serve as the source document (e.g., completion of outcome scales), then the PMO should review the CRF to ensure that site personnel will be able to complete the CRF in a practical fashion and consistent with the timelines for the study protocol.

PROVIDING MEDICAL INPUT TO REGULATORY/ETHICS AND IRBS

The PMO can assist the project team by playing a role in advising the study staff on issues related to informed consent and ethical practice. The PMO has two unique contributions to make. First, the PMO can help the individual principal investigator with the discussions and negotiations with the IRB. Second, the PMO can assist the principal investigator in anticipating the difficulties that may occur in the consenting process.

Many neuroemergency clinical trials take place at large medical centers. As a result, the use of a local IRB is much more common than in other neurological studies, in which a central IRB may facilitate the creation of an appropriate consent form and oversee the ethical issues in the study. When a local IRB is handling approval of both the content and wording in the consent and ethical issues related to the protocol, the IRB might need substantial supportive documentation regarding the underlying scientific basis of the project. The PMO can work with the local site principal investigator to help understand the informational needs of the IRB and be in a position to provide materials that may make the project more understandable to the IRB. When specific negotiations regarding wording in the consent form occur, the PMO can help relieve some of the burden from the principal investigator and the sponsor by making suggestions that would be acceptable to the sponsor and potentially acceptable to the IRB. The sponsor's regulatory team generally refers these difficult questions to the PMO when they have difficulty understanding specific medical concepts, language, or procedures.

In addition to assisting with negotiations with the IRB, the PMO should be in a position to consider issues surrounding the process of obtaining informed consent that may be unique to the clinical setting or patient's condition. It is important to recognize that once the regulatory parameters have been set, the study site faces the daunting task of imparting a large amount of information to potential subjects and to the subject's family in a situation of intense emotional upheaval. The comfort level of each the study personnel at delivering this information in a sensitive manner may be unique to the site. It is important to identify questions ahead of time. Is the local consent form appropriately designed to be patient friendly? Are the study personnel sufficiently trained and comfortable to discuss the consent with shocked and grieving relatives? Are there any unique regulatory difficulties imposed by the local IRB or the state legislature? How can the local principal investigator and their staff think these difficulties through and manage them in an appropriate manner? If the consenting process is difficult, does it cause study personnel to become apathetic to

patient recruitment? If so, what does the local principal investigator need to do to overcome this?

The PMO is well positioned to listen to communications from the site and to become proactive in helping the principal investigator recognize difficulties and think through problems to minimize unnecessary obstacles to subject identification and recruitment. It is extremely important for the study personnel to support the ethics of the project and to understand the potential contribution of the project to medical advancement in the treatment of stroke. Interaction with regulatory groups and IRB/ethic committees and difficult or negative experiences with potential subject families can be detrimental to morale. The principal investigator may or may not recognize when morale problems have led to identification/recruitment problems. The PMO can help the principal investigator recognize these difficulties and make suggestions to assist in working through these issues with the staff.

THERAPEUTIC TRAINING FOR SPONSOR'S STAFF

Therapeutic specialized training is an important component of training for all participating staff of the sponsor. This can begin during protocol development but needs to be in place to aid in the site identification and selection process. The PMO can assist with these educational efforts that are intrinsic to study startup and site selection and site maintenance. Varying levels of training can be performed. Often times a more general overview of the therapeutic area and the intricate components required to be present at a site can be done early in the training process. When final site selection is completed, appropriately designed training programs grow in importance. The PMO can provide education to study monitors regarding the specifics of the acute neurological disease/injury, the potential risks associated with the study drug that might affect the disease process. This educational process should be ongoing and will need to address specific questions over the course of the study.

The PMO may also be required to provide specific training at the investigator's meeting regarding unique aspects of the study, including diagnostic procedures, study drug delivery, and the reporting of AEs. In this capacity, the PMO will come to be recognized as a resource for the study. Training opportunities at the investigator's meeting range from protocol presentation to safety considerations and training on the various outcome scales that may be used in the specific study. Ongoing training for the sites may continue via teleconference after the project has been initiated. This training may include clarification of study-specific questions that arise during protocol performance or further elaboration of information related to the study instruments or processes as new employees take on study related roles at the sites during the conduct of the project.

SITE SELECTION

The site selection process demands consideration of many complex factors to assemble the most appropriate array of research settings. Site selection should not be dictated primarily by the perceived needs of the sponsor representatives or by the external consultants (e.g., steering committee). The PMO should have input into this process so that the proper criteria are developed for assessing the types of sites required and the relative personnel competencies required as well as to objectively evaluate the ability of any particular site to providing appropriate patients in the time frames required for the conduct of the trial. This needs to take into consideration not only the site's capabilities, but knowledge of the local country regulatory/ethics committee perspectives,

the use of centralized ethical review (e.g., Germany), and where these may improve or impede study startup. Often, the PMO may have local knowledge of individual investigator's capabilities, and this will place the PMO in a unique perspective to assist the study team in selecting particular investigators who are able to meet the specific needs of the project. In some situations, the PMO may also directly contact the investigator to review the details of the project to aid the investigator in determining their interest and abilities to pursue the project. The PMO may also have insights into whether the related site personnel, facilities, and the available patient pool are suitable for the specific project.

MEDICAL SUPPORT AT INVESTIGATOR'S MEETING

The PMO can play an important role at the investigator's meeting as a result of their knowledge of the specific therapeutic area, the specifics of the protocol, and the specific needs of the sponsor. They should already be familiar with the unique aspects of the study sites, special concerns that have been raised by regulatory and IRB/ethics, as well as the unique issues raised during discussions with study sites. This is an opportunity to provide consistent interpretation of the protocol and its execution and to demonstrate their ability to be a resource to the study site during the conduct of the study.

MANAGING EMERGING SITE ISSUES DURING ENROLLMENT

To optimize patient outcomes in the setting of acute neuroemergencies, the investigator/clinical specialist has to respond to the medical situation promptly and efficiently. In a like manner, the PMO must be responsive to the sites if the clinical trial sites are going to achieve timely recruitment of subjects and accurate data collection. The PMO must be completely proficient with the specific protocol and aware of issues that are likely to lead to protocol deviations. The PMO should also be aware of the types of subjects that are likely to contribute to AEs. The PMO must be ready to act as decisively in defense of the protocol's integrity as the investigator/clinical specialist acts in recruiting the patients and appropriately assessing their neurological function.

Site inquiries most often occur in the context of a medical crisis, and site personnel must be adequately prepared to appropriately query and request clarification only where necessary. Small variations in protocol procedures may profoundly affect the ability of the sites to get the job done in accordance with the prescribed timelines and thus may have significant impact on the ability of the site to perform the clinical trial. The PMO must be ready to identify which expectations influence the feasibility of the project and communicate the nature of these difficulties to the sponsor. When compromise is not possible, the PMO must be prepared to help the sites understand what kind of resources they may need to make protocol execution occur within their unique environments. Similarly, the PMO can assist the sponsor to understand the difficulties that may occur in the partnership with the site.

The nature of neuroemergency studies is that the enrollment and data collection occur rapidly after symptom onset. Appropriate medically trained personnel, in particular the PMO, must be prepared to be available 24 hours a day, 7 days a week. Many medical consultants (e.g., steering committee members, external consultants) are reluctant to accept this level of day-to-day responsibility. Site investigators may need to have a designated method of obtaining assistance during both the early enrollment phase and at critical time points for the study. The PMO or other

designated personnel must be intimately familiar with the protocol and with protocol interpretations that have been agreed on with the sponsor. Similarities in such interpretation are critical for the success of the study. Consideration should be given to the use of 24-hour call centers, which can appropriately use PMO resources to address the issues while taking into consideration time zone and language differences. An option is to have some of the more standard responses available in a question/answer log form that might be available on a secure website, but this will not usually be the most efficient manner of addressing concerns in the acute neurology trial. Available PMOs are the best manner of dealing with the unique study site concerns in an efficient manner.

As an example, many of the acute stroke projects share many similarities; however, nuances in protocol design may make a project more difficult in certain settings and easier in others. Neuroemergency trials typically have extremely tight protocol timelines. The sequences of tests and scheduled events require that the study staff (e.g., study coordinator and investigator) perform study procedures within minutes of patient arrival. The sequence of evaluations and procedures in these studies is always extremely rigid. Careful consideration of the timing of such events, the staffing required to provide appropriate adherence to these requirements, and the appropriate training to ensure that the study staff can adhere to the protocol and obtain the necessary trial data are areas where the PMO can provide significant input. When the site needs an answer to a question about the protocol, procedures, safety, neurological rating scales, or the study drug, they generally require an immediate answer to facilitate either subject recruitment or the appropriate data collection during the subject visit. The experienced PMO is aware of this and works with the sponsor to reduce ambiguity and to increase the efficiency of the study staff with these procedures and with the clinical scales and ratings used in patient assessment.

The PMO may also provide the requisite medical support for a call center that has been identified to provide clinical support for the study sites and their staff. This is of particular importance for issues regarding inclusion and exclusion that require immediate response. Such a call center should have 24-hour, 7 days per week coverage by appropriately trained personnel. Specific processes should be in place to ensure appropriate documentation is created for these calls in accordance with the sponsor's standard operating procedures. All responsible PMOs should be familiar with the protocol and the specific procedures to be followed in responding to call center requests. The use of such a call center ensures that critical questions that require immediate responses are appropriate triaged and addressed by appropriately trained personnel.

MANAGING SITE INITIATION PROCESS AND BOOSTER ACTIVITIES

The PMO works intensively with the site and with the sponsor during subject identification, screening, and recruitment to help solve problems that arise during the protocol execution. These problems may include ambiguities in interpretation of the protocol or complexities in workflow. The PMO should have a repertoire of potential solutions that have worked in other situations.

The PMO can also be a resource during the study by serving as a point person during booster calls with study sites. Teleconferences during study conduct can continue to expand the knowledge at the study site regarding issues surrounding recruitment and can continue to provide clarification of study protocol issues. The establishment of "best practices" for

protocol execution is an important component of these teleconferences.

GUIDING SAE AND CRF COMPLETION ISSUES

The occurrence of multiple AEs is a frequent issue in acute neurology trials. Patients enrolled into these trials are seriously ill and as such AEs occur quite frequently; in particular, SAEs are relatively frequent. The reporting of both AEs and SAEs in these patients, because frequent new medical events qualify, can become a major component of the study staff responsibility. A significant number of these events are reportable but may simply represent events that are related to the disease or to the patient's general medical condition rather than result from the study drug or device under study. However, all such events should be reported, unless prior agreement is reached with the relevant regulatory authorities, which is typically based on data from earlier trials with the same study agent and in the similar populations. As a result, attribution of causality may be very difficult. Frequently, investigators categorize the event as possibly related to the investigational agent (drug, biologic, or device), but this is determined mostly by the lack of a clear alternative mechanism and in the setting of temporal proximity. The PMO can be a resource to the principal investigator to provide information that can assist them in their decision making. In acute neuroemergency trials in which the disease process maybe ongoing or the clinical sequela may continue to develop, both identifying AEs and classifying their relationship to study drug may be insightful.

CODING OF MEDICAL EVENTS FOR AES AND SAES

The increased demands of appropriately understanding and accurately reporting AEs make the appropriate coding of such events critically important. The sponsor's internal safety department may not be adequately equipped, given the complexity of verbatim reporting, to properly code to MedRA preferred terms. The accuracy of such coding is considered critically important to appreciate properly significant trends in complications secondary to the study manipulations. External medical consultants like the PMO can provide a therapeutically knowledgeable resource that can assist the sponsor's staff or their designee in properly coding these events and in creating the appropriate trending reports to aid in the identification of ongoing safety issues during the study conduct and in appropriately describing these events in subsequent study reports and/or publications.

WRITING/CONTRIBUTING TO THE CLINICAL TRIAL REPORT

Each sponsor has their specific methods for creating a clinical trial report, but the common elements include an adequate description of the study conduct and the pertinent results both from an efficacy and safety perspective. In the case of significant PMO involvement during the conduct of the study, the PMO then is in the unique position to appreciate the nature of the trial and assist the sponsor's staff in designing an analysis plan and in interpreting the study results. The medical expertise they bring to the study implementation may contribute to a comprehensive clinical trial report that will appropriately convey the outcomes of the specific study.

SUPPLEMENTING SPONSOR'S STAFF AT REGULATORY MEETINGS

There are many points of interaction with the relevant authorities within each region of the world. The PMO may be uniquely suited to provide appropriate therapeutic specialization to aid the sponsor in these interactions. Such interactions can take the form of written communications with the authorities regarding the protocol, such as patient selection procedures, clinical procedures to be performed, clinical or surrogate endpoint selection, and/or methods for endpoint analyses. The PMO may be able to provide the therapeutic expertise to make significant contributions to these interactions. At time of completion of significant study phases (e.g., end of phase II or at time of NDA or BLA submission), the PMO may provide important expertise to aid in discussions and negotiations with the regulatory authorities, which may often times be difficult negotiations.

SUMMARY

The PMO can be a significant asset to any clinical trial but is of particular importance in the setting of neuroemergency trials. The support and guidance of the PMO can provide important information to the sponsor's staff and the study site's research team. This can result in significant savings in time and reduction in unnecessary stress, which may lead to improvements in both study quality and speed of completion. In addition, an effective PMO helps to build bridges with study sites and provides the important continuation or linkage into future work. The selection of a PMO or a group of PMOs to assist with the project provides important supplemental skills to the sponsor's medical team and in this manner leaves appropriate time for the sponsor's team to devote attention to other aspects of the drug development process.

13

Ethical Considerations in Neuroemergency Clinical Trials

Wayne M. Alves

Ethical considerations in neuroemergency clinical trials have evolved in keeping with current ethical guidances that take their origin from the Nuremberg Code stemming from the Nuremberg War Trials and the Declaration of Helsinki of the World Medical Association (1–3). Modern neuroemergency clinical trials are highly complex activities that are only distantly related to the so-called experimentation that might be performed in clinical practice by an individual clinician (4,5). Neuroethics challenges of clinical research in neuroscience populations have attracted vigorous debate (6–10). Research ethics considerations in neuroemergency clinical trials, however, play themselves out in the context of testing an evolving treatment hypothesis in a clinical trial. The inherent difficulty in obtaining informed consent from neuroemergency patients remains an active topic of discussion (11–17).

The evolution of a novel therapeutic hypothesis follows a stereotypical path. Typically, a therapeutic observation originates at the patient's bedside, and anecdotal case reports are quickly replaced by larger uncontrolled case series. Large prospective data series, using state-of-the-practice data collection tools, help to more clearly define the medical need and the likely outcomes of patients treated within current best clinical practices. The gold standard for the adoption of new therapeutic interventions is that they first should be subjected to adequate and well-controlled clinical trials to prove both efficacy and safety claims.

Numerous individuals and organizations are affected by ethical considerations in the context of clinical trials. The patient or volunteer subject, the investigators and their staff, the institutions where the research is conducted, the study sponsor, contract research organizations, and regulatory agency staff all have stakes in the ethical conduct of human research. Key ethical principles therefore underlie all clinical trials activities:

- "Respect for persons" guides the informed consent process;
- "Beneficence" leads us to make a benefit assessment of receiving the treatment;
- "Nonmalfeasance" leads us to make an appropriate assessment of risk; and
- "Justice" leads to specific concerns over access to clinical trials and the selection of subjects (1,2).

THE CONCEPT OF RISK

Risk is the dominant ethical concern in the risk-to-benefit consideration (1,2,18). Risk can range from "minimal" to "unacceptable," and sponsors and investigators are required to define the risks associated with the procedures and conduct of a proposed clinical trial. Ideally, this process should start in the early study planning stages at the point the protocol is being developed and should include a consideration of the nonclinical and clinical evaluations performed with the proposed treatment to date. A clinical investigator's brochure serves to codify the nonclinical and clinical knowledge base underlying a new chemical entity or biologic and provide a reasoned assessment of the balance of risks and benefits. Expert and nonexpert opinion may also be of value, especially in relation to pathological findings or side effects experience in earlier phases of development.

Risk must be understood to include any potential physical, mental, social, and economic harm to study participants. It should always be remembered that many neuroemergency patients are vulnerable subjects, and at a minimum the principles for protection of vulnerable populations should be carefully weighed (18–20). Conceptually, risk is the amount or degree of potential damage, its duration, and the extent of irreversibility due to exposure to the proposed novel treatment intervention (1). Models for assessing risk–harm relationships, which include both subjective and objective aspects of harm, can help to provide an estimate of the likelihood and extent of harm that could occur as a result of study participation (1,2). The value of involving the patient (or related persons) in the assessment of risk–harm relationships has been explored (21,22).

The process of risk assessment in acute neuroscience populations is confounded by the inherent risks associated with very complex disease and clinical conditions (1,2).The catastrophic nature or potential of neuroemergencies can make some of the risks associated with treatment seem rather pale in comparison, especially when the side-effect profile of the treatment is fairly good. For example, the mechanisms of damage in stroke and trauma are essentially few in number but often occur in combination (see the chapters in section 1 of this volume, especially Chapters 1 and 4). Also, risks associated with standard treatments are inherent features of the management of neuroemergency patients. Consider for example the risk of vasospasm associated with subarachnoid hemorrhage due to aneurysmal rupture (see Chapter 2). Rescue therapies such as hypertensive, hypervolemic, and hemodilutional (HHH) therapy are effective but carry considerable risks in their own right. Generally, the health utilities of patients are not known and in effect surrogate judgments are often substituted (1,2).

In everyday clinical trials practice, good clinical practice guidelines serve to remind participants of the ethical considerations and responsibilities of both sponsors and investigators in experimental research in humans. Good clinical practice is the gold standard for the design, conduct, performance, monitoring, auditing, recording, analysis, and reporting of clinical trials (23). An investigator's responsibilities in the clinical trials setting relate to ethical considerations. The investigator must follow federal legislation or directives regarding human subjects protect (such as 21 CFR Part 312, 21 CFR Part 812, or the 2001 European Directive on clinical trials). Table 13.1 lists selected ethical responsibilities of the clinical investigator adapted from various regulatory sources (23–25).

THE CONCEPT OF BENEFIT

The term "benefit" carries a societal connotation and reference to the potential direct personal benefit to the individual

TABLE 13.1   Clinical investigator's ethical responsibilities

Safeguard the rights and welfare of each research subject, and ensure that the subject's rights and welfare take precedence over the goals and requirements of the research.

Comply with the standards and requirements stipulated in regulations and research ethics documents and protect the rights and welfare of human subjects involved in research.

Comply with all other national, state, and local laws or regulations that may provide additional protection for human subjects.

Accept the final authority and decisions of the IRB, including directives to terminate participation in designated research activities.

Promptly report to the IRB proposed changes in the research conducted.

Do not initiate changes in the research without prior IRB review and approval, except where necessary to eliminate apparent immediate hazards to subjects.

Report immediately to the IRB any unanticipated problems in research that involve risks to subjects or others.

Seek, document, and maintain records of informed consent from each subject or the subject's legally authorized representative as required.

Cooperate in the IRB responsibility for initial and continuing review, record keeping, reporting, and certification.

Source: Adapted from Code of Federal Regulations 21 CFR 312.60.

study participant. In the context of experimental therapeutics, the direct benefit to the patient is generally unknown and typically considered unlikely in the specific clinical trial in which patients are participating. This creates an ethical conundrum as modern evolving standards, such as the 2001 European Union Directive on Clinical Trials, create potential conflicts with the clinical equipoise supporting the investigation of a novel treatment (26–30). Further, study participation in placebo or nontreatment controlled trials also means a pre-planned proportion of subjects will not receive other than the current standard of care at least until they complete the double-blind portion of the study and qualify for an open-label extension study, assuming one is offered (31).

The standard of care for neuro-emergency populations is complex and often idiosyncratic. Individual practice variations in hospital-based settings (e.g., emergency departments and neurological intensive care units) make for a plethora of examinations, drugs, and supportive interventions. Treatment decisions are often idiosyncratic, and there are few gold standards. Consequently, subjective definitions and perceptions are very important in guiding treatment decisions. Because modern clinical trials in neuroemergency populations are large, they take time to conduct and analyze and there is a danger that the rate of change in standard or background management could change during the clinical trial's duration.

Although reasoned ethical discussions of the potential benefits of a novel treatment often conclude that patients seldom directly benefit much from participation in specific clinical trials, clinician-scientists counter that the improved intensity and quality of care due to increased clinical surveillance during the trial often results in reduced mortality and morbidity of subjects as an unanticipated outcome of clinical trial participation. The oft stated shibboleth that patients who are in a clinical trial do considerably better than the natural history of the disease confirms this belief. Data are not systematically available to document this assertion. In addition, the placebo effect is also a well-established, albeit debated, phenomenon (32,33).

The issue of personal benefit to the clinician-investigator is also controversial, and disclosure of financial interest in a novel treatment or sponsor is now a regulated phenomenon. Regulations require the investigator to divulge relationships to sponsors and other potential sources of conflict of interest. Journal editors and professional associations are increasingly managing conflict of interest considerations. In addition, there is the conflict of interest inherent in the clinical and scientific biases of the investigators themselves that must be acknowledged and honestly confronted. The moral context of medicine and emphasis on dedication to patients and the advancement of medical science help to temper this (1,2).

STUDY DESIGN CONSIDERATIONS

Poorly designed, conducted, or reported trials are arguably unethical regardless of the degree of risk (1,2,34–36). Many institutions have developed research governance policies to ensure that the scientific quality of the proposed research is consistent with current clinical, scientific, and ethical standards (34,35,37,38). The need for the clinical trial, the recruitment and screening of subjects to be enrolled, whether a placebo treatment group is required, the use of active control groups, and procedures for controlling bias and deception have ethical, as well as scientific and clinical, considerations such as sufficient statistical power and adequate sample sizes (32,38,39). Similar considerations are relevant for early stopping of a study for either safety or efficacy reasons.

EVALUATION OF A THERAPEUTIC HYPOTHESIS

The randomized, double-blind, controlled, clinical trial is the gold standard for the acceptance of a new therapeutic hypothesis as proven. The regulatory definition of an "adequate and well controlled" clinical trial is presented in Table 13.2. The quintessential feature of a randomized clinical trial is the mechanisms used for avoidance of selection or treatment assignment bias. The achievement of "balance" in baseline characteristics and potentially confounding factors across treatment groups has proven difficult to achieve in neuroscience populations. Small apparently inconsequential imbalances may lead to failed trials. Randomization procedures can also pose an ethical dilemma,

TABLE 13.2   Elements of an adequate and well-controlled clinical trial

---

Clear statement of the objectives
Valid design and appropriate control
Proper selection of patients having the disease or condition
Proper treatment assignments (randomization)
Adequate measures to minimize bias (w.r.t. subjects, observers and analysis of data)
Well-designed and reliable methods of measuring responses
Data adequate to assess the effects of the drug

Source: Code of Federal Regulations 21 CFR 314.126 (a)–(e).

especially when placebo control is used. The "equipoise principle" demands all patients receive at least the standard of care. Clinical equipoise is a situation in which there is genuine clinical uncertainty regarding the relative merit of a novel treatment. The search is for a treatment superior to standard of care (19,40). This ethical calculation may differ in the context of diseases for which there are numerous proven treatments available to the clinician.

The emergence of treatment guidelines, for example, the guideline of the American or European Brain Injury Consortia (see Chapter 4), helps to provide a context for good scientific practice and practical ethics. The existence of these guidelines is not sufficient in their own right. There is also a need to confirm adherence to treatment guidelines. The American Brain Injury Consortium Technical Center provides an outstanding example of how to do this in the context of clinical trials.

Events during a clinical trial, such as the emergence of unexpected serious adverse events, may disturb the clinical equipoise present at the start of the study. Achieving an early signal of efficacy can also disturb equipoise. These situations have led to methodologies for early stopping rules and to adaptive study designs to ensure earliest possible detection of confirmation of success or failure of a therapeutic hypothesis. In practical terms, as side effects and serious adverse events are monitored, there may be a need to unmask treatment for clinical reasons or to conduct interim investigations for safety or efficacy. This is generally done by an independent data monitoring committee or data safety and monitoring board (see Chapter 11 for a full discussion of the role of the data safety and monitoring board).

USE OF PLACEBO CONTROL SUBJECTS

Arguments that the use of placebo control subjects may be unethical have arisen in the context of clinical trials (32,33). The need to control experimental and/or participant (both subject and investigator) bias creates the need for randomization. The "placebo" control subject is the gold standard for comparison, even in the presence of proven competing drugs because assay sensitivity is an important consideration. Control subjects are used in nonrandomized designs as well. A requirement for current standard of care as background treatment in a clinical trial is complicated in neuroemergency populations because, as mentioned above, treatment decisions can be largely idiosyncratic. This requires that we ensure balance in assignment to treatment arms. Add-on trial designs are used in which all patients receive the same standard of care and some are assigned to the novel treatment and others are assigned to placebo (or active) comparator.

STATUTORY BASIS OF CURRENT CLINICAL TRIALS REGULATIONS

Key federal legislation that is the statutory basis for today's regulations on drug, biologics, and device development have generally stemmed from unfortunate ethical lapses or wider concern over the safety of the consumer. Table 13.3 lists some key

milestones in U.S. regulatory history that illustrate the interrelationship between wider societal concerns and the legislation meant to address them. For example, in 1937 the use of diethylene glycol as a solvent in the elixir sulfanilamide led to over 100 deaths, largely of children. The 1938 Food, Drug and Cosmetics (FD&C) Act was enacted and established the proof of safety requirement for marketed products. Similarly, the tragic results stemming from the use of the sleeping pill thalidomide in the 1950 and 1960s stimulated the 1962 Kefauver-Harris Amendments to the 1938 FD&C Act, providing the statutory basis for requiring premarket testing for both safety and efficacy.

GENERAL ROLE OF THE INSTITUTIONAL REVIEW BOARD

Institutional review boards (IRBs) must consist of at least five or more individuals, including one unaffiliated with the institution. The selection and composition is specified in the 21 CFR Part 56. IRBs are authorized to invite individuals with special competencies to participate on an ad-hoc basis. An IRB review and approval must occur before study procedures are performed and includes review of the informed consent forms accompanying the study protocol. IRBs also monitor, through their reporting mechanisms, the activities of researchers at their respective institutions.

All investigational studies of novel treatments must undergo IRB review. This includes studies that are wholly investigational (i.e., randomized trials), those studies with treatment elements included (e.g., open-label safety or continuation studies), studies primarily intended for treatment (treatment protocols), and single patient protocols (novel ideas, compassionate use protocols). Table 13.4 lists the general criteria IRBs use in determining approval. The IRB may approve, require

TABLE 13.3   Drug and biologics regulatory milestones

19th Century	Rise of Ethical Pharmaceuticals companies in the United States
1820	*U.S. Pharmacopeia* established as national compendium of standard drugs
1846	Lewis Caleb Beck's *Adulteration of Various Substances Used in Medicine and the Arts* addresses problems with nation's largely imported drug supply
1848	*Drug Importation Act* stops entry of adulterated drugs into the United States
1902	Creation of the Drug Laboratory, Bureau of Chemistry (precursor to FDA)
1905	Samuel Hopkins Adams "The Great American Fraud" in *Collier's Magazine* exposes the patent medicine industry
	American Medical Association Council on Pharmacy and Chemistry launches a voluntary evaluation of manufacturer claims
1906	Upton Sinclair's *The Jungle* exposes unsafe practices in the Food-packing industry *Food and Drugs Act* June 30, 1906
1907	*Biologics Control Act* mandating purity of serums and vaccines
1912	*Sherley Amendment* brings therapeutics claims within the jurisdiction of the Food and Drugs Act of 1906
1930	Food and Drug Administration named (formerly the Food, Drug, and Insecticide Administration in the Bureau of Chemistry)
1937	107 die (mostly children) when poisonous solvent (diethylene glycol) is used in the elixir preparation sulfanilamide
1938	*Food, Drug, and Cosmetics Act (FD&C Act)* expands the FDA's jurisdiction over regulation of drugs, requiring proven safety
1962	Sleeping pill *thalidomide* found to have cause over 10,000 birth defects and unknown numbers of miscarriages and fetal effects
	Kefauver-Harris Amendments to FD&C Act of 1938 (Bill S1552 Oct. 10, 1962) requires premarket testing and approval providing evidence of both safety and efficacy
	World Health Organization Resolution proposing international standards for product development 24 May 1962
1970	Definition of *"substantial evidence"* for efficacy and safety published in *Federal Register* May 8, 1970
	FDA requires first *package insert* to provide information on risks and benefits of oral contraceptives
1990	Initiation of *International Conference on Harmonization* to define common global standards for drug development, including a common technical document for submissions for market approval
1991	FDA regulations to accelerate review of drugs for life-threatening diseases
1992	*Prescription Drug User Fee Act*
1997	*Food and Drug Administration Modernization Act* provides for accelerated review of new drug application and controls for off-label advertising

Sources: Adapted from Center for Drug Evaluation and Research, *FDA Time Line: Chronology of Drug Regulation in the United States* (www.cder.gov); *Milestones of FDA History* (www.fda.gov/oc/history); International Conference on Harmonization (www.ich.org).

modifications to secure approval, or disapprove a protocol. As part of their review they may request information cited in the regulations (or their own requirements), require documentation of informed consent, or may waive consent; they notify the investigator and the institution of their decisions in writing. The investigator must be provided with the reasons for disapproval and be allowed to respond. Continuing review typically occurs annually, but the IRB is free to require more frequent review for projects it deems appropriate.

OVERVIEW OF HUMAN SUBJECTS PROTECTION REGULATIONS

Today, the IRB or research ethics committee serves to moderate the problems and issues generated by modern clinical trial designs and to promote resolution of ethical conflicts associated with clinical

TABLE 13.4 General criteria for IRB approval of clinical trial protocols

Risks are minimized.

Risks are reasonable in relation to anticipated benefits and findings.

Selection of subjects is equitable.

Informed consent is sought.

Informed consent is appropriately documented.

The research plan makes provisions for monitoring data collected to ensure subject safety.

Adequate provisions are made for privacy and data confidentiality (recently enhanced by HIPAA legislation and regulations).

Additional concerns and safeguards for "vulnerable" populations.

HIPAA, Health Insurance Portability and Accountability Act of 1966.

Source: Code of Federal Regulations 45 CFR (Subpart A) 46.111.

experiments in acute neuroscience populations. The IRB and ethics committee conduct their activities under the statutes encoded in national regulations or law and widely accepted codes of ethics. Local norms and customs also influence IRB or ethics committee behavior and decisions (41). The clinician-trialist should consider and justify the proposed trial in their local practice setting before proceeding, a process that essentially is an evaluation of the risk-to-benefit ratio of the proposed therapeutic intervention. Clinical trial design issues, as well as informed consent issues, weigh heavy in this ethical calculation.

The Office for Human Research Protections (OHRP) institutional review board guidebook provides basic information for each IRB as well as the investigator and his or her staff. In the United States, human subjects' research protections are required under the Department of Health and Human Services (HHS) as set forth in the Code of Federal Regulation 45 CFR 46. Subpart A of the HHS regulations constitutes the Federal Policy (Common Rule) for the Protection of Human Subjects, whereby institutions provide assurance they will comply with the HHS human subjects regulations (45 CFR 46.103(a)).

The IRB should determine that the risks to subjects are reasonable in relation to anticipated benefits (21 CFR 56.111(a)(2)) and that the consent document contains an adequate description of the study procedures (21 CFR 50.25(a)(1)) as well as the risks (21 CFR 50.25(a)(2)) and benefits (21 CFR 50.25(a)(3)). Although there is no specific regulatory requirement to do so, a clinical investigator's brochure is usually submitted to the IRB. There are regulatory requirements for submission of information that normally is included in the investigator's brochure.

In the United States, all research must be reviewed and approved by an IRB designated under an OHRP-approved assurance. Except where the IRB specifically approves a waiver in accordance with HHS regulations, no investigator may involve a human being as a subject in research unless the investigator has obtained the legally effective informed consent of the subject or the subject's legally authorized representative. The meaning of "legally effective" and "legally authorized" is determined in part by applicable state laws.

There are various regulatory checks and balances to ensure adequate protection of human subjects. The sponsor must assure the U.S. Food and Drug Administration (FDA) that a study will be conducted in compliance with the informed consent and IRB regulations (21 CFR parts 50 and 56, see 21 CFR 312.23(a)(1) (iv)). It is not a sponsor's obligation to determine IRB compliance with the regulations. Rather, sponsors rely on the clinical investigator, who assures the sponsor on form FDA-1572 for drugs and biologics, or the investigator agreement for devices, that the study will be reviewed by an IRB.

An IRB notifies an investigator in writing of its decision to approve, disapprove, or request modifications in a proposed research activity (21 CFR 56.109(e)). A copy of this correspondence is then provided to the sponsor. The IRB is responsible for ensuring that informed consent documents include the extent to which the confidentiality of medical records will be maintained (21 CFR 50.25(a)(5)). In addition, the FDA requires sponsors (or the research monitors hired by them) to monitor the accuracy of the data submitted to FDA in accordance with applicable regulatory requirements.

The extent to which confidentiality of subject-related information is maintained may affect a subject's decision to participate in a clinical investigation. These data are in the possession of the clinical investigator, and each subject must be advised during the informed consent process of the extent to which confidentiality of records identifying the subject will be maintained and of the possibility that the FDA and the sponsor or its designee may inspect the records. Although FDA access to medical records is a regulatory requirement, subject names are not usually requested by FDA unless the study records of particular individuals require a more detailed study of the cases or unless there is reason to believe that the study records do not represent actual cases or actual results obtained.

INFORMED CONSENT PROCEDURES

In general, informed consent procedures must be legally effective and sufficient to allow the potential subject to consider whether or not to participate and should minimize the possibility of coercion or undue influence. Information should be provided in understandable language, and there should be no exculpatory language that either waives or appears to waive the subject's legal rights or releases or appears to release investigator, sponsor, or institution from liability of negligence. Failures to properly obtain informed consent are well known and have hurt the ability to get some populations in society to participate, for example, the deliberate infection of children in the Willowbrook experiments

or the distrust the African-American community has in light of the Tuskegee syphilis experiment. Several excellent websites provide detailed reviews of major cases in the history of research ethics (see Table 13.8).

To ensure adequate informed consent, patients should know and understand the reasons for the treatment and their participation and the benefits, risks, and alternatives to participation. But there are no guarantees in attempts to meet these standards. Modern medical interventions are complicated and highly technical. As such, the onus is on the investigator to ensure adherence to this standard. The necessary elements of an informed consent form are listed in Table 13.5 and involve elements of information, comprehension, and volunteerism. In neuroemergencies the ability to obtained informed consent from the patient is typically compromised by their disease, making it important to ensure that a legally authorized representative is appropriately consulted (11–17).

Informed consent is documented by the use of a written consent form preapproved by the IRB and signed by the person obtaining the consent and the subject or the subject's legally authorized representative after ample opportunity is provided to read and understand content. A copy of the informed consent form is given to the person signing the form on behalf of the subject. If information regarding the study, its conduct, and associated risks and benefits are provided orally, a short form written consent document can be used, stating that the elements of informed consent required by 45 CFR 46.116 have been presented orally to the subject or the subject's legally authorized representative. An independent witness is required when this method is used to obtain consent. The independent witness must be someone with no direct involvement with the study. If a written summary of the information provided orally is provided, only the

TABLE 13.5 Elements of the informed consent form

Basic elements:

- A statement that the study involves research
- An explanation of the purposes of the research
- The expected duration of the subject's participation
- A description of the procedures to be followed
- Identification of any procedures that are experimental
- A description of any reasonably foreseeable risks or discomforts to the subject
- A description of any benefits to the subject or to others that may reasonably be expected from the research
- A disclosure of appropriate alternative procedures or courses of treatment, if any, that might be advantageous to the subject
- A statement describing the extent, if any, to which confidentiality of records identifying the subject will be maintained
- For research involving more than minimal risk, an explanation as to whether any compensation and an explanation as to whether any medical treatments are available, if injury occurs and, if so, what they consist of, or where further information may be obtained
- An explanation of whom to contact for answers to pertinent questions about the research and research subjects' rights, and whom to contact in the event of a research-related injury to the subject
- A statement that participation is voluntary, refusal to participate will involve no penalty or loss of benefits to which the subject is otherwise entitled, and the subject may discontinue participation at any time without penalty or loss of benefits, to which the subject is otherwise entitled

Additional elements when appropriate:

- A statement that the particular treatment or procedure may involve risks to the subject (or to the embryo or fetus, if the subject is or may become pregnant), which are currently unforeseeable
- Anticipated circumstances under which the subject's participation may be terminated by the investigator without regard to the subject's consent
- Any additional costs to the subject that may result from participation in the research
- The consequences of a subject's decision to withdraw from the research and procedures for orderly termination of participation by the subject
- A statement that significant new findings developed during the course of the research, which may relate to the subject's willingness to continue participation, will be provided to the subject
- The approximate number of subjects involved in the study

Source: Code of Federal Regulations, 45 CFR 46.116. See also, 21 CFR 50.25.

short consent form itself is to be signed by the subject or the representative. The witness signs both the short consent form and a copy of the summary, and the person actually obtaining consent also signs a copy of the summary. A copy of the summary is given to the subject or the representative, in addition to a copy of the short consent form.

"Vulnerable" populations have been an explicit consideration in the regulations and guidelines surrounding the informed consent process. Children, prisoners, pregnant women, mentally disabled persons, and economically or educational disadvantaged persons are considered vulnerable populations. In general, neuroemergency

populations should also be considered vulnerable, especially during the acute period of their disease. Often, for a considerable time during the lengthy recovery characteristic of neuroemergencies, the patient remains vulnerable (16,17).

WAIVER OF CONSENT

An IRB may waive the requirement for the investigator to obtain a signed consent form for some or all subjects. The requirements for waived consent include the following:

- The patient is in a life-threatening situation.

- It is impossible to obtain consent from the patient.
- There is not sufficient time to obtain consent from the patient's legal representative.
- There is no therapy available that provides a greater chance of saving the patient's life.

There must be certification by the investigator that these conditions are met and by another physician not participating in the study. This certification must occur before administration of the test article.

A waiver of informed consent may also be granted if the IRB finds either of the following:

1. The only record linking the subject and the research would be the consent document, and the "principal risk" would be potential harm resulting from a breach of confidentiality. Each subject will be asked whether the subject wants documentation linking the subject with the research, and the subject's wishes will govern.
2. The research presents "no more than minimal risk" of harm to subjects and involves no procedures for which written consent is normally required outside of the research context.

In cases in which the consent documentation requirement is waived, the IRB may require the investigator to provide subjects with a written statement regarding the research. It should be noted that once waiver is given, the investigator does not obtain consent at a later date.

EMERGENCY RESEARCH CONSENT WAIVER

It is typically the case in neuroemergencies that research cannot practicably be carried out under the conventional informed consent process, for example, in the case of interventions developed for out-of-hospital cardiac arrest or traumatic brain injury. In these cases, an emergency research consent waiver may be granted for a class of research consisting of activities, each of which meet strictly limited conditions detailed under 21 CFR 50, and the requirements for exception from informed consent for emergency research detailed in 21 CFR 50.24 have been met. The IRB responsible for the review, approval, and continuing review of the research must approve both the research and a waiver of informed consent. The emergency research consent waiver regulations require the IRB to find that the human subjects are in a life-threatening situation; available treatments are unproven or unsatisfactory; and the collection of valid scientific evidence, which may include evidence obtained through randomized placebo-controlled investigations, is necessary to determine the safety and effectiveness of particular interventions. The emergency waiver rules are well codified, and the guidance provided by OHRP spells out how to implement appropriate procedures to be compliant with the regulations, including public disclosure of the research and its findings.

RESEARCH INVOLVING CHILDREN

Specific requirements for protection of children that are subjects in research studies are provided in the Code of Federal Regulations 45 CFR 46 Subpart D—Additional DHHS Protections for Children Involved as Subjects in Research. Subpart D of the HHS regulations requires additional protections for research involving children: "Children" are persons who have not attained the legal age for consent in the jurisdiction in which the research is conducted (45 CFR 46.402(d)). The IRB must find that the research activity represents one of four permissible categories of research and that adequate provisions are made for soliciting the assent of the children and the permission of each child's parents or guardian (45 CFR 46.404–408). Children who are wards of the state or any other agency, institution, or entity can be included in research only under certain conditions (45 CFR 46.409).

The IRB shall determine that adequate provisions are made for soliciting the assent of the children, when in the judgment of the IRB the children are capable of providing assent. If the IRB determines that the capability of some or all of the children is so limited they cannot reasonably be consulted or that the intervention or procedure involved in the research holds out a prospect of direct benefit that is important to the health or well-being of the children and is available only in the context of the research, the assent of the children is not a necessary condition for proceeding with the research. Even where the IRB determines that the subjects are capable of assenting, the IRB may still waive the assent requirement under circumstances, in which consent may be waived in accord with section 46.116 of Subpart A.

The IRB may find that the permission of one parent is sufficient for research to be conducted under sections 46.404 or 46.405. Where research is covered by sections 46.406 and 46.407 and permission is to be obtained from parents, both parents must give their permission, unless one parent is deceased, unknown, incompetent, or not reasonably available, or when only one parent has legal responsibility for the care and custody of the child.

If the IRB determines that a research protocol is designed for conditions or for a subject population, for which parental or guardian permission is not a reasonable requirement to protect the subjects (e.g., neglected or abused children), it may waive the consent requirements provided that an appropriate mechanism for protecting the children who will participate as subjects in the research is substituted and provided further that the waiver is not inconsistent with federal, state, or local law.

OTHER CONSENT MECHANISMS

Mentally incapacitated subjects are the norm in acute neuroscience populations, and the use of surrogates or legally authorized representatives is necessary. The investigator must be cautious because the decision to allow the mentally impaired subject to participate in the proposed study is not a best interest decision on the part of a proxy. Is it known whether the subject have wanted to participate in such a study. In the absence of prior written documentation, there is inherent difficulty in knowing what a person would want, as was illustrated in the recent Terri Schiavo case in the United States (13,42).

DEFERRED CONSENT

The concept of "deferred consent" arose in the context of the brain resuscitation trials pioneered by Dr. Peter Safer of the University of Pittsburgh (43–49). Out-of-hospital cardiac arrest wherein the brain is in imminent danger of irreversible damage is perhaps the *sine qua non* for this model for consent. The underlying idea is to immediately initiate the experimental treatment and seek informed consent as soon as it can be practically obtained from the patient or a legally authorized representative. The Office for Protection from Research Risk (OPRR, now OHRP) issued a decision that deferred consent does not meet HHS regulations for prospective informed consent (50). In their decision, OPRR legal staff indicated that waived consent would be potentially justifiable in the situation of a neuroemergency if the general conditions for waiver described earlier are met.

Deficiencies in consent forms must be guarded against. An FDA audit of 1521

TABLE 13.6   Data audit findings

Inadequate consent forms	52%
Named contacts	45%
Description of procedures	34%
Confidentiality statement	28%
Compensation/treatment for injury	29%
Alternative procedures/treatment	15%
Inadequate drug accountability	29%
Protocol nonadherence	26%
Record inaccuracy	20%
Records nonavailability	3%
IRB not kept informed	8%
Prohibited concomitant treatment	4%
Other deficiencies	10%

Source: Adapted from ref. 51.

regulatory submissions reviewed by FDA staff indicated that inadequate consent forms was a reason for an nonfileable decision in half the cases reviewed (52%) (51). Table 13.6 is adapted from a report from the OPRR and provides an interesting view of why regulatory submissions are not initially accepted by the agency.

IRB KNOWLEDGE OF LOCAL CONDITIONS

HHS regulations at 45 CFR 46.107 require that IRBs be knowledgeable about the local research context: The IRB must be sufficiently qualified through the experience and expertise and diversity of its members, including race, gender, cultural background, and sensitivity, to such issues as community attitudes to promote respect for its advice and counsel. The IRB must be able to evaluate research in terms of institutional commitments and regulations, applicable law, and standards of professional conduct and practice.

FINANCIAL COMPENSATION ISSUES

There are numerous issues regarding the role of financial compensation in the

clinical trial process. The financial interest the investigators may have in the product under development or the extent they are compensated for their participation, the compensation of subjects for their participation in the study, and charging the participant for the investigative treatment are all issues of concern.

Investigator Disclosure

Currently, sponsors of a study must submit to the FDA, as part of the marketing application, disclosures completely and accurately describing any financial arrangement entered into between the sponsor of the study and the clinical investigators involved in the conduct of the clinical trial (see 21 CFR Part 54). Of special interest is whether the value of the compensation to the clinical investigator for conducting the study is influenced by the outcome of the study. The investigators must disclose any significant payments of other sorts from the sponsor (grants to fund ongoing research, compensation in the form of equipment, retainers for ongoing consultation, or honoraria). The investigators must also disclose any proprietary interest they or their immediate family have in the tested product and significant equity interests in the sponsor. If there are disclosed arrangements, interests, or payments, the investigators must disclose steps taken to minimize the potential for bias.

Payment to Research Subjects

It is not uncommon for research subjects to be paid for their participation in studies, especially in the early phases of investigational drug, biologic, or device development where there is a greater reliance on healthy volunteers. Payments to research subjects for participation in studies generally is not considered a benefit; it is a recruitment incentive. Financial incentives are often used when health benefits to subjects are remote or nonexistent. The amount and schedule of all payments is presented to the IRB at the time of initial review. The IRB should review both the amount of payment and the proposed method and timing of disbursement to ensure that neither is coercive or present undue influence on the subject (21 CFR 50.20).

Credit for payment accrues as the study progresses and is not be contingent upon whether the subject completes the entire study. Unless it creates undue inconvenience or a coercive practice, payment to subjects who withdraw from the study may be made at the time they would have completed the study (or completed a phase of the study) had they not discontinued early. For example, in a study lasting only a few days, an IRB may find it permissible to allow a single payment date at the end of the study, even to subjects who had withdrawn before that date.

Although the entire payment should not be contingent upon completion of the entire study, payment of a small proportion as an incentive for completion of the study is acceptable to the FDA, providing that such incentive is not coercive. The IRB should determine that the amount paid as a bonus for completion is reasonable and not so large as to unduly induce subjects to stay in the study when they would otherwise have withdrawn. All information concerning payment, including the amount and schedule of payment(s), must be set forth in the informed consent document.

Charging for Investigational Drugs and Biologics

In general, we do not charge participants in clinical trials for the costs of providing the treatment. In addition, it is typical to provide continued access to study medications for patients who respond to treatment, however response is defined. This is usually accomplished through an open-label extension study open to those who complete a double-blind study. It is reasonable to require patients who enter the open-label extension to have completed the prior double-blind study. To not do so would make it difficult to know whether the patient is truly a "responder" or might create a situation making it difficult to interpret the findings of the double-blind study (i.e., risk of failed trial).

The investigational new drug (IND) regulations (21 CFR 312.7(d)) permit a sponsor to charge for an investigational drug or biologic that has not been approved for marketing only under very specific conditions. The charge should not exceed an amount that is necessary to recover the costs associated with the manufacture, research, development, and handling of the investigational drug or biologic. FDA may withdraw authorization to charge if the Agency finds that the conditions underlying the authorization are no longer satisfied.

The sponsor may not charge for an investigational drug or biologic in a clinical trial under an IND without the Agency's prior written approval. In requesting such approval, the sponsor must explain why a charge is necessary, that is, why providing the product without charge should not be considered part of the normal cost of conducting a clinical trial (21 CFR 312.7(d)(1)).

At times the provision of experimental therapies on a compassionate use basis is desired. The mechanism for doing so is the treatment protocol or treatment IND typically held by the treating clinician. The sponsor or investigator may charge for an investigational drug or biologic for a treatment use under a treatment protocol or treatment IND, as outlined in 21 CFR 312.34 and 312.35, provided

1. There is adequate enrollment in the ongoing clinical investigations under the authorized IND;

TABLE 13.7   Key research ethics documents

Oath of Hippocrates, *Hippocratic Writings*, translated by J. Chadwick and W. N. Mann, Penguin Books, 1950.

The Nuremberg Code, *Trials of War Criminals before the Nuremberg Military Tribunals under Control Council Law No. 10*. Nuremberg, October 1946–April 1949. Washington, DC: U.S. G.P.O, 1949–1953.

The Belmont Report, *Ethical Principles and Guidelines for the Protection of Human Subject of Research*. National Commission or the Protection of Human Subjects of Biomedical and Behavioral Research. Washington, DC, 1979

Federal Policy for Protection of Human Subjects, Public Health Services (PHS) Act 45 CFR 46

FDA Regulations 21 CFR Parts 50 and 56, 54, 312 and 314, 601, 812, 814

   Human Subject Protection (Informed Consent) (21 CFR Part 50)

   Additional Safeguards for Children in Clinical Investigations of FDA-Regulated Products (Interim Rule) (21 CFR Part 50, subpart D)

   Financial Disclosure by Clinical Investigators (21 CFR Part 54)

   Institutional Review Boards (21 CFR Part 56)

IOM Report: Responsible Research: A Systems Approach to Protecting Research Participants. Institute of Medicine, National Academies of Sciences, Oct. 3, 2002.

The Privacy Rule, A Health Insurance Portability and Accountability Act of 1996 (HIPAA)

ASSERT, "A Standard for the Scientific and Ethical Review of Trials" and

CONSORT, "Consolidated Standards for Reporting Trials". Mann H. Research ethics committees and public dissemination of clinical trial results. *The Lancet*, Aug 3, 2002: 406–408.

TABLE 13.8   Websites of interest for good clinical practice and research ethics

Website	URL
FDA websites	
Bioresearch Monitoring Information System File: Clinical Investigators, CROs and IRBs from FDA 1571 and 1572s	www.fda.gov/cder/foi/special/bmis/index.htm
CDER Center for Drug Evaluation and Research Guidance Documents	www.fda.gov/cder/guidance/index.htm
CBER Center for Biologics Evaluation and Research	www.fda.gov/cber
CDRH Device Advice	www.fda.gov/cdrh/devadvice/
CDRH Bioresearch Monitoring	www.fda.gov/cdrh/comp/bimo.html
Clinical Investigator Disqualifications Proceedings	www.fda.gov/foi/clinicaldis/
Expedited Safety Reporting Requirements Oct. 7, 1997 Federal Register Final rule	www.fda.gov/cder/regulatory/
FDA Debarred Persons List	www.fda.gov/ora/compliance_ref/debar/
FDA Disqualified/Restricted/Assurances Lists for Clinical Investigators	www.fda.gov/ora/compliance_ref/bimo/ dis_res_assur.htm
FDA Letters Providing Clinical Investigators with Notice of Initiation of Disqualification Proceedings and Opportunity to Explain	www.fda.gov/foi/nidpoe/default.html
FDA Modernization Act of 1997	www.fda.gov/cder/guidance/105–115.htm
CDRH Guidance	www.fda.gov/cdrh/modact/modguid.html
CDER-Related Documents	www.fda.gov/cder/fdama/
FD&C Act	www.fda.gov/opacom/laws/fdcact/fdctoc.htm
	www.fda.gov/cder/guidance/guidance.htm# International Conference on Harmonisation
International Conference on Harmonisation	
Investigational Device Exemptions (IDE) Policies and Procedures	www.fda.gov/cdrh/ode/idepolcy.pdf
Laws Enforced by FDA	www.fda.gov/opacom/laws/lawtoc.htm
MedWatch	www.fda.gov/medwatch/
Pediatric Medicine Page	www.fda.gov/cder/pediatric/
Warning letters	www.fda.gov/foi/warning.htm

(Continued)

TABLE 13.8 *Continued*

Website	URL
Non-FDA websites	
EMEA European Medicines Agency	http://www.emea.eu.int/
Clinical Trials Registry	www.clinicaltrials.gov
Government Printing Office (Federal Register, Code of Federal Regulations, Congressional Record)	www.access.gpo.gov/su_docs/
Institute of Medicine of the National Academy of Sciences	www.iom.edu/IOM/IOMHome.nsf/Pages/ human+research+protections
National Bioethics Advisory Commission (The charter for this commission has expired and the site is archived at this URL.)	http://bioethics.georgetown.edu/nbac/
National Human Research Protections Advisory Committee	Ohrp.osophs.dhhs.gov/nhrpac/nhrpac.htm
Office for Human Research Protections	Ohrp.osophs.dhhs.gov
OHRP IRB Guidebook	Ohrp.osophs.dhhs.gov/irb/irb_guidebook.htm
PHS List of Investigators Subject to Administrative Action: Research ethics websites	silk.nih.gov/public/cbz1bje.@www.orilist.html
Clinical research ethics training by NIH Clinical Center via the National Cancer Institute website. Certificate that documents training is provided	http://www.cc.nih.gov/researchers/training/ crt.shtml or http://cancer.gov/clinicaltrials/learning
NIH online training, *Introduction to the Responsible Conduct of Research*	http://researchethics.od.nih.gov
Central Office for Research Ethics Committees, United Kingdom	www.corec.org.uk
Family Health International site on international clinical research	http://www.fhi.org/en/RH/trainmat/ ethicscuri/index.html
Genome project site with useful information on ethical issues relating to genotyping research	www.genome.gov/PolicyEthics/
Site of Public Responsibility in Medicine and Research and the Applied Research Ethics National Association	www.primr.org
General bioethics site	www.bioethics.com/researchethics
Ethica, an international research ethics site	www.hf.uib.no

NIH, National Institutes of Health.

Source: FDA Information Sheets. Guidance for Institutional Review Boards and Clinical Investigators, 1998 Update (Revised May 10, 2001) and internet search by author.

2. Charging does not constitute commercial marketing of a new drug for which a marketing application has not been approved;
3. The drug or biologic is not being commercially promoted or advertised;
4. The sponsor is actively pursuing marketing approval with due diligence.

The FDA must be notified in writing before commencing any such charges for drug supply. Authorization for charging goes into effect automatically 30 days after receipt of the information by FDA, unless FDA notifies the sponsor to the contrary (21 CFR 312.7(d)(2)).

IMPORTANT RESOURCES FOR RESEARCH ETHICS TRAINING

The research ethics environment is constantly evolving, and stakeholders in the clinical development process must strive to monitor and assimilate new ideas and interpretations of relevant ethical issues. Numerous recent initiatives in international research agenda and research ethics training in the case of neuroemergency populations offer considerable encouragement in achieving harmonization with evolving regulatory guidances (4,52–55). Table 13.7 lists some of the key research ethics documents that should serve as a starting point for newer investigators and research staff. Seasoned investigators and their research staff should be familiar with these documents as part of their research ethics training. Most research institutions require this training as part of their policies on human subjects investigations. Table 13.8 provides some useful research ethics websites as well as excellent on-line research ethics training and certification. Researchers should become familiar with the content of regulatory websites. Useful historical materials on ethical issues in pharmaceutical development are available on these sites, and on-line training provides a mechanism for self-certifying research ethics training.

References

1. Alves WA, Macciocchi SN. Ethical considerations in clinical neuroscience. Current concepts in neuroclinical trials. Stroke 1996;27:1903–1909.
2. Macciocchi SN, Alves WA. Ethical considerations in neuroclinical trials. Neurosurg Rev 1997;20: 161–170.
3. World Medical Association. Declaration of Helsinki: Ethical Principles for Medical Research Involving Human Subjects. Ferney-Voltaire: France: The World Medical Association, June 1964 (Last Amended October, 2000; Last Clarification, 2004).
4. Barsan WG, Pancioli AM, Conwit RA. Executive summary of the National Institute of Neurological Disorders and Stroke conference on Emergency Neurologic Clinical Trials Network. Ann Emerg Med 2004;44:407–412.
5. National Commission for the Protection of Human Subjects of Biomedical and Behavioral Research. The Belmont Report: Ethical Principles and Guidelines for the Protection of Human Subjects of Research. Washington, D.C.: Department of Health Education and Welfare, April 18, 1979.
6. Farah MJ. Neuroethics: the practical and the philosophical. Trends Cogn Sci 2005;9:34–40.
7. Illes J, Raffin TA. Neuroethics: an emerging new discipline in the study of brain and cognition. Brain Cogn 2002;50:341–344.
8. Kulynych J. Legal and ethical issues in neuro-imaging research: human subjects protection, medical privacy, and the public communication of research results. Brain Cogn 2002;50:345–357.
9. Northoff G, Witzel J, Bogerts B. Neuroethics—a future discipline? Nervenarzt 2005.
10. Sententia W. Neuroethical considerations: cognitive liberty and converging technologies for improving human cognition. Ann N Y Acad Sci 2004;1013:221–228.
11. Cooper J. The law and ethics of the use of experimental medication in patients incapable of expressing consent: between a rock and a hard place. Med Law 2000;19:189–195.
12. Foex BA. The problem of informed consent in emergency medicine research. Emerg Med J 2001;18:198–204.
13. Shaul RZ. Potato, potato, proxy consent, permission—just don't call the whole thing off. Crit Care 2005;9:123–124. Epub 2005 Jan 7.
14. Tu JV, Willison DJ, Silver FL, Fang J, Richards JA, Laupacis A, Kapral MK. Investigators in the Registry of the Canadian Stroke Network. Impracticability of informed consent in the Registry of the Canadian Stroke Network. N Engl J Med 2004;350:1414–421.
15. Vanpee D, Gillet JB, Dupuis M. Clinical trials in an emergency setting: implications from the fifth version of the Declaration of Helsinki. J Emerg Med 2004;26:127–131.
16. Winslade WJ, Tovino SA. Research with brain-injured subjects. J Head Trauma Rehabil 2004;19:513–515.
17. Winslade WJ. Research on minimally conscious patients: innovation or exploitation? J Head Trauma Rehabil 2004;19:178–179.
18. Office for Human Research Protections. Protecting Human Research Subjects: Institutional Review Board Guidebook. Department of Health and Human Resources. June 21, 2001.
19. Ijichi S, Ijichi N. The scientific establishment of a new therapeutic intervention for developmental conditions: practical and ethical principles. Childs Nerv Syst 2003;19:711–715.
20. Neill SJ. Research with children: a critical review of the guidelines. J Child Health Care 2005;9: 46–58.
21. Koops L, Lindley RI. Thrombolysis for acute ischaemic stroke: consumer involvement in design of new randomised controlled trial. Br Med J 2002;325:415.
22. Nurock S. Patients may be less risk averse than committees. Br Med J 2005;330:471–472.
23. International Conference on Harmonization. E6: Good Clinical Practice: Consolidated Guideline. Issued 5/9/1997.
24. Code of Federal Regulations 21 CFR 50, 50.24, 50.25(a)(1)–(5), 56, 56.108(e), 56.111(a)(2), 312.7 (d1)–(d2), 312.23(a)(1)(iv), 312.34–35, 312.60, 314.126.
25. Code of Federal Regulations 45 CFR 46 (Subpart A), 46.103(a), 46.107, 46.109, 46.111, 46.116, 46.402(d), 46.404–409.
26. European Medicines Agency. The European Union Directive 2001/20/EC, 4 April 2001.

Published in Official Journal of the European Communities, 1 May 2001, L121, pp. 34–44.

27. Gennery B. Academic clinical research in the new regulatory environment. Clin Med 2005; 5:39–41.

28. Kompanje EJ, Maas AI, Dippel DW. Clinical trials on medical products for human use in the case of acutely incapacitated patients within the fields of neurology and neurosurgery; implications of the new European legislation. Tijdschr Geneeskd 2003;147:1585–1589.

29. Stocchetti N, Dearden M, Karimi A, Lapierre F, Maas A, Murray GD, Ohman J, Persson L, Servadei F, Trojanowski T, Unterberg A. New European directive on clinical trials: implications for traumatic head injury research. Intensive Care Med 2004;30:517–518. Epub 2004 Jan 16.

30. Warlow C. Over-regulation of clinical research: a threat to public health. Clin Med 2005;5:33–58.

31. Grady C. The challenge of assuring continued post-trial access to beneficial treatment. Yale J Health Policy Law Ethics 2005;5:425–435.

32. La Vaque TJ, Rossiter T. The ethical use of placebo controls in clinical research: the Declaration of Helsinki. Appl Psychophysiol Biofeedback 2001; 26:23–37; discussion 61–65.

33. Rothman KJ, Michels KB. The continuing unethical use of placebo controls. N Engl J Med 1994;331:394–398.

34. Guc MO. Ethics in publication. Acta Neurochir 2002;83(suppl):101–104.

35. Hoeksema HL, Troost J, Grobbee DE, Wiersinga WM, van Wijmen FC, Klasen EC. A case of fraud in a neurological pharmaceutical clinical trial. Ned Tijdschr Geneeskd 2003;147:1372–1377 [in Dutch].

36. Tuech JJ, Pessaux, P, Moutel G, Thoma V, Schraub S, Herve, C Methodological quality and reporting of ethical requirements in phase III cancer trials. J Med Ethics 2005;31:251–255.

37. Kansu E, Ruacan S. Research ethics and scientific misconduct in biomedical research. Acta Neurochir 2002;83(suppl):11–15.

38. Walsh MK, McNeil JJ, Breen KJ. Improving the governance of health research. Med J Aust 2005; 182:468–471.

39. Bacchetti P, Wolf LE, Segal MR, McCulloch CE. Ethics and sample size. Am J Epidemiol 2005; 161:105–110.

40. Freedman B. Equipoise and the ethics of clinical research. N Engl J Med 1987;317:141–145.

41. Rikkert MG, Lauque S, Frolich L, Vellas B, Dekkers W. The practice of obtaining approval from medical research ethics committees: a comparison within 12 European countries for a descriptive study on acetylcholinesterase inhibitors in Alzheimer's dementia. Eur J Neurol 2005;12:212–217.

42. Emanuel EJ, Emanuel LL. Proxy decision making for incompetent patients. JAMA 1992; 267: 2067–2071.

43. Abramson NS, Safar P. Deferred consent: use in clinical resuscitation research. Brain Resuscitation Clinical Trial II Study Group. Ann Emerg Med 1990;19:781–784.

44. Beauchamp TL, Fost N, Robertson JA. The ambiguities of "deferred consent." IRB 1980; 2:6–9.

45. Fost N, Robertson JA. Deferring consent with incompetent patients in an intensive care unit. IRB Rev Hum Subj Res 1980;2:5–6.

46. Levine RJ. Research in emergency situations. The role of deferred consent. JAMA 1995; 273:1300–1302.

47. Levine RJ. Deferred consent. Control Clin Trials 1991;12:546–550.

48. Miller BL. The ethics of cardiac arrest research. Ann Emerg Med 1993;22:118–124.

49. Miller BL. Philosophical, ethical, and legal aspects of resuscitation medicine. I. Deferred consent and justification of resuscitation research. Crit Care Med 1988;16:1059–1062.

50. Office for Protection from Research Risks. OPRR Reports: Informed Consent Requirements in Emergency Research, October 31, 1996.

51. Turner G, Lisook AB, Delman DP. FDA's conduct, review, and evaluation of inspections of clinical investigators. Drug Inf J 1987;21:117–125.

52. Arnold LK, Razzak J. Research agendas in global emergency medicine. Emerg Med Clin North Am 2005;23:231–257.

53. Hyder AA, Dawson L. Defining standard of care in the developing world: the intersection of international research ethics and health systems analysis. Dev World Bioeth 2005;5:142–152.

54. Rivera R, Borasky D, Rice R, Carayon F. Many worlds, one ethic: design and development of a global research ethics training curriculum. Developing World Bioeth 2005;5:169–175.

55. Worrall BB, Chen DT, Meschia JF. Ethical and methodological issues in pedigree stroke research. Stroke 2001;32:1242–1249.

14

Industry Perspective on Drug Development

Thomas C. Wessel and Christopher Gallen

Stroke, head trauma, and spinal cord injury share common biological mechanisms that, if left unopposed, can lead to irreversible cell injury and death. Over the last 20 years, we have gained a more complete understanding of the early biochemical steps involved in cell injury, from what seem to be relatively easily described phenomena, such as the influx of calcium into cells, excitatory amino acid release, and oxygen radical formation to more complex, often delayed, changes brought about by active gene transcription. Several attractive pharmacological targets have been identified in the ischemic cascade and cell death pathways, and our understanding of the temporal sequence of events in the expanding penumbra has nurtured continuing belief in neuroprotective strategies for stroke, head trauma, and spinal cord injury. Yet the string of failures in recent neuroprotective trials and the medical community's low level of thrombolysis utilization for acute stroke give pause to drug developers who considering entering this treacherous area.

SELECTION OF PHARMACOLOGICAL TARGETS AND NECROSIS VERSUS APOPTOSIS

A number of factors can trigger destructive reactions within the brain and spinal cord, including calcium flux into cells, excitatory amino acid release, formation of nitric oxide and free radicals, lipid peroxidation, inflammation and immune responses (1,2). The time frame in which individual steps unfold in this chain of events ranges from seconds after ischemia onset (for calcium influx, cell depolarization, nitric oxide, and radical formation) to days (for delayed gene transcription and inflammation). Detailed descriptions of the time course and overlapping steps in necrotic and apoptotic cell mechanisms in stroke and head trauma are reviewed in elsewhere in this handbook and in the literature (3). The focus of research in ischemic injury in stroke and head trauma has historically been on mechanisms related to necrosis, not apoptosis.

Although the principal biochemical and histological differences between necrosis and apoptosis have been characterized in detail and are widely accepted (4), it is now increasingly clear that a continuum of necrotic and apoptotic cell death exists in most central nervous system (CNS) injuries. The trigger for entry into either of these cell death pathways is presumed to be related to the severity of the initial insult (5) and may be determined by the degree of ATP depletion (6). From the drug development perspective, it seems a daunting task to influence this intricate system by attacking a single element that may only shift the likelihood of pushing a cell toward necrotic or apoptotic cell death (7).

A number of agents have been explored in head trauma and spinal cord injury models, including calcium channel blockers, glutamate receptor antagonists, corticosteroids and antioxidants, opioid receptor antagonists, and magnesium administration, along with anti-inflammatory and immunomodulatory treatments (8). Essentially, the same agents have been used for ischemic stroke and have been reviewed in detail (9). Although some of these agents have resulted in modest improvements in outcome after acute spinal cord injury in humans (10), neuroprotection studies in head injury (11) and stroke (12) have been uniformly disappointing.

ANIMAL MODELS

Current pharmaceutical discovery is heavily dependent on various disease models, yet fundamental questions have been raised about the adequacy of these models to predict human outcomes in clinical stroke trials (13). A great deal of work has gone into characterizing different animal models for stroke (14) and head trauma. But it is important to realize that for any animal model to be relevant to the development of treatment for a disease, at least three logical conditions should be met (15) (but rarely are): First, one should understand the nature of the *animal model* in detail (i.e., the pathology and the mechanisms and systems biology of the animal model); second, one should understand the nature of the *human disease* in detail (i.e., the pathology and the mechanisms and systems biology of the human disease); and third, one should be sure that the animal model and the target disease in humans are congruent in all important respects, because if they are not congruent in some important aspect, predictions made with the animal model will very likely be wrong. For neurotrauma and some stroke models, a plausible case can be made that a reasonable congruity between animal model and human disease exists. Head trauma models can mimic the process that induces lesions in humans, within the limitations of cross-species differences in reaction to trauma, and various methods of stroke induction, from ligation to introduction of clots or microspheres to global ischemia, can induce lesions similar to those observed in human stroke.

CONTROLLED CONDITIONS OF THE ANIMAL EXPERIMENT VERSUS CLINICAL CHAOS

Over the last 20 years, much experience has been gained in standardizing animal models of ischemic stroke and head trauma, especially with respect to experimental conditions and time windows. These techniques provide relatively uniform reproducible lesions in young animals achieved under constant experimental conditions (e.g., body temperature, glucose), and precise application of drug is possible in a genetically homogeneous cohort at a defined time point. This, however, is in striking contrast to the highly variable conditions at play during the acute phase of actual stroke or head trauma, where the constellation of physiological derangements

TABLE 14.1 Differences between preclinical models and clinical reality

	Preclinical models	Clinical reality
Application of neuroprotective agent		
Time window	Short	Long
Plasma concentration	High	Low
Penetration into brain	Usually established	Often unknown
Dose ranging	Usually carefully tested	Often neglected
Experimental conditions		
Temperature	Controlled during stroke, sometimes postinjury	variable
Blood pressure	Usually controlled	variable
Oxygenation		
Blood glucose	Normal	variable
Electrolytes	Normal	Often abnormal
Drugs other than test agent present	None	Frequent
Subject characteristics		
Species	Inbred albino rats, gerbils	Humans
Genetic variation	Small	Large
Age	Young, same age	Elderly, broad range
Sex	male	Both
Lesion site	Standardized gray matter (range of lesion sizes)	Anywhere, mostly combined gray and white matter injury
Penumbra duration	2 hours?	Unknown
Comorbidities	none	Many
Other concomitant condition	none	Frequent: infection, dehydration, stress, drug/ethanol
Outcome measures		
Primary endpoint	lesion size	Functional outcome and/or disability
Secondary endpoints	Occasionally motor test	Occasionally lesion size
Timepoints measured	Focal: 2 days Global: 7 days	1 month, 3 months

differs from patient to patient: Differences in brain temperature, variability in blood pressure and viscosity, oxygenation, electrolyte and glucose disturbances, and the presence of alcohol or other drugs (to name a few) all contribute to variable patient outcomes. Furthermore, considerable individual differences exist in gene expression and hormonal status that may contribute profoundly to head trauma (16) and stroke outcome (17) (e.g, apolipoprotein E expression [18], homocysteine levels [19], and estradiol [20]). Table 14.1 summarizes the most obvious differences between conditions in animal models and the clinical reality that stroke researchers confront.

TIME WINDOW, GLUCOSE, TEMPERATURE, AND BRAIN COMPOSITION

One of the most striking areas of discrepancy between the animal model and human condition is probably related to the time window for intervention in neuroprotective or rheological/antithrombotic trials for stroke. As reviewed by Kidwell and colleagues (21), during the period from 1980 to 1999 the median time window allowed for enrollment decreased by half in each decade, from 48 to 12 hours. This, of course, does not come close to the 2-hour therapeutic window most commonly used with rodents, and most neuroprotective trials include patients for up to 6 hours (or longer, considering the number of patients with protocol violations enrolled).

In terms of physiological derangements that are known to have profound consequences for stroke outcome and are carefully controlled in animal experiments but commonly neglected in clinical trials, just two are mentioned here: body temperature and blood glucose. Compelling evidence suggests that the animal model and the human patient behave similarly on both of these variables. Well-documented examples have shown how seemingly minor increases in body temperature can have similar consequences in animal experiments (22,23) and in humans (24,25), even in relation to the admission body temperature in acute stroke (26). Similar congruity between the animal model (27) and stroke outcome in humans has been shown for elevated glucose, both in terms of acute (28) and persistent delayed hyperglycemia (29). As DeBow and colleagues noted (30), the vast majority of animal studies used young adult rodents, mostly males (98% in their review), most studies (60%) do not assess functional outcome, and survival times were often short (48 hours [66%] for focal ischemia and 7 days [80%] for global ischemia). Furthermore, although rectal temperature during ischemia induction was recorded in more than 60% of the animal experiments, only 32.6% of ischemia studies in humans measured temperature after surgery.

As pointed out by Cheng and colleagues (31), another discrepancy between animal models and human conditions lies in the difference in brain composition: Approximately 90% of brain tissue in rodents is gray matter, whereas the gray matter-to-white matter ratio in humans is about 1:1, indicating humans' much greater likelihood of having concomitant gray and white matter injury (32), which could negate or obscure benefit from a neuroprotective agent that has differential effects on gray and white matter. In addition, animal models do not account for the negative impact of preexisting comorbidities, some of which strongly influence stroke outcome (33).

UNMET MEDICAL NEEDS AND DRUG DEVELOPMENT

The unmet medical need for effective treatments in stroke and head trauma as well as other catastrophic neurological emergencies and more insidious neurological problems is immense. Beyond the toll in suffering and loss for patients and their families, the societal costs of inadequate current therapies are enormous. Individual academic researchers may allocate their resources to neuroemergency indications based on many complex and competing reasons: humanitarian motives, love of science, continuing focus on an area of expertise, or other personal reasons. Industrial enterprises also base their allocation of resources on complex internal and external reasons but must, of necessity, do so in an economically remunerative manner. The costs of pharmaceutical research are very high for neuroemergency indications oriented toward winning regulatory and commercial acceptance of novel neurotherapeutic agents. The chances of success are generally small, particularly in the areas of stroke and head injury; indeed, the low success rates of past efforts are an important reason why the current unmet medical need is so high. Over the course of time, the return from the small number of therapeutics that succeed must cover the direct costs of development and capital and provide a sufficient economic return to warrant continuing investment. The key factors influencing the economic desirability of a program are the market's size and severity of need, the real competitive advantages conferred by the new therapeutic, its patent life ("marketing life") and potential price, and the number of competitors and number of providers who need to be aware of the product.

RISKS TO NEUROTHERAPEUTIC DRUG DEVELOPERS

Ideas for potential novel neurotherapeutics surface fairly often. Although assessments of the underlying disease mechanism and initial animal data may help prioritize the most promising drugs, in most cases the outcome of the experiments—preclinical or clinical—is difficult to predict. Indeed, if one knew beforehand whether an effort would succeed or fail, there would be no failures. Yet failures greatly outnumber successes. Because the failures come together with the process that generates the successes and incur costs that must be recouped for the investigative enterprise to continue, the cost of all the successful and unsuccessful efforts must ultimately be carried by the successful products. This is also true for the pharmaceutical industry as a whole: The returns from successful developments pay for the investigations into therapies that fail, and, in the long run, investment in areas that consistently fail will dry up. Conversely, the more successful and profitable efforts are, the more resources are available to apply to the successful area and explore different indications. In some cases, investigations into mechanisms shown to be important by the success of therapies based on them have been derided as "me-too" therapies, but such new therapies may provide valuable marketplace choices. Moreover, as we look at the different patterns of gene expression that results from "me-too" compounds, it seems likely that we will start to think differentially about these drugs in the future as we begin

to better understand the complexity of their other secondary effects.

ECONOMICS OF CNS DRUG DEVELOPMENT

CNS diseases pose immense burdens on patients, their families, the health care system, and society as a whole. The unmet medical and human needs are great and drive the need for significant investments in therapeutics development. Overall estimates of the CNS market size depend on definitions, that is, whether CNS is considered to include neurological, psychiatric, pain, anesthetic, or drug addiction and abuse agents in total or as subsets. Broadly defined, sales of CNS therapeutics comprise approximately 15% of total pharmaceutical sales, approximately $30 billion worldwide. These costs are a small fraction of the societal costs incurred by the underlying CNS diseases, because most available CNS therapeutics provide symptomatic treatment and are not curative. For example, the estimated annual economic costs of anxiety disorders, depression, and schizophrenia are $47 billion, $44 billion, and $33 billion per year, respectively (34). The societal costs of stroke and head trauma are 54 and 48 billion/year. As a general rule, the cost of medications is about one-tenth of the overall cost of health care. Pharmaceutical expenditures focus on those areas where the therapies to date have proved of sufficient utility to win acceptance in large markets. Consequently, about two-thirds of current CNS therapeutics are for psychiatric indications, despite the fact that neurological emergencies are quite common.

Effective CNS pharmaceutical agents have the potential to provide a huge benefit to patients and economies. Present market size numbers reflect the current market but vastly understate the potential market. First, because many important CNS disorders have no curative treatment at all, development of a treatment would create an entirely new market together with a corresponding benefit. Second, many disorders have only ameliorative therapies that either have limited efficacy or are associated with significant side effects that strongly restrict their use. More effective or better-tolerated safer therapies have the potential to dramatically increase utilization and therefore market size. Third, markets currently served by workable but suboptimal generic therapies have the potential to grow dramatically with the introduction of more effective or safer therapeutic agents. Certainly any new therapies would incur a direct health care cost, but because the cost of pharmaceuticals is actually only a small fraction of the overall cost of health care, effective therapies have a strong potential to reduce overall health care expenditures and to ameliorate the economic impact of disease on individual patients. In assessing the need for any prospective therapeutic agent economically, the number of potential patients who could be served by a safe and effective therapy, the seriousness of the illness, and the disability entailed all must be considered (and all of which affect the potential benefit and hence the price of a therapy).

An important concern arises from such assessments, which are much easier in areas such as epilepsy, where there are many therapeutics of varying utility to compare, but are much more difficult in areas such as stroke or head injury, where there are few or no specific therapies for many clinical situations. Such uncertainty usually results in too-conservative market estimation and hence in underestimation of the worth of a potential therapeutic. On more than one occasion in recent decades, once a therapy was actually available on the market and once patients and clinicians became aware of it, the market value for the therapy turned out to be an order of magnitude greater than originally estimated. This is certainly not always the case and counterexamples exist, but there is little doubt that lack of knowledge about the opportunity in a given market contributes to a pattern of under-investment. Overall, the large size of the potential medical impact of effective neurotherapeutic agents is a strong factor in favor of investment in neurotherapeutics.

FACTORS INFLUENCING ECONOMIC FEASIBILITY IN CNS DRUG DEVELOPMENT

Once some estimate of the potential unmet medical need is available, a few general factors control whether development of a particular therapeutic is economically feasible. The key economic determinants of risk are success rate, development time, and development cost. The challenge is steep in general therapeutics development but far less so than in neurotherapeutics development. In general, for every 5000 to 10,000 compounds screened, about 250 will enter preclinical testing; of those, 5 will enter clinical testing and 1 will win approval by the U.S. Food and Drug Administration (FDA) (35). Overall, only about 11% of new active substances entering clinical development are predicted to reach the market (36). The success rates for neurotherapeutic agents, however, are far lower than average. The relative difficulty of neurotherapeutics development is illustrated by a comparison of the chance of compounds that are initiated into human testing to progress to eventual marketing across therapeutic areas: anti-infectives, 33%; cardiovascular, 6%; anticancer, 6%; and nervous system, 1–2% (36). The low success rates seen with nervous system compounds represent a value that has fallen over time. In the past, the success rate for nervous system compounds was similar to that of anticancer compounds. Furthermore, neurotherapeutic compounds fail late in development, during phase III pivotal trials, far more often than other compound categories. For most non-CNS indications, animal models are more predictive of efficacy as screens and early data in humans are more predictive of efficacy in later trials. The chance that CNS compounds initiated into pivotal trials will subsequently progress to eventual marketing across therapeutic areas is much lower than for other therapeutic areas: anti-infectives, 75%; cardiovascular, 43%; anticancer; 32%; and nervous system, 14% (36). Because these large pivotal programs cost about three times as much as the combined cost of early development phase I and II trials, this pattern of late failure acts as a powerful disincentive to CNS therapeutics developers. Further, of the pharmaceuticals that do win regulatory approval, only one in three will produce revenues that match or exceed development costs (37).

In summary, the revenue stream and profits produced by that one pharmaceutical will have to pay for the tens of thousands of antecedent compounds produced, screened, and rejected along the way. The very high failure rates for CNS compounds and their propensity to expensive late-stage failures are factors that weigh heavily against investments in neurotherapeutic development.

COSTS OF DEVELOPMENT

Both fixed and variable components factor into overall development costs. Expensive discovery resources, both human and material, are needed to explore mechanisms, probe interventions molecularly, develop therapeutic molecules and delivery mechanisms, and assess toxicology and other parameters in detail. Testing of potential therapies in human drug development is governed by strict international regulations, including broadly accepted principles of good clinical practice. Human testing involves extensive efforts to conceive development plans and protocols, conduct detailed testing in thousands of patients, capture and analyze all data in robust and reliable systems, and write extensive reports and regulatory submissions. Cost estimates for pharmaceutical development vary widely but were estimated by the Tufts Center to be approximately $802 million in 2000 (38). This estimate included an average out-of-pocket cost per new drug of $403 million plus costs of capital; it was further noted that capitalized postapproval development costs raise the overall pre- and post-approval cost to $877 million (39), a number that includes out-of-pocket pre-clinical and clinical expenses and costs of capital for preclinical and clinical expenditures for the expenses of both project failures and successes. These costs have consistently been driven upward by the progressive increase in the number of clinical trials required for regulatory approval: from 30 in the period from 1977 to 1980 to 68 in the period from 1994 to 1995. Similarly, the number of patients per new drug application has increased from 1576 in the period from 1977 to 1980 to 4237 in the period from 1994 to 1995 (40).

The $802 million figure is likely conservative. As a cross-check, one could examine the research and development (R&D) budgets of most major pharmaceutical companies, divide by the portion of the $403 million related to direct expenses rather than costs of capitalization, and get a number of predicted compounds several times higher than the average yearly number of new chemical entities registered by that company. Even if the $802 million was considered as fully loaded costs (including costs of capital) and divided into the research and development budgets of the larger companies, the resulting number would be much higher than the average number of new chemical entities registered by those companies per year. In 2000, for example, 11 major pharmaceutical companies had R&D budgets greater than

$2 billion per year, yet predictions based on current pipelines and success rate estimates suggest launch rates averaging 1.3 new active substances per year over the past 6 years (41). As already noted, these numbers are of all the more concern when coupled with the realization that the R&D costs are such that only 3 in 10 marketed drugs produce revenues that exceed or match their development costs (42).

In addition to direct investment, the length of development time is a key parameter governing the developer's economic risk because long development times increase the cost of capital and reduce the time available to recoup investment. Any resource invested in any economic endeavor can be compared against other alternative investments for rate of return. Every investment requires capital, and capital has a cost per year that should at least exceed what the investor could have gotten for the capital had they invested it in a safer venue for a comparable time period. Because interest is exponential, the more years before the capital is recovered, the greater the sum required to justify the alternative investment. Studies by the Tufts Center for the Study of Drug Development indicated that reducing the total development time by half will reduce total costs by 29% (43). Even more relevantly, once a pharmaceutical agent's patent expires and the agent is subject to generic competition, the profit produced is not remotely enough to pay for the costs of development of the pharmaceutical. Hence, the costs of development must be recouped while the product is on patent. Longer development times directly reduce the number of profitable years remaining for any therapeutic. In the United States, it takes 10 to 15 years to move a new drug from discovery through regulatory approval, leaving on average about 5 years to recoup the costs of development, capital, and a profit to attract subsequent investment. Those 15 years include approximately 6 years for the

discovery/preclinical phase, 5 years for the clinical phase, and 1 to 2 years for the FDA approval process, and the length of any of these phases can easily mushroom. In a patent life of 20 years, that leaves only 5 to 10 years to recoup the fully loaded costs of development and capital.

Development times are driven both by the length and the number of trials serially needed to prove efficacy. For any given trial, length is determined by a series of segments: idea to protocol approval, protocol approval to first patient in, first patient in to last patient in (enrollment), last patient in to last patient out (treatment period), last patient's last visit to database lock, database lock to key statistics, and key statistics to final study report. Across therapeutic areas, several of these segments are roughly equivalent and depend more on the developer's operational excellence than on anything specifically related to the therapeutic area: protocol approval to first patient in, last patient's last visit to database lock, database lock to key statistics, and key statistics to final study report. The time to first patient in can be lengthy in studies looking for rare patients, but overall across the industry this time period is generally similar from trial to trial. The two key variable areas are enrollment and treatment period. Enrollment can turn to a significant degree on two factors: the real accessibility of the patient population (i.e., how many patients there are who meet the trial's inclusion and exclusion criteria) and how well organized and effective the investigators are in reaching, informing, and enrolling those patients. Most protocols with very poor enrollment times can trace that failure to unrealistic inclusion and exclusion criteria. The tendency to try to perfect the trial by limiting enrollment to a narrowly defined group of patients is usually rooted in a desire to reduce sample size by limiting variance, but this can lead to trials that fail to enroll or produce results that do not generalize to the population one actually seeks to treat. The treatment period is a time component that typically separates CNS trials from most other areas. Like oncology trials, CNS trials often require prolonged patient follow-up either to document full recovery or to detect relapse of longer term adverse effects. This issue is not a problem for acute stroke or head-injury therapeutics but can be an issue for stroke prevention trials.

INDUSTRIAL INVESTMENT AND PRODUCTIVITY

Pharmaceutical R&D investments are high and growing geometrically. In 2002, members of the Pharmaceutical Research and Manufacturers of America spent approximately $32 billion on pharmaceutical R&D, which represents about a 15-fold rise over the past 20 years and exceeds the National Institutes of Health (NIH) budget of $24 billion. This massive increase in expenditure has not produced anything like a corresponding increase in output—on average, the number of new therapeutics per year did not grow by half during the period when funding increased 15-fold. NIH funding is critically important for the general advance of health sciences and should not be underestimated, but a 2001 report by the NIH indicated that when specific links to pharmaceutical developments were assessed for 47 drugs with U.S. sales of $500 million or more per year, only 4 drugs had been developed in part with NIH-funded technologies (44). Governmental expenditures in basic research are critical in generating the observations, concepts, and methods for understanding pathophysiological phenomena and their substrates, but the complexity of the disease process and of pharmacology is such that fundamental insight does not in itself guarantee successful therapies. Some rational concepts based on basic research fail to be significant in the clinic, whereas others end up requiring many attempts to eventually produce a favorable intervention. The result is the need for massive investment above and beyond government expenditures.

Domestic U.S. pharmaceutical R&D expenditures exceed those of any other major industrial sector, even high-investment sectors such as computer software and services and the electrical, electronics, or aerospace industries. Company-financed research in products affecting the central nervous system and sense organs was estimated at $7.3 billion in 2001, significantly exceeding the $3.9 billion expenditure for agents acting on the cardiovascular system and roughly equal to the combined $7.4 billion expenditure on products affecting neoplasms, the endocrine system, and metabolic diseases (45). More than 80% of larger pharmaceutical companies are developing agents to treat CNS disorders. They focus primarily on larger more well-defined indications such as depression, schizophrenia, or multiple sclerosis or on underserved indications in which the medical need is high, such as dementias, brain tumors, or substance use disorders.

DECREASED RESEARCH PRODUCTIVITY

Geometrical increases in R&D expenditures notwithstanding, the overall productivity of pharmaceutical research in producing new chemical entities has not increased in proportion to the investment. Over the past 20 years, pharmaceutical research has increased about 1500%, but the number of approvals of new therapeutic agents has been relatively small, rising from approximately 20 per year in the 1980s to approximately 30 per year in the 1990s and currently (46). This failure of productivity has been partially ameliorated by the wider markets opened by globalization and regulatory developments that facilitate simultaneous global registrations for new drug treatments. Looking forward, however, rapid growth from enhanced market penetration seems unlikely now that the pharmaceutical industry is very global and present in virtually every economically significant market. Growth from incremental population growth and growth of health care resource allocation as societies become more wealthy and health conscious can continue, but this is unlikely to occur at a pace matching the geometrical growth of pharmaceutical R&D expenditures. Further, over the last few decades the pressures of productivity challenges have been exacerbated by both the competition from "fast followers" and the challenge to generate new therapeutics fast enough to replace those that go off patent. Patent expiration and fast followers are a huge challenge to innovator companies, in some cases engendering crisis and the risk of economic failure. The ability of new technologies to rapidly close discovery gaps once a promising new target for development is proven greatly facilitates the ability of fast-follower companies to exploit the discoveries of innovator companies. This process is reflected in the progressively shorter period of marketing exclusivity and market share enjoyed by an innovator company before competitors match the initial breakthrough process. The huge economic importance of patent expiration is driven by the fact that generic production has taken an increasing share of the U.S. prescription pharmaceutical market (47). Once the patents constraining the generic use of a compound expire, the compounds are typically produced by a manufacturer that has small costs to recover ($1 million to $2 million for bioequivalence studies) and is not required to meet the significant costs involved in the development of next-generation pharmaceuticals. In a sense, the compound passes into the patrimony of humankind, where it is sold at markedly reduced prices that do not cover the costs of further research and development.

Many major pharmaceutical companies have recognized the need to double or triple their discovery output to maintain current profitability and growth in the face of generic competition (48). High throughput screening has been portrayed as a major and massive source of new compounds (49,50). Yet analyses suggest that actual utilization is only 2% to 7% of installed capacity (51) and likely is not the rate-limiting step. Later issues such as biometabolism and compound toxicology are more important limitations to discovery output and are being managed by industrializing the screening process.

In recent years, price increases played an important role in covering the gap between geometrically rising investments and slowly rising production of new therapies, but price increases are unlikely to play a significant role moving forward in compensating for the productivity challenge to the pharmaceutical industry. The economic rationale underlying price increases was the fact that pharmaceuticals as a small part of the overall health care cost had the potential to replace expensive medical therapies (e.g., H2 blockers replacing gastric resection for the treatment of peptic ulcers). Because most of the expenses associated with pharmaceuticals derive from R&D and marketing costs rather than the specific cost of goods production, prices tend to be driven by the relative benefit conferred, not the cost of unit production (although, in general, the cost of goods is much higher for protein therapeutics than for small molecule therapeutics and can constitute a significant portion of the total price). Although this logic persists, the development of therapies for many indications has resulted in pharmaceutical costs rising, as a percentage of health care costs, to a level where price pressures are high. Health care expenditures are a major budget item worldwide. In 1997, health care costs as a percentage of gross domestic product were higher in the United States than for other major industrialized nations. Pharmaceutical costs are a small percentage of overall health care costs, about 8% in the United

States. On average, pharmaceutical costs are similar to the average telephone bill but are of concern because they disproportionately affect vulnerable segments of the population such as the elderly. In most of the world, pricing is tightly controlled by governments at levels that are, in aggregate, not compatible with sustaining current worldwide pharmaceutical R&D expenditures. Such prices tend to cause pharmaceutical research to shift outside the borders of the countries with lower pricing (52) and to decrease availability of therapeutics to patients by slower introduction of new therapies (53). It can be argued that pharmaceuticals may actually decrease overall health care costs by reducing larger expenditure items such as hospitalization. But the key strategic implication of this pressure for developers of neurotherapeutics and other drugs is the fact that increased R&D costs are unlikely to be covered by price increases, and therefore the need to dramatically improve productivity is inescapable.

The scientific challenges for developing effective neuroprotective agents or agents to facilitate recovery from major CNS insults are formidable, as are the economic challenges for all of CNS development. However, the unmet medical need for effective therapies is huge, as is the societal cost of the lack of adequate therapies. The emerging sophistication of our science together with the continuing efforts of academicians and industry will eventually prevail over both the scientific and economic hurdles.

References

1. McIntosh TK. Neurochemical sequelae of traumatic brain injury: therapeutic implications. Cerebrovasc Brain Metab Rev 1994;6:109–162.
2. Fisher M. Characterizing the target of acute stroke therapy. Stroke 1997;28:866–872.
3. Yakovlev AG, Faden AI. Mechanisms of neural cell death: implications for development of neuroprotective treatment strategies. NeuroRx 2004;1:5–16.
4. Bredesen DE. Neural apoptosis. Ann Neurol 1995;38:839–851.
5. Bonfoco E, Krainc D, Ankarcrona M, Nicotera P, Lipton SA. Apoptosis and necrosis: two distinct events induced, respectively, by mild and intense insults with N-methyl-D-aspartate or nitric oxide/superoxide in cortical cell cultures. Proc Natl Acad Sci USA 1995;92:7162–7166.
6. Eguchi Y, Shimizu S, Tsujimoto Y. Intracellular ATP levels determine cell death fate by apoptosis or necrosis. Cancer Res 1997;57:1835–1840.
7. Pohl D, Bittigau P, Ishimaru MJ, Stadthaus D, Hubner C, Olney JW et al. N-methyl-D-aspartate antagonists and apoptotic cell death triggered by head trauma in developing rat brain. Proc Natl Acad Sci USA 1999;96:2508–2513.
8. Faden AI. Pharmacological treatment of central nervous system trauma. Pharmacol Toxicol 1996;78:12–17.
9. Labiche LA, Grotta JC. Clinical trials for cytoprotection in stroke. NeuroRx 2004;1:46–70.
10. Bracken MB, Holford TR. Effects of timing of methylprednisolone or naloxone administration on recovery of segmental and long-tract neurological function in NASCIS 2. J Neurosurg 1993;79:500–507.
11. Doppenberg EM, Choi SC, Bullock R. Clinical trials in traumatic brain injury: lessons for the future. J Neurosurg Anesthesiol 2004; 16:87–94.
12. Labiche LA, Grotta JC. Clinical trials for cytoprotection in stroke. NeuroRx 2004; 1: 46–70.
13. Lees KR. Neuroprotection is unlikely to be effective in humans using current trial designs: an opposing view. Stroke 2002;33:308–309.
14. Recommendations for Standards Regarding Preclinical Neuroprotective and Restorative Drug Development. Stroke Therapy Academic Industry Roundtable (STAIR). Stroke 1999;30:2752–2758.
15. Horrobin DF. Modern biomedical research: an internally self-consistent universe with little contact with medical reality? Nat Rev Drug Discov 2003;2:151–154.
16. Teasdale GM, Graham DI. Craniocerebral trauma: protection and retrieval of the neuronal population after injury. Neurosurgery 1998;43: 723–737.
17. Meschia JF, Brott TG, Brown RD Jr, Crook RJ, Frankel M, Hardy J, Merino JG, Rich SS, Silliman S, Worrall BB. Ischemic Stroke Genetics Study. The Ischemic Stroke Genetics Study (ISGS) Protocol. BMC Neurol 2003;8:4.
18. Chiang MF, Chang JG, Hu CJ. Association between apolipoprotein E genotype and outcome of traumatic brain injury. Acta Neurochir (Wien) 2003;145:649–653.
19. Wald DS, Law M, Morris JK. Homocysteine and cardiovascular disease: evidence on causality from a meta-analysis. Br Med J 2002;325:1202.
20. Roof RL, Hall ED. Estrogen-related gender difference in survival rate and cortical blood

flow after impact-acceleration head injury in rats. J Neurotrauma 2000;17:1155–1169.

21. Kidwell CS, Liebeskind DS, Starkman S, Saver JL. Trends in acute ischemic stroke trial through the 20th century. Stroke 2001;32:1349–1359.

22. Wass TC, Lanier WL, Hofer RE, Scheithauer BW, Andrews AG. Temperature changes of .1oC alter functional neurologic outcome and histopathology in a canine model of complete cerebral ischemia. Anesthesiology 1995;83:325–335.

23. Kim Y, Busto R, Dietrich D, Kraydieh S, Ginsburg M. Delayed postischemic hyperthermia in awake rats worsens the histopathologic outcome of transient focal cerebral ischemia. Stroke 1996;27:2274–2281.

24. Castillo J, Martinez F, Leira R, Prieto R, Lema M, Noya M. Mortality and morbidity of acute cerebral infarction related to temperature and basal analytic parameters. Cerebrovasc Dis 1994;4: 66–71.

25. Reith J, Jorgensen H, Pedersen P, et al. Body temperature in acute stroke: relation to stroke severity, infarct size, mortality, and outcome. Lancet 1996;347:422–425.

26. Kammersgaard LP, Jorgensen HS, Rungby JA, Reith J, Nakayama H, Weber UJ, Houth J, Olsen TS. Admission body temperature predicts long-term mortality after acute stroke: the Copenhagen Stroke Study. Stroke 2002; 33:1759–1762.

27. Pulsinelli WA, Waldman S, Rawlinson D, Plum F. Moderate hyperglycemia augments ischemic brain damage: a neuropathologic study in the rat. Neurology 1982;32:1239–1246.

28. Parsons MW, Barber PA, Desmond PM, Baird TA, Darby DG, Byrnes G, Tress BM, Davis SM. Acute hyperglycemia adversely affects stroke outcome: a magnetic resonance imaging and spectroscopy study. Ann Neurol 2002;52:20–28.

29. Baird TA, Parsons MW, Phanh T, Butcher KS, Desmond PM, Tress BM, Colman PG, Chambers BR, Davis SM. Persistent poststroke hyperglycemia is independently associated with infarct expansion and worse clinical outcome. Stroke 2003;34:2208–2214.

30. DeBow SB, Clark DL, MacLellan CL, Colbourne F. Incomplete assessment of experimental cytoprotectants in rodent ischemia studies. Can J Neurol Sci 2003;30:368–374.

31. Cheng YD, Al-Khoury L, and Zivin JA. Neuroprotection for ischemic stroke: two decades of success and failure. NeuroRx 2004;1:36–45.

32. Dewar D, Yam P, McCulloch J. Drug development for stroke: importance of protecting cerebral white matter. Eur J Pharmacol 1999;375:47–50.

33. Demchuk AM, Buchan AM. Predictors of stroke outcome. Neurol Clin 2001;18:455–473.

34. PhRMA. New Medicines in Development for Mental Illness. 2000.

35. PhRMA. The Value of Medicines. p. 3.

36. Benjamin GA, Lumley CE. Industry Success Rates 2003. CMR, 2003.

37. Grabowski H, Vernon J, DiMasi J. Returns on research and development for 1990's new drug introductions. Pharmacoeconomics 2002;20 (suppl 3):11–29.

38. DiMasi JA, Hansen RW, Grabowski HG. The price of innovation: new estimates of drug development costs. J Health Econ 2003;22:151–185.

39. Tufts Center for the Study of Drug Development Impact Report, 5(3) May/June 2003.

40. Peck CC. Drug development: improving the process. Food Drug Law J 1997;52:163–167.

41. Ogg MS, van den Haak MA, Halliday RG. Activities of the International Pharmaceutical Industry in 2000 Pharmaceutical Investment and Output, CMR International (available at http://www.cmr.org/index.htm).

42. Grabowski H, Vernon J. Returns to R&D on new drug introductions in the 1980s. J Health Econ 1994;13:383–406.

43. Tufts Center for the Study of Drug Development, Outlook 2003, R&D Efficiency, 2003.

44. National Institutes of Health. A Plan to Ensure Taxpayer Interests are Protected. NIH Response to the Conference Report Request for a Plan to Ensure Taxpayers' Interests are Protected. http://ww.nih.gov/news/070101wyden.htm 1, 2001.

45. Pharma Annual Survey, 2001. Pharmaceutical Industry Primer 2001. PhRMA, 2001:5.

46. PhRMA Annual Survey, 2001. U.S. FDA. Global Market Research & Analysis, 2001.

47. IMS Health, 2001. In: Pharmaceutical Industry Primer 2001, PhRMA, 2001:8.

48. Peakman, Franks TS, White C, Beggs M. Delivering the power of discovery in large pharmaceutical organizations Drug Discovery Today 2003;8:203–221.

49. Beggs M. HTS—where next? Drug Disc World 2000;2:25–30.

50. Beggs M, Long AC. High throughput genomics and drug discovery-parallel universes or a continuum? Drug Disc World 2002; 3:75–80.

51. Lin L. Betaseron. Dev Biol Stand 1998;96:97–104.

52. U.S. International Trade Commission. Global Competitiveness of U.S. Advanced Technology Manufacturing Industries: Pharmaceuticals. Report to the Senate Finance Committee. Washington, DC: ITC, September 1991.

53. IMS Health. U.S. Outpaces Europe in Growth of Emerging Biotech Market: Industry Perspective. http://www.imshealth.com/ims/portal/front/articleC/0,2777,6599_40183890_–40053441,00.htm 17, January 2003.

15

Regulatory Perspective

Lisa L. Travis

In this chapter we provide a pragmatic approach to the interpretation of U.S. Food and Drug Administration (FDA) regulations and their relationship to research, development, and marketing of neuro-emergency pharmaceutical products. The development of neuroemergency treatments presents specific challenges that are best dealt with through thoughtful interpretation and implementation of regulatory guidelines. Drug Development in the United States is heavily governed and regulated by the FDA; therefore, important strategies related to pharmaceutical regulatory processes are discussed. Neurological indications such as traumatic brain injury (TBI) and intracerebral hemorrhage (ICH) fall into a special category of serious and life-threatening conditions for which few treatment options are available. FDA prefers to work closely with sponsors of promising development programs for serious conditions to help maximize the possibility of success and therefore encourages close and early communication in all aspects of these programs. It is therefore important to be aware of the strengths and limitations of available options to help expedite the development process for promising treatments where there is an unmet medical need and also to comprehend important nuances of these regulatory mechanisms.

INVESTIGATIONAL NEW DRUG APPLICATIONS

To ensure regulatory success, the sponsor must recognize the value of honest communication with the FDA. Sponsors communicate with the FDA primarily through telephone contacts, meetings, and formal filings. The primary and most common mechanism for sponsors to communicate with the FDA during drug development is the investigational new drug application (IND) process, which is legally required when the sponsor wants to move from testing the product in animals to testing in humans. Each document filed to the sponsor's IND becomes a part of the official record and is considered a formal message to the FDA. Communications to the IND build a story that starts when a sponsor expresses intent to dose a patient and may continue beyond the time a product is marketed. Neuroemergency indications fall into a special category of unmet medical need where close and early communications with the FDA are encouraged and mechanisms are provided to facilitate the relationship between the FDA and sponsors at the "pre-IND" stage.

There are two categories of INDs, commercial and investigator sponsored. Commercial INDs are sponsored by a company that financially supports the conduct of clinical trials and may involve multiple clinical sites, whereas investigator-sponsored INDs are generally financed and conducted by a physician at one study site. In the latter case, the investigator is the sponsor of the IND and assumes all legal responsibility for the conduct of the study and under whose direction the investigational drug is administered or dispensed (1). For both IND categories, sponsors are not allowed to proceed with proposed clinical studies until a 30-day waiting or "exemption" period has been completed.

When a drug is commercially available, an investigator may choose to initiate an IND for a new use without notifying the commercial sponsor in the event they are willing to assume responsibility for clinical supply costs. For unapproved products, investigator-sponsored INDs must include a letter of permission from the marketing application holder authorizing the applicant to cross-reference manufacturing information in their file. Data from trials conducted under investigator-sponsored INDs may prove to be useful in the overall drug development process as long as the integrity of the data is retained. Data from investigator-sponsored studies cannot contribute to a statistically significant cohort, but these data may provide supporting evidence to proceed directly to phase III. It may be appropriate for neuroemergency development programs to build on phase II data from an investigator-sponsored study and proceed directly to phase III. When considering fast track development (see Fast Track Development Programs, later text), front-loaded planning is critical to a program's success. Investigator-initiated studies must be carefully monitored and designed so that potential sponsors, and later the FDA, are confident an optimal dose is found in phase II.

A sponsor may file an IND for a new drug once they establish by nonhuman *in vitro* and *in vivo* methods the product is reasonably safe for initial use in humans and exhibits pharmacological activity that justifies commercial development. For a marketed product or a product already under investigation in another indication, a new IND is opened for the additional indication. Some sponsors start a development program for a new indication under an existing IND, but this is not advisable. Ultimately, indications should be broken out by IND. Many neuroemergency trials start in phase II, but if the sponsor initiates a new program for a drug that is first tested in humans, the sponsor is always required to conduct new preclinical studies designed to provide necessary evidence to support the safety of administering the compound to humans.

Regulatory requirements in ICH and TBI include animal toxicity studies and the demonstration of pharmacological activity in at least two rodent models and in nonhuman primates if feasible (2). The pharmacology of the drug must be characterized and the sponsor must demonstrate its mechanism. For TBI, sponsors should elucidate blood–brain transport activity and concentrations required to induce activity in the brain. Behavioral and histological data in animals provide a rationale to support the use of the drug (3). Additional animal studies are important to model the efficacy of the drug in humans, particularly for the fast track programs described later in this chapter. Accelerated programs require thorough preplanning, and to maximize the associated time savings, sponsors should conduct critical analyses at the preclinical stage so animal studies are predictive of human efficacy outcomes. Clinical programs are expensive, so predictive models are desirable when designing phase II clinical programs, particularly when the development timeline is collapsed in programs regulated under 21 CFR 312 Subpart E (see Subpart E Programs, later text).

In stroke, different nonclinical paradigms are used to support the use of neuroprotective drugs and stroke recovery drugs. During a 1999 stroke therapy academic roundtable by the University of Massachusetts, recommendations for standards regarding preclinical studies in stroke were discussed. For neuroprotective drugs, the following assessments were recommended before initiating clinical trials (4):

- Correlate an adequate dose-response curve with serum levels defining at least the minimally effective and maximally tolerated doses in at least one species (typically rat);
- Perform time window studies showing benefit when therapy is initiated at delayed time points after stroke onset in animal models;
- Demonstrate reproducible treatment effects in randomized blinded animal studies;
- Describe outcome measures including both infarct volume and functional assessment in both acute and long-term phase animal studies;
- Conduct initial studies in species such as rodents subjected to permanent occlusion models (unless reperfusion is necessary for drug effect);
- Consider preclinical studies in non-human primates for novel drugs.

Stroke recovery models predictive of efficacy have not been standardized but may include any number of sensorimotor, behavioral, and cognitive tests predictive of functional outcome. Route of administration must also be rationalized in animal models (5).

After an IND is filed, the sponsor then focuses on collecting data and information necessary to establish that the product will not expose humans to unreasonable risks when used in limited early-stage clinical studies. Procedures and requirements for the submission to, and review by, the FDA of INDs are described in Title 21 of the Code of Federal Regulations (CFR) Section 312. The basic format and content of an IND is outlined in Form FDA 1571 (6). Considerations to be made

when proceeding to the IND stage in neuroemergency trials are as follows (7):

- First in human neuroemergency studies are normally phase II trials to confirm the drug's mechanism of action and to establish optimal dose (for Subpart E programs, these must be established in phase I).
- Phase II trials must focus on the population of interest. Head trauma patients are heterogenous, so inclusion criteria must be carefully focused. Phase II and III trials compare active and placebo groups.
- The success of any clinical trial is dependent on the selection of appropriate endpoints. Therefore, sponsors should discuss all aspects of trial design at a pre-IND meeting. Before presenting trial designs to the Agency, all aspects of phase II trial design must be carefully defined and substantiated.

An emergency-use IND may be submitted to the FDA for the emergency use of an unapproved investigational drug or biologic in cases where a subject does not meet the criteria of a current protocol. At times, when an approved study protocol is not in place, the drug or biologic may be made available for the emergency use under the company's IND. In the case of ICH, TBI, or acute stroke, it is impossible to prospectively define whether a subject will be eligible for treatment, thereby rendering an emergency-use IND impractical.

Treatment INDs, described in 21 CFR 312.35, are used to make promising new drugs available to desperately ill patients as early in the drug development process as possible and before marketing of the product begins. The FDA will permit an investigational drug to be used under a treatment IND if there is preliminary evidence of drug efficacy and the drug is intended to treat a serious or life-threatening disease or if there is no comparable alternative drug or therapy available to treat that stage of the disease in the intended patient population. An immediately life-threatening disease is defined as a stage of a disease in which there is a reasonable likelihood that death will occur within a matter of months or in which premature death is likely without early treatment. Patients enrolled in treatment INDs are not eligible to be enrolled in pivotal clinical trials, which should be well underway or completed at the same time. The available patient pool for neuroemergency studies is rather limited; therefore, companies most often opt not to pursue treatment INDs and rather focus their efforts recruiting patients into controlled clinical trials. To date, there have been 39 treatment INDs that have been allowed to proceed by the FDA. Of these 39, 13 were for cancer indications and 11 were for HIV/AIDS indications (8).

When the FDA does not believe or cannot confirm that a study can be conducted without unreasonable risk to patients, the study is placed on "clinical hold" (9). Clinical hold as described in 21 CFR 312.42 is a mechanism by which the FDA delays a proposed clinical investigation or suspends an ongoing investigation. In general, the Agency initiates a clinical hold when it believes (10)

- There is an unforeseen or demonstrated risk to human subjects;
- Clinical investigators are not qualified to conduct the study;
- The investigator brochure is inadequate;
- The IND does not contain sufficient information to allow the FDA to conduct a risk assessment;
- The study includes women but involves reproductive risk (special exceptions are described in 21 CFR 312.42 (b)(5)(a–c)).

The FDA may also impose a clinical hold for studies involving an exception from informed consent when any of the requirements for the exception are no longer met (see Waiver from Informed Consent Requirements, later text). Most of these deficiencies can be avoided by careful planning and measuring the FDA's expectations based on requirements for products in the same class or therapeutic indication. After a clinical hold is issued, a sponsor must address each deficiency before the hold is lifted (see 21 CFR 312.42). Sponsors improve their chances of avoiding clinical holds or resuming stalled clinical studies by effectively communicating their plans to the FDA.

COMMUNICATING WITH THE FDA

Effective communication with the FDA is best achieved by fostering a nonadversarial relationship with the Agency, and this requires a commitment of appropriate time and resources to focus on defining a plan and executing against that plan. Sponsors are asked to communicate their "general investigational plan" in their original IND, and then sometimes find themselves changing the plan without discussing these changes with the FDA. The FDA operates in a data-driven domain where it is best to be forthright with plans to avoid surprises. Sponsors should examine in advance the implications of each decision and be willing to examine each plan from all angles so that if one aspect of a development program fails, there is a contingency plan. A solid relationship with the FDA is built on forthright communication. Agency meetings are an excellent form of communication and are encouraged for development programs that are intended to address an unmet medical need, and close early communication helps to prevent future mistakes. Meetings with the FDA fall into one of three categories described in Table 15.1.

Before filing an IND, sponsors may request a pre-IND meeting with the FDA. Pre-IND meetings commonly are conducted to reach agreement on the design of animal studies needed to support human clinical testing. However, pre-IND

TABLE 15.1 Categorical FDA meeting types

Meeting type	Description	Meeting timing	Background package timing
Type A	A meeting that is necessary for an otherwise stalled drug development program to proceed (e.g., to address an issue that has resulted in a clinical hold).	Meeting held within 30 days of request	Background package submitted at least 2 weeks before meeting
Type B	A pre-IND, end of phase I meeting (for Subpart E, Subpart H, or similar products, e.g., a product and indication designated as fast track program), end of phase II meeting, or a pre-BLA or NDA meeting. One meeting of each type is usually requested.	Meeting held within 60 days of request	Background package submitted at least 1 month before meeting
Type C	Any other type of meeting (e.g., cost recovery, facility design, or general product issues meeting).	Meeting held within 75 days of request	Background package submitted at least 1 month before meeting

Source: Manual of Standard Operating Procedures and Policies: Scheduling and Conduct of Regulatory Review Meetings with Sponsors and Applicants. Food and Drug Administration website. Available at http://www.fda.gov/cber/regsopp/81011.htm. Accessed on February 2, 2005.

meetings are also appropriate for new development programs for an established or investigational drug where the product is already being studied under an existing IND and the sponsor intends to initiate a phase II study. In the absence of a plan to file an IND, sponsors may discuss general drug development issues with the FDA in a type C meeting.

For Subpart E programs (see Subpart E Programs, later text), end of phase I meetings are appropriate after completion of early phase I studies to review phase I data and reach agreement on plans for pivotal phase II studies to support licensure (11). Information necessary to support marketing approval is also discussed. For Subpart E programs, an end of phase I meeting replaces a traditional end of phase II meeting. End of phase II meetings are scheduled to review phase II data to determine whether it is safe to proceed to phase III, to evaluate plans for the phase III program and protocols, and to identify any additional information necessary to support a marketing application for the uses

under investigation (12). Sponsors of neuroemergency studies should never proceed to phase III in the absence of an end of phase II meeting. End of phase II and pre-IND meetings are necessary should a sponsor pursue fast track designation (see Fast Track Development Programs, later text); therefore, sponsors should always express their intent to eventually request fast track designation, thereby establishing important regulatory criteria for their development program.

When phase III pivotal clinical trials are subject to discussion at an end of phase II meeting and data from these studies form the primary basis of an efficacy claim, sponsors may request "special protocol assessment" from the FDA. Such assessment is particularly useful to secure the FDA buy-in on the design, conduct, and analysis of key protocols. In the absence of broad and recent precedence for the approval of neuroemergency treatments, it is important to assess the FDA's current expectations for approval. The FDA recommends that a sponsor submit clinical protocols for assessment no later than 90 days before the anticipated start of a study and after having met with the review division (13). The Agency may provide their assessment in the absence of a meeting, but only when they are already familiar with the developmental context of a proposed clinical trial. Protocols for additional pivotal studies initiated after a phase III program is already underway or protocols for supplemental indications for an approved product may fall into this category. Assessment requests for protocols of neuroemergency studies should include a cohort of information necessary to assist the FDA in their evaluation. In addition to and separate from the proposed study protocol, sponsors must include focused questions describing issues specific to the protocol design, study conduct, goals, and data analysis. Projected regulatory outcomes should be described including proposed language for labeling that is supported by the described study. Other supportive information may include but is not limited to a copy of the proposed statistical analysis plan, imaging charter associated with endpoint assessment, and unique pages of the subject case report form.

The FDA has a 45-day review time frame within which they issue their comments in a special protocol assessment letter. Comments are primarily based on questions posed by the sponsor, supportive data, information provided in the assessment request, and relevant FDA policies and guidance. For neuroemergency programs, pivotal phase III studies should always be discussed at an end of phase II meeting; therefore, sponsors of these studies should already be poised to take advantage of this review mechanism. Special protocol assessment is also available for carcinogenicity protocols and certain stability protocols. Procedures for submitting a request for special protocol assessment in addition to the Agency review process are described in detail in the FDA's Guidance for Industry: Special Protocol Assessment (14).

Neuroemergency medicine is a field devoid of treatment options, where few products reach the approval stage. Nonetheless, promising treatments eventually proceed through the approval process. In the recent past, sponsors have requested, and been granted, pre–biologics license application (BLA) or pre-NDA meetings in the absence of unblinded phase III data. Often the FDA and sponsors have met to discuss a promising program only to discover the product failed in phase III. The FDA has become wiser in the process and now rarely grants meetings unless the sponsor provides positive data from unblinded studies. It is clearly in the best interest of both parties to meet with a full set of data. The sponsor generally drives the discussion points during a pre-BLA/NDA meeting through focused questions submitted in a premeeting background package. Therefore, the general meeting topics outlined in the FDA guidelines for pre-BLA and pre-NDA meetings are flexible (15,16). For programs with marketed products, meetings focus on clinical and statistical issues, but for new investigational products, meetings also focus on manufacturing issues.

A sponsor may request a separate pre-NDA or pre-BLA meeting to focus only on manufacturing issues, thereby maximizing time spent in a clinical/statistical focused meeting. Two meetings, one focused on manufacturing issues and another focused on all remaining data, may be scheduled on the same day or sequential days to accommodate sponsor travel schedules. Separate meetings must be requested in advance at the time of the original meeting request with a description of proposed topics for each. To hold a productive meeting, sponsors should provide the Agency with as much information as possible in advance of the meeting. Guidelines state that pre-BLA/NDA meetings are held to describe general information that will be submitted in a marketing application, to discuss preliminary efficacy results and methods for final statistical analyses, to discuss the proposed format for data in a planned marketing application, to identify pivotal studies, and to discuss any major outstanding issues (17). To ensure productive discussion, it is critical to provide the FDA with a useful information package that includes the following:

- Outline of completed nonclinical and clinical studies and overview of all studies to be submitted in the marketing application, including proposed size of safety and efficacy database;
- Brief discussion of preliminary unblinded results from phase III pivotal trials;
- Proposed indication statement for labeling;
- Overview of plan to integrate safety and efficacy analyses.

If a separate manufacturing meeting is not required, basic manufacturing information should be provided including names and addresses of the manufacturer and distributor and a summary of release tests and criteria.

As with all documentation provided to the FDA, the quality of a background package is greatly improved with adequate review from all necessary functional areas as well as outside experts. The success of an FDA meeting is dependent on the FDA's ability to read and understand the background package. Therefore, sponsors should always include tables of contents with each submission, use clear consistent pagination, label each section with tabs, and provide clear concise summaries of information. When citing published literature, full copies (not just abstracts) of the referenced material must always be provided along with the package. When cross-referencing another submission, sponsors should be sure to provide summary information of the reference in the background

package to minimize the need for the FDA to search for necessary information. Each review division at the FDA handles hundreds of documents, so background packages must be self-contained with all information necessary for the review. Sponsors should not assume the FDA has any prior knowledge of their development program, so packages should be written as if to a naive audience.

ROLE OF THE CORE DATA SHEET

The language contained in a product label provides an important public message describing a product and its intended use. Production of preferred labeling information early in a development program directly supports decision-making processes related to clinical trial design, implementation, and analysis. When starting a development program, regardless of its stage, an internal "company core data sheet" (CCDS) describing language for required essential labeling information provides an overall picture of potential risks versus benefit and, over time, brings the larger picture into focus as development progresses. The CCDS is an internal company document used by a sponsor to map essential language to propose or "target" labeling information for different regions. In turn, preferred labeling language in conjunction with the CDS is then used to drive the generation of early or late stage product development plans and position strategy for discussions with the FDA.

The language used in the CCDS is a representation of available data, whereas language in regional target labeling represents the sponsor's prospective goals. The CDS includes clinical data such as therapeutic information, posology, contraindications, special warnings and precautions, adverse effects, overdose, pharmacokinetic (PK) and pharmacodynamic (PD), and a summary of safety experience. Target labels describing prefer-red language are useful to drive the design of clinical studies in support of development programs and often are developed in consideration with competitor information and market positioning. The FDA provides specific guidance on developing the clinical sections of product labels in the FDA Guidance for Industry, "Clinical Studies Section of Labeling for Prescription Drugs and Biologics: Format and Content" (18).

When developing a target label in the United States, sponsors should carefully review and consider competitor approval information. Competitor information is available through the Freedom of Information Act at http://www.fda.gov/cber/efoi/approve.htm for biologics and http://www.fda.gov/cder/da/da.htm for drugs. For example, in stroke, sponsors should examine the FDA's clinical review of Genentech's Alteplase, available at http://www.fda.gov/cder/biologics/products/altegen061896.htm. In programs for which there is no standard of care, all sponsors should evaluate product approval information as well as advisory committee transcripts for therapies in the same product class. Class assessments are particularly useful when evaluating the FDA's position on safety standards and allow sponsors to examine safety trends and make program impact assessments. Complete packages of information including live taped videos of advisory committee meeting are available at http://www.FDAlive.com.

Regardless of the program development stage, CDS and target label information provide a useful reference for evaluating programs goals in support of program decisions and discussions with regulatory agencies (Fig. 15.1).

FAST TRACK DEVELOPMENT PROGRAMS

The FDA's fast track program was first described in the FDA Modernization Act (FDAMA) of 1997 (section 506, 21 U.S.C. 356) as an initiative to facilitate the development of products intended to treat serious and life-threatening diseases where no alternative therapies exist. When FDAMA was first legislated, sponsors had high hopes for this program and little understanding of how the FDA would interpret and implement its nuances. The FDA first described the program in the Guidance for Industry: Fast Track Drug Development Programs: Designation, Development, and Application Review, issued September 1998. For the 2004 fiscal year, the Center for Drug Evaluation and Research (CDER) received 87 requests for fast track designation, 45 of which were granted. The median response time by the FDA was 50 days (19). Programs for neuroemergency indications have received fast track designation, none of which has made it to the market to date (20,21).

As with some pieces of legislation, fast track has evolved to become somewhat different than what was expected, but in many ways has proven itself to be useful. The guidance was updated in July 2004 to reflect the Agency's current thinking based on 6 years of experience with the program. To maximize the opportunities offered within the fast track legislation specifically for therapies intended to treat neuroemergency indications, it is essential to fully understand what it is and how to best use it.

The guidance references three programs referred to as "fast track programs" but not automatically conferred along with fast track designation. The three programs referenced in the guidance are Accelerated Approval (21 CFR 314 Subpart H for drugs; 21 CFR 601 Subpart E for biologics), Priority Review (CBER SOP 8405; CDER MAP 6020.3), and Subpart E (21 CFR 312 Subpart E for drugs; 21 CFR 601 Subpart H for biologics), sometimes referred to in the Industry as accelerated development. Table 15.2 compares each program. Fast track designation provides for increased

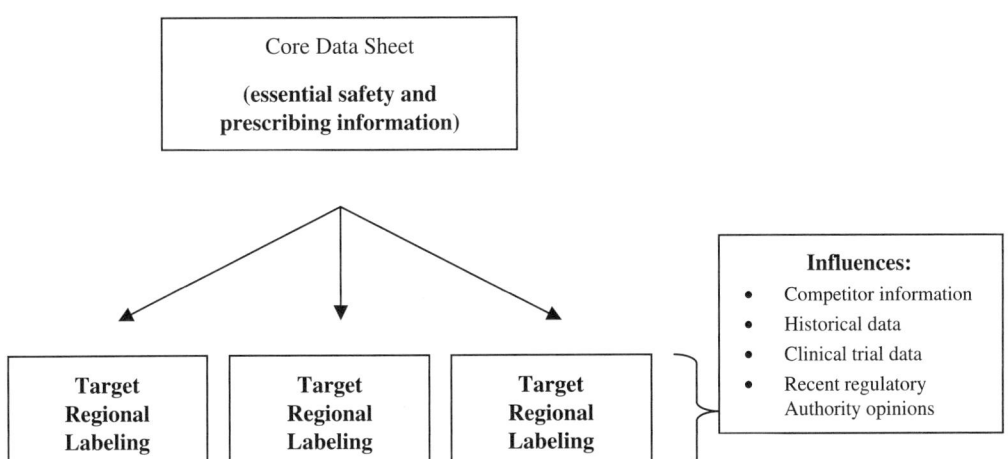

FIGURE 15.1 Relationship between clinical sections in the core data sheet and region-specific labeling.

communication with the Agency and provides a mechanism for submitting pieces of a license application separately (e.g., "rolling review"). It does not automatically confer rights to the other programs, which must be warranted separately. Fast track designation provides a mechanism by which the FDA acknowledges they will participate in specific meetings such as pre-IND, end of phase II, and pre-NDA or pre-BLA meetings. Some FDA divisions traditionally have been less amenable to agreeing to frequent meetings with sponsors. To be fair, the FDA is under a deluge of meeting requests, some of which may not be necessary. Products for ICH, TBI, and acute stroke are often reviewed in the Division of Neuropharmacological Drug Products (or Neuropharm Division), which is inundated with meeting requests. It is often easier to schedule meetings for fast track products, so the designation can be quite useful in this regard. The utility of rolling review is questionable. Technically, the Agency is not required to review a license application until all the pieces are submitted, and there is no proof that rolling review results in rapid approval.

Fast track is a designation that is given to a product and corresponding development program that has the potential to address an unmet medical need. The fast track program is centered around building an excellent working relationship with the FDA to bring successful products to market. A sponsor may request fast track designation at any time during a development program, but a program cannot be designated without hard evidence that a product will actually work in the intended population. For example, if an applicant is able to demonstrate potential through nonclinical studies, specifically pharmacologic and animal model data (22), these studies can be presented to the FDA at a pre-IND meeting and fast track can be incorporated into pre-IND discussions. In cases where there are no predictive animal models, it is impossible to obtain designation at this stage of development, so designation can only be made on the basis of well-controlled trials. Practically, neuro-emergency trials cannot be carried out as active comparator trials, so a placebo control is necessary. Theoretical evidence is not enough to sway the Agency, and as the guidance points out, the program is intended for promising new therapies, the potential of which must be established.

Promising programs exploring treatments for stroke or head trauma theoretically meet the criteria for fast track designation as long as they demonstrate

TABLE 15.2   Comparative summary of the FDA's fast track programs

Program	Criteria	How does it accelerate development?	Important considerations	How to apply
Fast Track Designation (Section 112 of FDAMA)	Development program has potential to address unmet medical need for serious or life-threatening condition.	Separate program that is built on close and early communication with the FDA throughout development program. Provides for "rolling review," which allows the sponsor to submit pieces of license application one at a time. The FDA has the option to review on an as-submitted basis, theoretically shortening time to approval once final piece has been submitted. Most fast track products qualify for priority review (not guaranteed).	Does not automatically confer faster marketing application review. Does not automatically confer accelerated approval or "Subpart E" development. Does not automatically mean the sponsor does not need two well-controlled pivotal trials to support licensure. The potential to address unmet medical need may be demonstrated via nonclinical data, indicating that a drug may be granted fast track designation before it has even reached the clinic based on its potential.	Requests for fast track designation are submitted in triplicate to the sponsor's IND as an IND amendment at any time during the development process, from the time of original IND submission to the marketing application. Such requests are provided as "stand-alone" documents, including copies of all literature references by which the product reviewer may evaluate the request.
Accelerated Approval (21 CFR 314 Subpart H for drugs; 21 CFR 601 Subpart E for biologics)	Claimed indication is normally serious or life-threatening illness in which product is superior relative to other treatment options.	Shortens development time by permitting approval based on early clinical endpoints or surrogate markers.	Accelerated approval does not confer a faster license application review time. Shortened review time, which is priority review, must be requested separately.	Discussion takes place during development program, but approval is conferred at the time of product approval. Discussions regarding use of early clinical endpoints or surrogate measures take place when negotiating appropriate trial design with the FDA at the end of phase II stage.

Priority Review (CBER SOP 8405; CDER MAP 6020.3)	In CDER: significant improvement in treatment, diagnosis, or prevention of nonserious or serious disease; in CBER, for a serious disease.	Priority review reduces marketing application review from 10 months (standard review) to 180 days or less.	Discussions should be incorporated early in the product development process to determine whether a product is eligible for priority review. At a minimum, sponsors should plan to add this item to the pre-NDA or pre-BLA meeting agenda.	Priority review is formally requested in writing at the time of BLA or NDA submission. This may be done as a stand-alone letter or as a separate white paper with accompanying rationale. The request may be submitted as an attachment to the license application cover letter.
Subpart E (21 CFR 312 Subpart E or 21 CFR 601 Subpart H)	Life-threatening and severely debilitating illnesses where there are no satisfactory treatment options.	Provides for shorter development by collapsing phase I and phase II requirements; sponsors proceed directly from phase I to phase II trials to support licensure. Phase II trials replace the requirement for larger phase III studies.	The sponsor must plan far enough ahead of time to incorporate dose-finding studies into the phase I trial (or other equivalent clinical trial) because they will not be allowed to proceed with a pivotal trial to support licensure until the optimal dose is found, and the test article is shown to be safe and well tolerated. Historical control subjects may only be used on a case-by-case basis, and must first be found acceptable to the FDA. A historical database must be scientifically sound and should be able to statistically support the pivotal clinical endpoints.	Subpart E development must be discussed at the pre-IND stage or at the time of requesting an end of phase I meeting.

TABLE 15.3   Fast track designated neuroemergency programs

Drug name	Indication	Product description	Fast track designation basis
Desmoteplase (Forest Laboratories, licensed from Paion, Frankfurt, Germany)	Acute ischemic stroke	Desmoteplase is a genetically engineered version of a clot-dissolving protein found in the saliva of the vampire bat *Desmodus rotundus*. It possesses a high fibrin selectivity, allowing it to dissolve a clot locally without affecting the blood coagulation system, which is though to potentially reduce the risk of intracranial bleeding.	Product shows potential to treat patients up to 9 hours past stroke onset in which current standard of care is 3 hours.*
Viprinex (Neurobiological Technologies)	Ischemic stroke	Viprinex is a thrombin-like enzyme obtained from the venom of the Malayan pit viper (*Agkistrodon rhodostoma*) and is highly specific for fibrinogen, producing anticoagulation by defibrinogenation. It has been reported to reduce the level of plasminogen activator inhibitor and may stimulate the release of tissue plasminogen activator from the endothelium.†	Product shows potential to treat patients up to 6 hours past stroke onset in which current standard of care is 3 hours.‡
Dexanabinol (Pharmos Corporation)	TBI	Dexanabinol is an anti-inflammatory compound thought to block synthesis of proinflammatory cytokines in the injured brain, slowing breakdown of the blood–brain barrier. It is also an *N*-methyl-D-aspartate receptor antagonist that may prevent the lethal massive influx of calcium ions into cells of the injured brain. As an antioxidant, it may help protect the brain by scavenging free radicals formed after injury.§	TBI is a serious often life-threatening condition for which no approved therapies exist. Dexanabinol showed potential in phase II to prevent secondary brain damage. This program failed in phase III (announced 12/20/2004).

* Press release issued by Forest Laboratories. Forest Laboratories Starts Confirmatory Study of Desmoteplase, A Novel Investigational Treatment for Acute, Ischemic Stroke. Feb 9, 2005. Available at http://ir.frx.com/ phoenix.zhtml?c=83198&p=irol-newsArticle&ID=672233&highlight. Accessed March 9, 2005.

† Viprinex Product Information Sheet. Available at http://www.rxmed.com/b.main/b2.pharmaceutical/ b2.1.monographs/CPS-%20Monographs/CPS-%20(General%20Monographs-%20V)/VIPRINEX.html. Accessed April 11, 2005.

‡ FDA Grants Fast Track Status to Viprinex. Yahoo Press Releases. Available online at http://biz.yahoo.com/ ap/050128/neurobiological_stroke_therapy_1.html. Accessed April 11, 2005.

§ Press release issued by Pharmos. Pharmos Receives FDA Fast Track Designation for Dexanabinol Accelerates New Drug Application Review for First Neuroprotective Brain Trauma Product: September 30, 2003. Available at http://www.biausa.org/Pages/BI_news.html. Accessed March 9, 2005.

potential to address unmet medical need (Table 15.3). On January 28, 2005, Neurobiological Technologies, Inc., in Richmond, California announced they received fast track designation for their product Viprinex for the treatment of acute nonhemorrhagic stroke (23). Viprinex is intended to decrease disability in stroke patients when administered within 6 hours of stroke onset. Tissue plasminogen activator, the only FDA-approved treatment for acute ischemic stroke, must be administered within 3 hours of onset, therefore leaving room for products such as Viprinex to receive fast track designation on the criteria of unmet medical need. Paion GmbH of Germany and Forest Labs of New York are collaborating on the development of a treatment, desmoteplase, for the treatment of acute ischemic stroke. The desmoteplase development program for stroke was granted fast track designation for its potential to treat acute ischemic stroke up to 9 hours after onset. Pharmos Corporation received fast track designation in September 2003 for dexanabinol in TBI, but the program was closed after failure of their phase III pivotal trial that was designed to determine whether dexanabinol helped patients regain their memory after TBI.

Experience has shown that sponsors are required to meet each of the criteria outlined in the FDA's fast track guidance to achieve designation. If all the criteria are met in the absence of predictive data, designation will not be granted (Fig. 15.2).

The FDA reserves the right to withdraw fast track designation at any point in a development program should any of these criteria cease to be met. The FDA recommends in the guidance that meeting packages for meetings after fast track designation "include a discussion of how accumulated data and study plans continue to demonstrate that the product and the development plan meet the criteria for fast track designation" (24).

ACCELERATED APPROVAL

Accelerated approval was legislated in 1992 to allow for earlier approval of drugs intended to treat serious or life-threatening illnesses; in 1997 it was codified as a fast track initiative under FDAMA. The FDA is obligated under the Federal Food, Drug, and Cosmetic Act to require sponsors to provide substantial evidence of safety and efficacy for a drug on the basis of adequate and well-controlled studies. To support claims under this paradigm, traditional safety and efficacy endpoints have been morbidity and mortality. However, for products intended to treat serious and/or life-threatening illnesses, the FDA may allow risk/benefit decisions to be based in part on evidence of clinical efficacy demonstrated by showing a positive effect on a surrogate endpoint (e.g., a biological laboratory finding) or a clinical endpoint other than survival or irreversible morbidity (25). Such endpoints can only be used in pivotal studies intended to demonstrate the usefulness of a drug that provides meaningful therapeutic benefit over existing treatments or where other treatment alternatives do not exist.

The use of surrogate endpoints may shorten a drug's overall time to market. However, marketing approval can be withdrawn if phase IV postmarketing studies fail to verify a drug's long-term clinical benefit. Accelerated approval is not related to priority review and does not confer a more rapid review and approval of a license application but rather allows for the shortening of an overall development program based on the evaluation of early clinical endpoints or surrogate markers. Accelerated approval is described under the FDA's fast track initiative; however, drug development programs without fast track designation may separately be granted accelerated approval. Likewise, development programs that qualify for fast track do not automatically qualify for accelerated approval.

Accelerated approval is something that is conferred at the time of product approval, but negotiations for approval based on a surrogate or endpoint reasonably likely to predict clinical benefit should be built into the drug development program as early as possible. Under the Accelerated

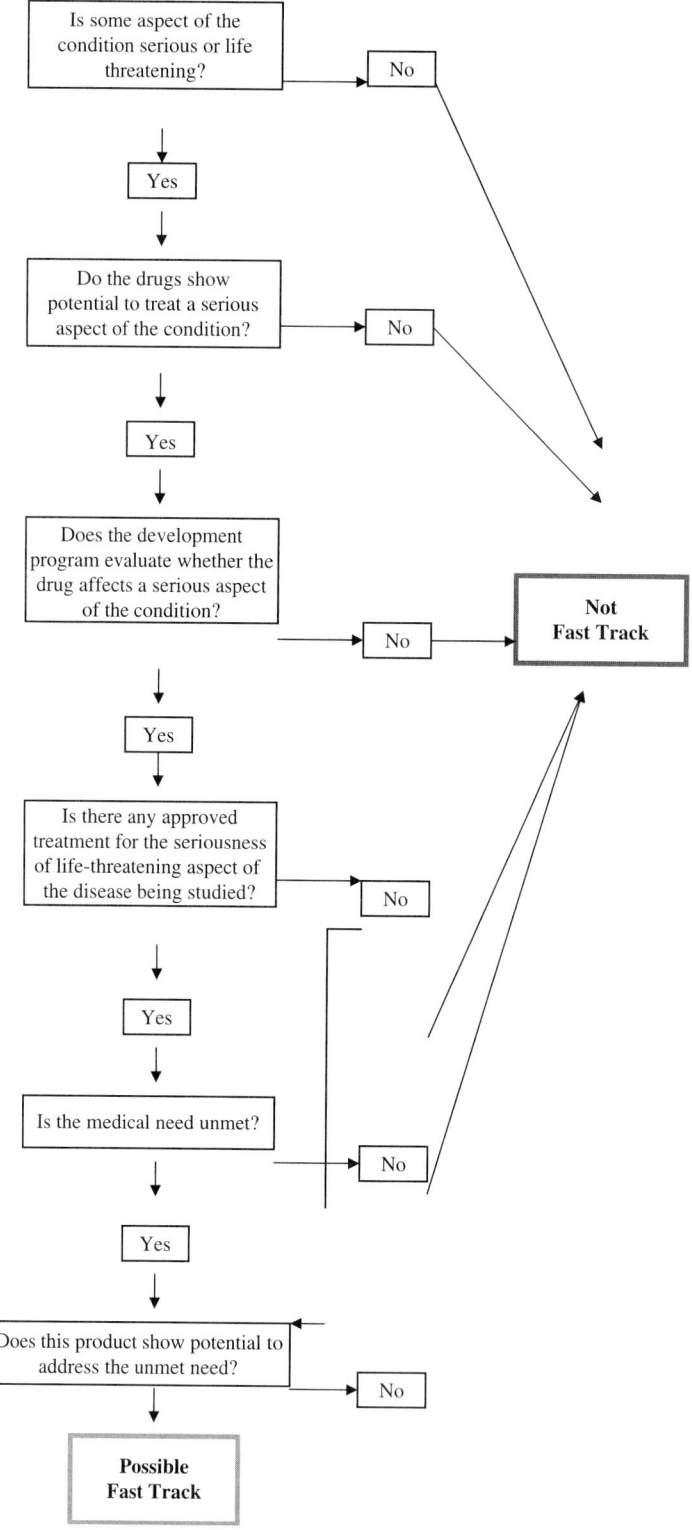

FIGURE 15.2   Fast track decision tree.

Approval Regulations (21 CFR 314 Subpart H), a surrogate must be reasonably likely to predict clinical benefit. Ideally, the surrogate does not just correlate with outcome, but the treatment actually affects the prediction made by the surrogate. During a May 2000 conference of the National Institute of Neurological Disorders and Stroke, researchers were encouraged to identify and validate surrogate endpoints for brain injury. Suitable surrogate endpoints are described as those ideally related to the biological processes underlying brain injury that would predict future status. In general, useful surrogates are quantifiable endpoints with a correlation to the degree of brain damage, associated with dynamic response with time, related to drug's mechanism of action, and they will also be technically feasible (26). Possible surrogates are described as intracranial pressure, therapy intensity level (SiVo$_2$), imaging, magnetic resonance spectroscopy (delayed ischemic hypersensitivity [DiH], membrane-bound tissue factor [MTF]), dialysis, and neurological worsening or improvement, but none of these meets all the defined criteria (27).

The choice of a surrogate marker for brain injury is problematic in that reliance on a surrogate marker is not necessarily predictive of successful clinical outcome (28). A treatment may affect the surrogate but not the disease mechanism and may negatively affect clinical outcome while there is a positive effect on a surrogate marker or early clinical endpoint. Surrogates are selected on a case-by-case basis because one surrogate that is appropriate to a particular drug or class of compounds may prove to be irrelevant to another. The FDA has explored surrogate markers such as intracranial pressure, biochemical markers, and imaging by magnetic resonance imaging (MRI) or computed tomography (CT). The Agency has focused on intracranial pressure as a short-term predictive effect, but to date sponsors have not shown a correlation with long-term clinical

benefit. Some research has correlated serum and cerebrospinal fluid concentrations of the calcium-binding protein S100B (S-100 protein, beta chain) with functional and morphological outcome in severe and moderate head injury, but these possible surrogates have not been accepted due to doubts regarding the ability to correlate the surrogate with clinical outcome (29).

As predictive as surrogate markers may be, early predictive endpoints sometimes lead to incorrect conclusions. When conducting stroke studies, there is no single predictive outcome measure and efficacy measures are designed based on the drug's mechanism of action. Stroke studies commonly integrate time-to-event or time-to-treatment analyses where endpoints of mortality, recurring thrombosis, or functional outcome are measures of success. Imaging endpoints such as MRI have not been considered acceptable surrogate markers to date but may be used to supplement primary efficacy data.

SUBPART E PROGRAMS

The Accelerated Approval provision, regulated under 21 CFR 314, Subpart H, is sometimes mistakenly confused with a separate initiative regulated under 21 CFR 312, Subpart E. This entirely separate initiative is sometimes referred to as "Subpart E" or as "Accelerated Development." During the IND stage, the fast track initiative described under 21 CFR 312 Subpart E allows for expediting product development where data from phase II studies are used to support licensure. Subpart E was established to expedite the development, evaluation, and marketing of new therapies intended to treat persons with life-threatening and severely debilitating illnesses, especially where no satisfactory alternative therapy exists. The term "life threatening" is defined as a disease or condition where the likelihood of death is high unless the course of the disease is interrupted and where the end-point of clinical trial analysis is survival. The term "severely debilitating" is defined as a disease or condition that causes major irreversible morbidity. Subpart E is built on the premise of close and early communication with the FDA through frequent meetings including pre-IND and end of phase I meetings. The end of phase I meeting takes the place of the end of phase II meeting as sponsors proceed directly from phase I to the conduct of pivotal trials to support licensure. The primary purpose of the end of phase I meeting described in 21 CFR Subpart E is "to review and reach agreement on the design of phase II controlled clinical trials, with the goal that such testing will be adequate to provide sufficient data on the drug's safety and effectiveness to support a decision on its approvability for marketing." The usual procedures for end of phase II meetings, including documentation of agreements reached, are also used. The FDA will conduct a risk-to-benefit assessment to determine whether benefits of a drug outweigh the risks.

Most accelerated approval programs are based on well-controlled phase II and phase III studies, and it is uncommon for a Subpart E program also to be approved on a surrogate endpoint under the accelerated approval regulations. In the absence of plausible surrogate markers for neuroemergency studies, a Subpart E program is still possible; however, data from phase II studies must be statistically significant (30).

PRIORITY REVIEW

Priority review was first described under FDAMA and provides for a 6-month review of a license application rather than the standard 10-month review. Priority review is ordinarily given to products that are intended to treat serious or life-threatening conditions and demonstrate the potential to address unmet medical needs for such conditions (31). Therefore, most products with fast track designation fit the prescribed criteria. A sponsor does not necessarily need to demonstrate benefit with a product in a head-to-head trial. Rather, if data are compelling and a case is made that the product is filling a void in the market, then priority review may be granted. Priority review is not automatically granted and sponsors must formally request it at the time of BLA or NDA submission. Sponsors should express their intentions to request priority review by the time of the pre-BLA or pre-NDA meeting. Rationale for the request can be discussed at that time; however, a formal request is submitted with the BLA or NDA, often as an attachment to the application cover letter in "white paper" format. The FDA meets internally to decide whether the application will be accepted or refused 45 days after filing. During this meeting, a decision is made whether to grant priority review and this is communicated to the sponsor shortly thereafter (Fig. 15.2).

WAIVER OF PROSPECTIVE INFORMED CONSENT

Normally, the FDA allows investigators to legally involve human beings as research subjects only after they have prospectively obtained informed consent from their subjects or the subject's legally authorized representative. To safeguard vulnerable patient populations from assuming undue risks and to minimize the possibility of coercion or undue influence, investigators may only obtain consent when the prospective subject is allowed sufficient opportunity to consider whether or not to participate in a clinical trial (32). Treatments of neuroemergency conditions must rely on established practices that produce less than satisfactory results in many patients. The American College of Emergency Physicians notes "we have an obligation to ensure that the American public receives the benefit of improvement in acute care medicine. Such improvement can only come through continued biomedical research into the causes and treatments for injury and illness" (33). Neuroemergency clinical trials differ from nonemergency trials in that physicians must respond instantaneously and make every attempt to administer lifesaving treatment in the absence of varied treatment options, often when the patient is incapable of prospectively providing informed consent and a legally authorized representative is not available to act on the patient's behalf.

Neuroemergency physicians, along with other emergency care practitioners, have a long history of discussions with regulatory authorities regarding options to waive prospective informed consent requirements for patients in emergency situations. In 1993, the concept of permitting exception from prospective informed consent for emergency research was first explored in a series of industry–FDA meetings. The FDA's rules were designed for single-use intervention and were not applicable to the totality of a clinical study. The FDA rules also allowed waiver of consent and institutional review board (IRB) approval based on assessing whether in a particular emergency-use setting (a) the subject is confronted by a life-threatening situation, (b) informed consent cannot be obtained, (c) there is no time to obtain consent from a legally authorized representative, and (d) no approved therapy is available in the therapeutic window (34). Department of Health and Human Services (DHHS) regulations did not waive IRB approval but did provide for a waiver on a patient-by-patient basis when research only involved a minimal risk to the subject and could not be carried out without the waiver (35). Researchers began to argue that "the severe conditions and treatments represented in emergency medicine were incompatible with this concept" (36) and that both the FDA and DHHS rules were

incompatible with neuroemergency trial standards.

Discussions ensued regarding the harmonizing of consent waiver policies between the FDA and the DHHS and their application to the emergency trial setting. These discussions prompted a congressional hearing in May 1994, and a draft proposal to amend the informed consent regulations was published in the *Federal Register* in September 1995 (60 FR 49086). After a notice and comment period and careful consideration, the final regulation was published in the *Federal Register* on October 2, 1996 (61 FR 51498). FDA standards are harmonized with the DHHS, who published its waiver criteria matching the requirements in the same *Federal Register* issue. In November 1996, federal regulations were finally amended to provide for an exception to the requirement for prospective informed consent when enrolling critically ill patients in clinical trials of emergency treatments, and these regulations are described in 21 CFR 50.24 (37). The Agency later issued a draft guidance on March 30, 2000 titled "Guidance for Institutional Review Boards, Clinical Investigators and Sponsors: Exception from Informed Consent Requirements for Emergency Research," that helped to clarify 21 CFR 50.24 by defining key aspects of the regulation (38). The FDA explained that the intent of the new regulation is to allow research on life-threatening conditions for which available treatments are unproven or unsatisfactory and where it is not possible to obtain informed consent, while establishing additional protections to provide for safe and ethical studies (39).

Investigators are required to obtain prospective informed consent unless the criteria described in 21 CFR 50.24 are met, yet practically speaking, interpretation of the exemption requirements has varied within the Agency. Obtaining prospective consent is deemed feasible unless both the investigator and a physician not otherwise participating in the study (this person may be a member or consultant to the IRB) certify in writing all the following (40):

- The human subject is confronted by a life-threatening situation necessitating use of the investigational drug;
- Available treatments are unsatisfactory;
- Prospective informed consent cannot be obtained because of an inability to communicate or obtain legally effective consent from the subject;
- There is insufficient time to obtain consent from a legally authorized representative;
- There is no reasonable way to prospectively identify the individuals likely to become eligible for participation in the investigation;
- Participation in the research may provide a direct benefit to the subject and the risks and possible benefit is supported by preclinical evidence;
- The clinical investigation could not practically be carried out without the waiver.

Investigators participating in trials where an exception from prospective consent has been granted by the FDA must attempt to contact the subject's family member within the therapeutic window who is not also the legally authorized representative to ask whether he or she objects to the subject's participation in the investigation. All attempts to contact a family member must be summarized by the investigator and related documentation must be made available to the applicable IRB at the time of protocol review (41).

As described in 21 CFR 50.24(d), protocols involving an exception to the standard requirements must be conducted under a separate IND, even when there is an existing IND for the same product. For products already under investigation in another indication, sponsors should file a protocol for emergency studies along with the request for waived consent and accompanying documentation described in FDA

Form 1571. Waivers are granted before the initiation of a study and therefore cannot be conferred while a study is in progress. The rules also require that before initiation of the study, the applicable IRB evaluates whether the proposed clinical trial is acceptable to members of the community where the trial will be conducted. To do so, the IRB must consult the community in a forum, providing opportunity for community members to learn about the proposed study and respond with their opinions about its acceptability to investigators or IRB representatives.

The FDA and DHHS rules do not define specific acceptable methods for performing community consultation, but one method is presented as an example. Baren and colleagues (42) from the Department of Emergency Medicine at the Hospital of the University of Pennsylvania in Philadelphia described a possible approach to community consultation for an emergency research study for the prophylaxis of posttraumatic seizures in children with severe closed head trauma. Parents of children being seen for minor traumatic injuries in three pediatric emergency departments were first asked whether they would participate in a study about informed consent. They were provided a description of a randomized placebo-controlled trial of phenytoin for the prophylaxis of posttraumatic seizures in children with severe closed head trauma and were then asked whether they would have consented for their own child's participation if their child had suffered such head injury. Sixty-six percent of parents (149/227) stated they would give consent for their child's participation. Other results are described. Baron and colleagues concluded that community consultation for determining the acceptability of an emergency research protocol could be obtained using interview techniques in the emergency department. This methodology provides an avenue by which investigators may gather data from a targeted community for presentation to

IRBs during the review of emergency research studies proposing a waiver of informed consent.

Neuroemergency trials are particularly difficult to conduct in the absence of an exception to prospective consent. Although it would appear that neuroemergency studies meet the necessary criteria, the FDA has shied away from granting exceptions for these trials. Experience has shown that once a trial has been initiated without consent exception, the FDA will not reconsider an exception. Additionally, the FDA has stated that if certain patients are able to consent whereas others cannot, two different patient populations are represented. It seems reasonable that if a trend toward a favorable risk-to-benefit profile is clinically demonstrated and many patients are lost to treatment due to their inability to consent, then an option for waived consent should be considered.

There is little question that neuroemergency trials meet criteria 1–6 described previously; however, there has been a gray area around 21 CFR 50.24(a)(4) requiring that the clinical investigation could not be practically carried out without the waiver. Although in some cases patients admitted to an emergency room are able to prospectively consent to a trial, more often these patients cannot speak, are unresponsive, and are not accompanied by a legally authorized representative. The FDA has taken a very conservative position regarding waiving the prospective consent, stating that if even one patient can be enrolled in the trial who is capable of consenting themselves, then waived consent cannot apply to the entire study. In these cases, the FDA anticipates the investigational study will eventually enroll. Although both consumers and the FDA had high hopes for this regulation, ethical debate and experience with failed clinical trials has prompted the FDA to take a more conservative approach.

The ethical debate around exception from informed consent rests on the argument that patient autonomy must be protected and vulnerable patients may be exploited. The standard of care for neuroemergency trials is ill defined with few available options. Therefore, one can argue that if a "hypothetical reasonable patient" faces a fatal outcome, they may be willing to assume potential risk associated with promising investigational treatment. Interview techniques may be used to assess whether, when faced with such a choice, a hypothetical reasonable patient would be more likely than not to choose the opportunity to receive an investigational treatment option. After consulting the community, the concern would shift to the potential harm due to an investigational treatment. Exception from informed consent requirements described in 21 CFR 50.24 state that "risks associated with the investigation are reasonable in relation to what is known about the medical condition of the potential class of subjects... and what is known about the risks and benefits of the proposed intervention or activity."

Neuroemergency trials are slow to enroll under conditions where waived consent is not optional, so is it ethical to disallow consent exception in these trials? It would seem reasonable that a sponsor should be allowed to proceed to incorporate waived consent in neuroemergency trials if the sponsor could demonstrate the following:

- Community consultation demonstrates the hypothetical reasonable patient would elect investigational treatment in a neuroemergency;
- Preliminary evidence of efficacy is available;
- Favorable risk-to-benefit ratio has been clinically demonstrated.

The regulations under 21 CFR 50.24(a)(3)(2) state that (in addition to other criteria) exception will be granted when evidence from animal studies supports the potential for the intervention to provide a direct benefit. Animal data have not always been predictive of clinical outcome in human studies, so it would be premature for the FDA to allow consent exception in the absence of predictive efficacy. But for data showing a favorable impact on efficacy in early human studies, there is ethical debate around disallowing an exemption for patients that would potentially benefit and for whom no other treatment options exist and where the risks-to-benefit ratio associated with the investigational drug is deemed reasonable.

In 1994, the DHHS granted a waiver from the prospective informed consent regulations for pegorgotein superoxide dismutase (PEG-SOD) for the treatment of severely head-injured patients. PEG-SOD, a free radical scavenger, was intended to prevent secondary injury and thereby improve the outcome of severe head injury (43). As described in 1996 in the *Journal of the American Medical Association*, a randomized placebo-controlled trial was conducted in 29 centers in 463 patients with a primary endpoint of Glasgow Outcome Scale at 3 months after brain injury (44). Secondary endpoints were mortality and Disability Rating Scale. Of the 463 patients enrolled, 162 received placebo, 149 received PEG-SOD 10,000 U/kg, and 152 received PEG-SOD 20,000 U/kg. The only statistically significant difference between the groups was a decreased incidence of adult respiratory distress syndrome in the 10,000 U/kg group as compared with the placebo group ($p < 0.015$). The primary and secondary endpoints were not statistically significant, prompting a move in the FDA toward a more conservative stance on granting waived consent.

One example of a high profile case in which the FDA has granted exception from informed consent is a clinical trial of PolyHeme, Northfield Laboratory's synthetic hemoglobin for the treatment of patients in hemorrhagic shock after traumatic injuries. The trial involves administering PolyHeme to patients in hemorrhagic

shock instead of saline at the scene of injury or during transport to the hospital (45). Public controversy over the Northfield study prompted further discussion about the ethics of dosing unconscious patients and waiving prospective informed consent requirements. Some bioethicists are concerned because the trial can continue even after the patient has been admitted to a hospital and that the criteria for the waiver, that satisfactory treatment is not available, is no longer being met once patients have access to blood. The current phase III protocol for PolyHeme allows for the trial to continue until 12 hours after admission to a hospital (46). On November 18, 2004, Northfield issued a statement saying that the FDA and the IRBs at 16 medical institutions (of 25 planned sites) had approved the research protocol, whereas 2 IRBs have not. In the midst of ethical debate, trials with PolyHeme continue to enroll and the FDA has stood by their decision to waive consent.

In general, the pendulum of waived consent has swung in a conservative direction; however, for programs where a positive preliminary safety and efficacy profile has been demonstrated, exception from prospective consent requirements should be considered and the future of waived consent studies should be reevaluated.

ORPHAN PRODUCT DESIGNATION

To obtain orphan drug designation (ODD), the disease indication for which a particular drug is being studied must first be classified as a rare disease, defined either as having a prevalence of 200,000 cases in the United States or, if the drug is a vaccine, diagnostic drug, or preventive drug, a condition with an incidence of less than 200,000 cases per year (47). As described in 21 CFR 360.20(a), sponsors may request ODD for a previously unapproved drug or for a new orphan indication for an approved drug. Sponsors of a drug previously approved as an orphan drug may request ODD for the subsequent drug for the same indication if the drug is chemically different or if they present a plausible hypothesis that the drug may be clinically superior to the first drug. Multiple sponsors may receive ODD for one drug for the same indication. Sponsors may also be eligible for designation if they are developing a product for which there is no reasonable expectation for product development costs to be recouped (48). Special incentives accompanying OPD are as follows:

- Waiver of the user fee (application fee) upon filing a marketing application (the fee for an application that includes clinical data is $672,000 in 2005) (49);
- Tax credits for clinical research undertaken by the sponsor to generate required data for market approval (50);
- Marketing exclusivity for an approved drug for 7 years after product approval (51);
- Protocol assistance from either CDER or the Center for Biologics Evaluation and Research (CBER) (52);
- Financial assistance from the Office of Orphan Products Development (OOPD), which provides up to $200,000 per year for phase I trials and up to $350,000 per year for phase II and III trials (53);
- For orphan development programs, a representative from OOPD will also attend any product-related FDA meetings at the sponsor's request.

When sponsors request designation based on the criteria of limited prevalence or incidence, the decision whether to grant designation hinges on the sponsor's case for a medically plausible population subset in addition to patient numbers. Dr. Marlene Haffner, Director of the Office of Orphan Products Management,

states that "problems may be encountered in determining what criteria the OPD reviewer should use to determine if a subset of a disease is medically plausible. Medically plausible subsets are groups of patients with special requirements differentiating them from a larger group" and "development of treatments for these medically plausible subsets of patients is unlikely without the benefits granted by the Orphan Drug Act" (54).

Also required is a discussion of the scientific rationale for use of the drug for the rare disease or condition along with supportive data from animal and human studies. Sponsors normally draw on supportive data available in an IND or published literature (55). When drawing from information already contained in a submission to the FDA (such as an IND), this information does not need to be repeated but may simply be cross-referenced to the submission and page. When literature is cited, full copies of each literature source must be provided. Additionally, copies of supportive literature to substantiate population incidence claims must be provided. Should the sponsor reference a page in a book, a copy of the chapter in which the reference is contained is sufficient. Certain neuroemergency indications may meet the criteria for orphan product designation. Incidence data for ICH and TBI in support of orphan designation are described.

Patients with TBI who experience long-term disability have already been established as a medically plausible orphan population. In August 2004, Pharmos Corp. was granted orphan drug designation for dexanabinol for the attenuation or amelioration of the long-term neurological sequelae associated with moderate and severe TBI, but to date no other products have achieved the same designation (56). According to the Centers for Disease Control and Prevention, reporting for 2004, each year in the United States an estimated 1.4 million people sustain a TBI. Of these, 235,000 are hospitalized and survive,

50,000 die, and 80,000 to 90,000 experience the onset of long-term or lifelong disability (57,58). There is an additional small but unknown proportion of all persons with TBI who are not hospitalized but may experience long-term disability. The Centers for Disease Control and Prevention estimates this number to be around 1 percent, which translates to an additional 10,000 persons with long-term disability. The overall conclusion is that between 80,000 and 90,000 persons become disabled each year from TBI, which meets the criteria for an orphan population (59).

Stroke is the interruption of blood supply to any part of the brain, normally caused by blood clots, whereas hemorrhagic stroke results from the rupture of an intracerebral vessel leading to the development of bleeding in the brain. Therefore, hemorrhagic stroke is thought to be a medically plausible subset of stroke in general. To date, no development programs have been granted ODD for the treatment of hemorrhagic stroke, even though clinical trials in the indication are ongoing. There is a question of whether products intended to treat conditions such as ICH also may also be useful in treating subarachnoid hemorrhage. This may be problematic for development programs of versatile products that address one indication rather than the other. Regardless, the numbers for hemorrhagic stroke in general meet the numeric criteria for designation. Stroke is the third leading cause of death and a major cause of serious disability in the United States. In 2002, the National Institute of Neurological Disorders and Stroke noted that more than 700,000 people have strokes each year, of which hemorrhagic strokes account for 15% and the remainder are ischemic (60,61). Additionally, 100,000 of these are recurrent cases (62). The conclusion is that around 90,0000 persons experience hemorrhagic stroke each year. Approximately 30,000 Americans experience subarachnoid hemorrhage each year (63). To date, one

product, nitroprusside, has been granted orphan designation for the treatment and prevention of cerebral vasospasm after subarachnoid hemorrhage (64).

The FDA provides helpful guidelines for submitting an application for ODD at http://www.fda.gov/orphan/designat/destips.htm. Products that have received orphan designation as well as orphan products that have received marketing approval are listed on the FDA's website at http://www.fda.gov/orphan/designat/list.htm.

IMAGING PROTOCOLS

When images are used as evidence of drug effect in support of a regulatory submission, consistent procedures must be used to ensure data are reproducible across multiple study sites. Imaging protocols or charters describe specified conditions under which images are obtained and are required. Imaging data are sometimes used in support of clinical outcomes in accelerated approval programs where efficacy evaluations are based primarily on the evaluations of such protocol images. Procedures for procuring both valuable and nonvaluable images, from both active and control groups, should be decided with the FDA before implementation. As with any clinical protocol, after submission, the FDA provides for a 30-day exemption period in which they may return comments; however, protocols may be implemented at risk to the sponsor within 30 days after submission. To minimize the need to change the conduct of a study, should an imaging component be integrated into clinical studies, the content of an imaging protocol should be discussed with the FDA along with potential trial design prior to submission of the protocol to the FDA.

Evaluation of protocol images by central readers should be completed before other images, such as nonprotocol images, and in some cases where large numbers of images are obtained or where image tapes are obtained, sponsors have implemented image selection procedures. This is generally discouraged because the selection of images can introduce selector bias. In cases where preselection is thought to be needed, the sponsor is encouraged to clearly identify and discuss the selection procedures with the appropriate agency division before their implementation.

Sponsors should prospectively describe in protocols of efficacy studies how missing images (in addition to images that are technically inadequate, uninterpretable, or show results that are indeterminate or intermediate) will be handled in the data analysis. For example, images may be missing from analyses for reasons such as patient withdrawal, image selection processes, technical challenges, or protocol violations. Sponsors may also incorporate analyses in a statistical analysis plan based on the principle of intention-to-treat but which are adapted to a diagnostic setting such as intention-to-image or intention-to diagnose (65). Analyses beyond intention-to-treat should be discussed and agreed to with the agency should the sponsor intend to use such data to support label claims. To eliminate bias, images must be read by a core laboratory that handles clinical trial data according to proscribed procedures defined in a protocol. Software used to acquire, store, and transmit information must be validated to rigorous standards.

HEALTH INSURANCE PORTABILITY AND ACCOUNTABILITY ACT PRIVACY REGULATIONS

In 1996, the DHHS implemented the Health Insurance Portability and Accountability Act (HIPAA), covering various topics from fraud and abuse control to revisions of criminal law. The area of HIPAA impacting pharmaceutical development is Title II, Subtitle F, administrative simplifications covering privacy and security rules. Subtitle F was effective April 2003 and fully enforced by the FDA April 20, 2005. The privacy and security rules provided special safeguards to protect the confidentiality of patient records. Persons conducting clinical research became liable for the misuse of "personally identifiable, non-public information" (66). Privacy is defined as implementing and maintaining controls over who is allowed to access information and security is the ability to prevent accidental or intentional disclosure, alteration, or destruction of information by unauthorized individuals (67).

Under the security rule, covered entities include any health care provider that electronically maintains or transmits health information relating to an individual. To comply, sponsors must assess the vulnerabilities of health data and then develop and implement relevant security measures. They must be able to identify and address known or suspected security incidents, such as attempted unauthorized access to or destruction of electronic protected health information (ePHI). They are required to have audit controls in place to monitor ePHI network activity. They must also implement measures to ensure that only authorized individuals or programs can access ePHI, and ensure that the individual or entity attempting to access a system is authentic. To do so, sponsors should develop appropriate standard operating procedures (SOPs) that describe methods for safeguarding the integrity, confidentiality, and availability of electronic data.

Sponsors are required to monitor security incidents against specifications described in SOPs and audit activities related to ePHI for compliance. Auditing processes must involve verification of everyone who is attempting to access the ePHI to provide assurance that only authorized users are able to access requested information.

Sponsor access control and user authentication mechanisms are often nonexistent in work environments incorporating MRI or CT, but security controls required by HIPAA may be adequately addressed by simply locking a room and providing only authorized staff with access to the equipment. These limited controls may be acceptable as long as they are described in company procedures that are implemented and enforced (68).

Some ways to protect image information include controlling viewing of images, removing patient information from plain site, implementing the proper security access to prevent patients from entering controlled areas, and making sure that patients do not overhear other patient information. From an imaging perspective, the software in CT and MRI machines is not editable; therefore, it is not easy for investigators to maintain total privacy of images at a site before transferring these images to a central laboratory. Images may be transferred electronically in an environment where patient identification cannot be completely protected as long as sponsors include language in an informed consent where the patient agrees to the manner by which data are transferred. Products such as MDInteractive (see www.mdinteractive.com) are available on the market to facilitate the transfer of CT and MRI data in a secure HIPAA-compliant environment.

LICENSE APPLICATION PROCESS

After a sponsor has shown under an IND that a product is safe and effective, they must compile supportive data in a license application for FDA review and approval before commercial distribution in the United States. License applications for new products are filed as an NDA or BLA. For new (or supplemental) indications for marketed products, sponsors file a supplemental NDA or BLA containing data from studies focused on the new indication. In 2001, CDER and CBER implemented an optional license application filing structure and formout based on International Conference on Harmonization guidelines describing a common application format, the common technical document (CTD). The structure and format of the CTD is described in International Conference on Harmonization M4 Guidance: The Common Technical Document (69). By late 2003, CTD submissions were becoming more common as the FDA worked to establish procedures for accepting electronic CTDs. Many sponsors have adapted to the new CTD format and continue to work toward implementation of electronic CTD requirements.

It is important to note that although the CTD format provides a different mechanism by which to present information, data requirements for product approval are unchanged. For new products, the FDA requires inclusion of physiochemical characterization, biological activity, manufacturing facility information, and nonclinical and clinical data. These data allow the Agency to review the application by making four principle determinations (70,71):

- Whether the benefit of the product outweighs the risks and whether it is safe and effective;
- Whether proposed labeling is appropriate;
- Whether manufacturing methods and quality controls are adequate to preserve the product's identity, strength, quality, and purity;
- Whether the manufacturing process is capable of producing a product consistent with specifications, good manufacturing practices, and relevant regulations.

The FDA provides an extensive description of the license application approval process in an interactive handbook available on their website (72). The description references CDER processes but is also applicable to CBER. The process flow, as described by the FDA, is provided as Figure 15.3. In general, after filing an NDA or BLA, the file undergoes a screening process during which the FDA evaluates the format of the file and determines whether the application is complete and includes necessary information to facilitate review of the file. For electronic license applications, this assessment also involves an evaluation of the electronic application

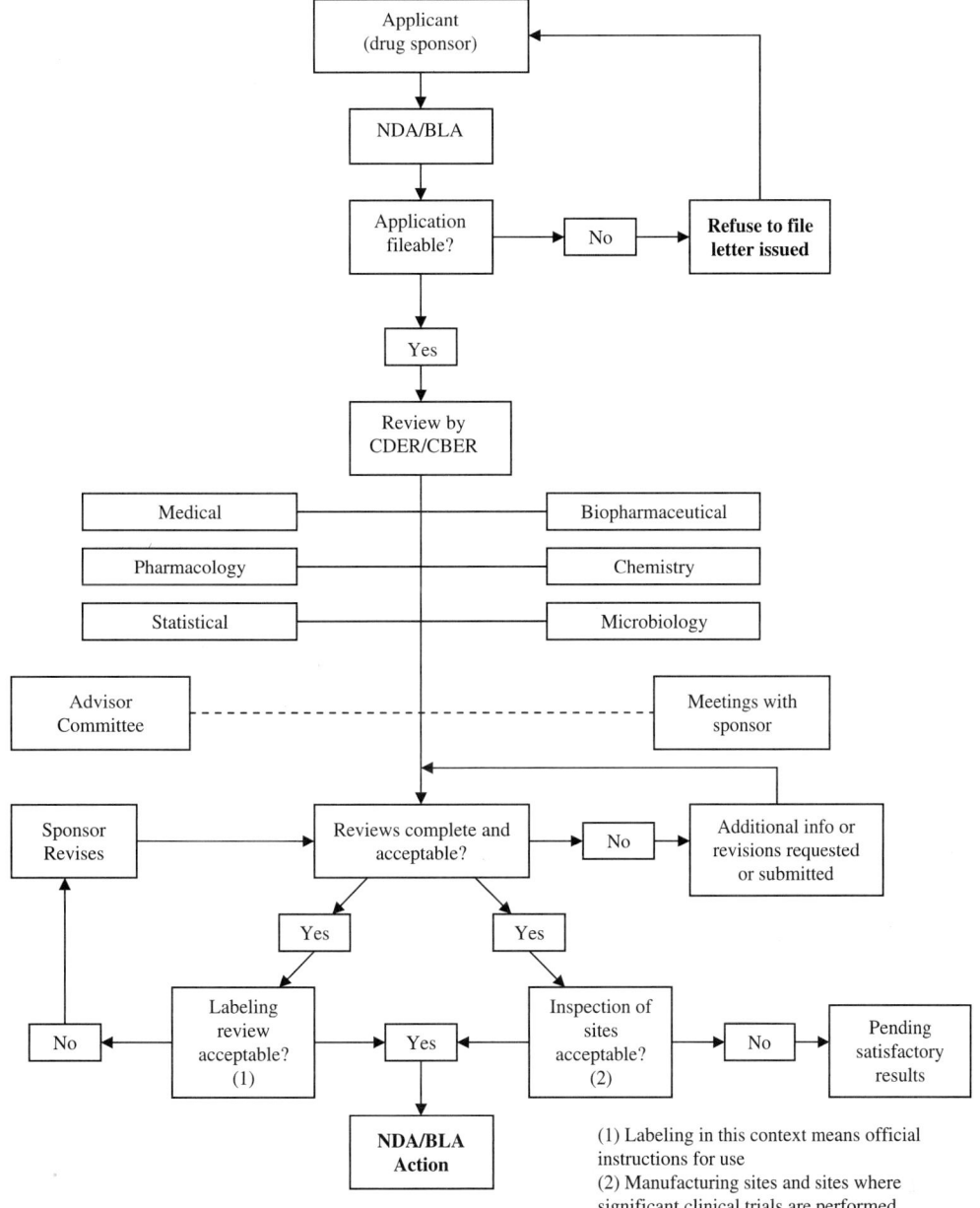

FIGURE 15.3 License application approval process. (From NDA Review Process. Food and Drug Administration website. Available at http://www.fda.gov/cder/handbook/nda.htm. Accessed on April 5, 2005.)

for compliance with electronic review requirements. Forty-five days after filing, the FDA holds an internal meeting to determine whether the file is acceptable for review, and within 60 days from filing, the FDA notifies the sponsor by issuance of a formal letter whether the file has been accepted. When an application is accepted for filing, the review focus depends on the nature of the application. For supplemental applications, chemistry reviewers may not be involved, and the review may focus entirely on clinical and nonclinical issues.

NDAs or BLAs are subject to a multidisciplinary review that includes clinical and statistical evaluations. In the United States the license application review process is extremely data driven, which means that it does focus on sponsor interpretation of clinical data. This differs from the review process in other regions such as Europe where reviews are driven by sponsor interpretation of data that must be supported by the data themselves. FDA clinicians and biostatisticians delve immediately into source clinical data provided electronically as data listings or SAS transport files. Clinical reviewers recreate each of the sponsor's analyses to support efficacy and safety claims and conduct sensitivity analyses to determine whether the data are robust. Only after manipulating source data and reaching their own conclusions will the FDA concur with sponsor claims.

During the review process, the FDA may consult with an advisory committee. The primary role of an advisory committee is to provide independent advice on a variety of issues such as whether a product should be approved or whether the FDA should implement a particular regulatory requirement. Advisory committees contribute to the quality of the FDA's product review process, thereby helping the Agency make sound decisions about new medical products and other public health issues. Often, advisory committees make strong recommendations, but it is ultimately the FDA's decision whether to approve a new drug or a new indication for an approved product. Each advisory committee advises a corresponding FDA review group, and for programs of neuroemergency indications, the main advisory committee of interest is the advisory committee of Peripheral and Central Nervous System Drugs (73).

Advance preparation for an advisory committee meeting is key to success. Effective preparation takes place during the development process, long before a license application is filed. For example, the first step is to know as much as possible about the members of the committee. Sponsors can assess how a member may potentially react by evaluating transcripts of prior committee meetings for review of products of the same indication or same drug class. Electronically searchable transcripts are available free of charge on the FDA's website, http://www.fda.gov. It is also useful to examine publications by that member to better understand a sense of their area of expertise. Sponsors should be aware of the current activities and issues of all patient advocacy groups for neuroemergency products such as the Brain Injury Association of America or the National Stroke Association. When a meeting is scheduled, preparation may be daunting. Therefore, although it is possible for larger companies with more resources to coordinate their own preparation activities, smaller companies may find it useful to engage the help of experienced consultants. An overview of preparation activities is provided by Harder and Perry (74).

Before product approval, labeling review and facility inspections must be complete. For approval of the label, each proposed statement in the label must be justified by data in the license application. Most often, label review takes place through several discussion rounds and multiple iterations of comments and changes between the sponsor and the FDA. Upon completion of all necessary actions, it is most often the division director who signs off on the action letter for approval of the product. Companies receive a letter by fax (followed by a hard-copy original) stating that the product is approved, not approved, or conditionally approved (approvable). Upon receipt of an approval letter, the sponsor is immediately free to market the product.

POSTAPPROVAL ISSUES: MARKETING, PATIENT REGISTRIES, ADVERTISING, AND PROMOTIONAL MATERIALS

After the FDA has granted approval to market a product, sponsors may be required to conduct an additional study(ies) or to gather further information to address unresolved questions about the product including safety, efficacy, or manufacture. Two different types of commitments are required: those that are conditional for approval and those that are necessary but not required for approval. One example of a commitment necessary for approval is an accelerated approval clinical benefit study to validate clinical benefit after approval based on a surrogate marker. Other post-marketing submissions include clinical data from studies initiated by the sponsor without a request from FDA and submissions to address other postmarketing commitments such as annual reports.

Sponsors may discuss planned postmarketing commitments as early as an end of phase II meeting and no later than the pre-NDA or pre-BLA meeting. Postmarketing commitments are agreed with the FDA before product approval and are officially documented in each product approval letter. Approval letters for products of the same indication and class as well as those of competitor products provide a guide to the FDA's thinking regarding possible postapproval commitments. For example, the product approval letter of 1996 for Genentech's alteplase for acute

ischemic stroke describes postapproval commitments as follows:

- To continue the ongoing study to evaluate the safety and efficacy of alteplase treatment in patients treated more than 3 hours after stroke onset;
- To conduct a prospectively designed uncontrolled clinical study in patients treated within 3 hours of stroke onset to identify patient characteristics that may be associated with an increased risk from treatment;
- To provide the electronic data set and additional safety analyses from The European Cooperative Acute Ischemic Stroke Study.

For accelerated approval products, FDA review and approval of advertising and promotional materials is required before distribution (75).

Product approval is normally based on a small subset of patients providing a representative sample of product safety in a broader population. Once a product is on the market, the FDA continues to follow it closely to assess product safety and associated risks. The Agency's postmarketing surveillance programs focus primarily on evaluating events that were not observed or recognized before approval and identifying adverse events that might occur because a product is not being used as anticipated. Postmarketing safety assessment and risk surveillance relies on two methods of adverse event reporting, the first of which is direct voluntary reporting by health professionals and consumers and the second of which is required reporting by sponsors. Required reporting is based on voluntary reports provided by manufacturers, health care professionals, and patients. Medical, statistical, and epidemiological experts within the Agency use these reports to provide and ongoing assessment of the product safety.

The Agency uses this approach to postmarketing risk assessment to evaluate the likelihood and seriousness of adverse events. The risk profile is progressively updated and weighed against demonstrated benefit. This evolving risk-to-benefit profile allows the FDA to make decisions about product safety and take necessary actions such as mandating an update of labeling information, requiring a Dear Health Care Professional letter, and sometimes reevaluating an approval decision (76).

CONCLUSION

It is important not only to be aware of the regulatory mechanisms governing the development of neuroemergency products, but also to understand the regulatory climate around these processes. Many procedures associated with the development of serious conditions provide a useful framework for the design and implementation of neuroemergency regulatory strategies. Sponsors considering the development of such programs should clearly define prospective goals through projected product labeling and identify regulatory strategies to support these objectives. Close and early dialogue with the FDA within the context of mechanisms such as pre-IND meetings is necessary to ensure the appropriate use of key strategies. Sponsors in the later stage of development may still be able to capitalize on mechanisms such as fast track, accelerated approval, priority review, special protocol assessment, and associated meetings with the FDA. Regardless of a program's development stage, sponsors should carefully consider how they will work toward effectively using the pre-IND, IND, and license application process to enter into a constructive collaboration with FDA.

References

1. Frequently Asked Questions on Drug Development and Investigational New Drug Applications. Food and Drug Administration website. Available at http://www.fda.gov/cder/about/smallbiz/faq.htm. Accessed February 5, 2005.

2. Investigational New Drug Application (IND) Application Process. Food and Drug Administration website. Available at http://www.fda.gov/cder/regulatory/applications/ind_page_1.htm. Accessed on February 20, 2005.

3. Clinical Trials in Head Injury. National Institute of Neurological Disorders and Stroke. Available at http://www.ninds.nih.gov/news_and_events/proceedings/headinjurywkshp.htm. Accessed February 20, 2005.

4. Recommendations for Standards Regarding Preclinical Neuroprotective and Restorative Drug Development. Stroke. 1999;30:2752. Available at http://stroke.ahajournals.org/cgi/content/full/30/12/2752. Accessed February 20, 2005.

5. Guidance for Industry, Single-Dose Acute Toxicity Testing for Pharmaceuticals. FDA website. August 1996. Available at http://www.fda.gov/cber/guidance/pt1.pdf. Accessed August 20, 2005.

6. Information for Sponsors-Investigators Submitting Investigational New Drug Applications. Food and Drug Administration website. Available at http://www.fda.gov/cder/forms/1571–1572-help.html. Accessed on April 11, 2005.

7. Hirschhorn W. Investigator Initiated Clinical Research, Planning, Developing, Conducting, Managing, and Succeeding. Presentation available at http://www.temple.edu/ovpr/oct/doc/investigator-Initiated%20Research.ppt. Accessed March 1, 2005.

8. Treatment Investigational New Drugs (IND) Allowed to Proceed, Product Index. Food and Drug Administration website, www.fda.gov/oashi/patrep/treatind.html. Accessed February 16, 2005.

9. Clinical Hold Decision. Food and Drug Administration website. Available at http://www.fda.gov/cder/handbook/clinhold.htm. Accessed April 1, 2005.

10. Title 21 Code of Federal Regulations Section 312.42. (b), Investigational New Drug Applications. Food and Drug Administration website. Available at http://www.accessdata.fda.gov/scripts/cdrh/cfdocs/cfcfr/CFRSearch.cfm. Accessed April 1, 2005.

11. Title 21 Code of Federal Regulations Section 312.82, Investigational New Drug Applications. Food and Drug Administration website. Available at http://www.accessdata.fda.gov/scripts/cdrh/cfdocs/cfcfr/CFRSearch.cfm. Accessed April 1, 2005.

12. Title 21 Code of Federal Regulations Section 312.47, Investigational New Drug Applications.

Food and Drug Administration website. Available at http://www.accessdata.fda.gov/scripts/cdrh/cfdocs/cfcfr/CFRSearch.cfm. Accessed April 1, 2005.

13. Guidance for Industry, Special Protocol Assessment. Food and Drug Administration website. May 2002. Available at http://www.fda.gov/cber/gdlns/protocol.pdf. Accessed May 12, 2005.

14. Guidance for Industry, Special Protocol Assessment. Food and Drug Administration website. May 2002. Available for http://www.fda.gov/cber/gdlns/protocol.pdf. Accessed May 12, 2005.

15. Guidance for Industry, Formal Meetings With Sponsors and Applicants for PDUFA Products. Food and Drug Administration website. February 2000. Available at http://www.fda.gov/cber/gdlns/mtpdufa.htm. Accessed February 5, 2005.

16. Guidance for Industry, IND Meetings for Human Drugs and Biologics Chemistry, Manufacturing, and Controls Information. Food and Drug Administration website. May 2001. Available at http://www.fda.gov/cder/guidance/3683fnl.htm. Accessed February 5, 2005.

17. Guidance for Industry Formal Meetings With Sponsors and Applicants for PDUFA Products. Food and Drug Administration website. February 2000. Available at http://www.fda.gov/cder/guidance/2125fnl.htm. Accessed February 5, 2005.

18. Draft Guidance for Industry Clinical Studies Section of Labeling for Prescription Drugs and Biologics—Content and Format. Food and Drug Administration website. July 2001. Available at http://www.fda.gov/cder/guidance/1890dft.htm. Accessed February 28, 2005.

19. CDER Response to Requests for Fast Track Designation, FY2004. Food and Drug Administration website. Available at http://www.fda.gov/cder/rdmt/internetftstats.htm. Accessed on April 11, 2005.

20. Henney J. International Conference on Surrogate Endpoints and Biomarkers National Institutes of Health. April 15, 1999. Food and Drug Administration website. Available at http://www.fda.gov/oc/speeches/surrogates8.html. Accessed on February 1, 2005.

21. Suydam L. Statement Before the Committee on Energy and Commerce, United States House of Representatives. May 3, 2001. Food and Drug Administration website. Available at http://www.fda.gov/ola/2001/fdama0503.html. Accessed February 1, 2005.

22. Guidance for Industry: Fast Track Drug Development Programs. July 2004. Food and Drug Administration website. Available at http://www.fda.gov/cder/guidance/5645fnl.htm. Accessed March 9, 2005.

23. Press Release: Neurobiological Technologies Receives FDA Fast-Track Status for Anti-Stroke Treatment in Development, Viprinex(TM) (ancrod). Yahoo Finance News. January 28, 2005. Available at http://biz.yahoo.com/prnews/050128/sff011_1.html. Accessed March 9, 2005.

24. Guidance for Industry: Fast Track Drug Development Programs. Section 5(B). July 2004. Food and Drug Administration website. Available at http://www.fda.gov/cder/guidance/5645fnl.htm. Accessed March 10, 2005.

25. Title 21 Code of Federal Regulations Section 314.510, Applications for FDA Approval to Market a New Drug. Food and Drug Administration website. Available at http://www.accessdata.fda.gov/scripts/cdrh/cfdocs/cfcfr/CFRSearch.cfm. Accessed March 15, 2005.

26. Clinical Trials in Head Injury Workshop. May 12–13, 2000. National Institute of Neurological Disorders and Stroke. Available at http://www.ninds.nih.gov/news_and_events/proceedings/headinjurywkshp.htm. Accessed on March 10, 2005.

27. Clinical Trials in Head Injury Workshop. May 12–13, 2000. National Institute of Neurological Disorders and Stroke. Available at http://www.ninds.nih.gov/news_and_events/proceedings/headinjurywkshp.htm. Accessed on March 10, 2005.

28. Tasker R. Pharmacological advance in the treatment of acute brain injury. Arch Dis Child 1999; 81:90–95. Available at http://adc.bmjjournals.com/cgi/content/full/81/1/90. Accessed on March 2, 2005.

29. Intracranial pressure monitoring in the management of penetrating brain injury. J Trauma 2001;51:S12–S15.

30. Guidance for Industry: Fast Track Drug Development Programs. Section 5(B). July 2004. Food and Drug Administration website. Available at http://www.fda.gov/cder/guidance/5645fnl.htm. Accessed March 10, 2005.

31. Center for Drug Evaluation and Research Manual of Policies and Procedures Priority Review Policy. Food and Drug Administration website. Available at http://www.fda.gov/cder/mapp/6020-3.pdf. Accessed on February 25, 2005.

32. Title 21 Code of Federal Regulations Section 50.20, General Requirements for Informed Consent. Food and Drug Administration website. Available at http://www.accessdata.fda.gov/scripts/cdrh/cfdocs/cfcfr/CFRSearch.cfm. Accessed March 15, 2005.

33. Statement of the American College of Emergency Physicians presented and the FDA/NIH public

forum. U.S. *Federal Register*: September 21, 1995:60(No.183), 49085–49103.

34. Title 21 Code of Federal Regulations Section 50, General Requirements for Informed Consent. Food and Drug Administration website. Available at http://www.accessdata.fda.gov/scripts/cdrh/cfdocs/cfcfr/CFRSearch.cfm. Accessed March 15, 2005.

35. Title 21 Code of Federal Regulations Section 50.23(a) Informed Consent of Human Subjects: Exceptions from General Requirements. Food and Drug Administration website. Available at http://www.accessdata.fda.gov/scripts/cdrh/cfdocs/cfcfr/CFRSearch.cfm. Accessed March 15, 2005.

36. American Medical Association Council on Ethical and Judicial Affairs. Waive of Informed Consent for Emergency Room Research. Report 1-A 97. American Medical Association. Available at http://www.ama-assn.org/ama1/pub/upload/mm/369/ceja_1a97.pdf. Accessed on March 4, 2005.

37. Title 21 Code of Federal Regulations Section 50.23(a) Informed Consent of Human Subjects: Exceptions from General Requirements. Food and Drug Administration website. Available at http://www.accessdata.fda.gov/scripts/cdrh/cfdocs/cfcfr/CFRSearch.cfm. Accessed March 15, 2005.

38. Draft Guidance for Institutional Review Boards, Investigators, and Sponsors: Exception from Informed Consent Requirements for Emergency Research. March 30, 2000. Food and Drug Administration website. Available at http://www.fda.gov/ora/compliance_ref/bimo/emrfinal.pdf. Accessed March 4, 2005.

39. FDA Information Sheet, Guidance for Institutional Review Boards and Clinical Investigators, Exception from Informed Consent for Studies Conducted in Emergency Settings: Regulatory Language and Excerpts from the Preamble, 1998. http://www.fda.gov/oc/ohrt/irbs/except.html.

40. Title 21 Code of Federal Regulations Section 50.24(a) Informed Consent of Human Subjects: Exceptions from General Requirements. Food and Drug Administration website. Available at http://www.accessdata.fda.gov/scripts/cdrh/cfdocs/cfcfr/CFRSearch.cfm. Accessed March 4, 2005.

41. Title 21 Code of Federal Regulations Section 50.24(a)(7)(5) Informed Consent of Human Subjects: Exceptions from General Requirements. Food and Drug Administration website. Available at http://www.accessdata.fda.gov/scripts/cdrh/cfdocs/cfcfr/CFRSearch.cfm. Accessed March 4, 2005.

42. Baren JM, Anicetti JP, Ledesma S, Biros MH, Mahabee-Gittens M, Lewis RJ. An approach to community consultation prior to initiating an emergency research study incorporating a waiver of informed consent. Acad Emerg Med 1999;6:1210–1215. Available at http://www.ncbi.nlm.nih.gov/entrez/query.fcgi?cmd=Retrieve&db=PubMed&dopt=Citation&list_uids=10609922. Accessed March 4, 2005.

43. http://www.liebertonline.com/doi/pdf/10.1089/089771502753754037.

44. Young B, Runge JW, Waxman KS, Harrington T, Wilberger J, Muizelaar JP, Boddy A, Kupiec JW. Effects of pegorgotein on neurologic outcome of patients with severe head injury. A multicenter, randomized controlled trial. JAMA 1996;276:538–543.

45. Robeznieks A. Blood product trial sparks informed consent debate: the FDA allows informed consent to be waived in emergency situations if other treatment is unavailable or unsatisfactory. Dec. 6, 2004. Available at http://www.ama-assn.org/amednews/2004/12/06/prsc1206.htm. Accessed on February 20, 2005.

46. PolyHeme® Phase III trial description. Northfield Labs Inc. website. Available at http://www.northfieldlabs.com/amb_trial.html. Accessed on February 20, 2005.

47. Title 21 Code of Federal Regulations Section 316(a) Orphan Drugs: General Provisions. Food and Drug Administration website. Available at http://www.accessdata.fda.gov/scripts/cdrh/cfdocs/cfcfr/CFRSearch.cfm?CFRPart=316. 316. Accessed March 15, 2005.

48. OOPD Program Overview, Food and Drug Administration website. Available at http://www.fda.gov/orphan/progovw.htm. Accessed March 15, 2005.

49. FDA Office of Financial Management User Fees for 2005. Food and Drug Administration website. Available at. http://www.fda.gov/oc/oms/ofm/userfees/userfees.htm. Accessed April 2, 2005.

50. Tax Credit for Testing Expenses for Drugs for Rare Diseases or Conditions. Food and Drug Administration website. Available at. http://www.fda.gov/orphan/taxcred.htm. Accessed on April 2, 2005.

51. Orphan Drug Act (As Amended), Section 527. Food and Drug Administration website. Available at http://www.fda.gov/orphan/oda.htm. Accessed on March 19, 2005.

52. Orphan Drug Act (As Amended), Section 525. Food and Drug Administration website. Available at http://www.fda.gov/orphan/oda.htm. Accessed on March 19, 2005.

53. Office of Orphan Product Designation Frequently Asked Questions. Food and Drug Administration website. Available at http://www.fda.gov/orphan/faq/index.htm. Accessed on March 19, 2005.

54. Haffner M. Orphan drugs: the United States experience. Drug Inform J April-June 1999. Available at http://www.findarticles.com/p/articles/mi_qa3899/is_199904/ai_n8844995. Accessed on March 27, 2005.

55. Title 21 Code of Federal Regulations Section 316.20. Orphan Drugs: Content and Format of a Request for Orphan-Drug Designation. Food and Drug Administration website. Available at http://www.accessdata.fda.gov/scripts/cdrh/cfdocs/cfcfr/CFRSearch.cfm?CFRPart=316.316. Accessed March 15, 2005.

56. Cumulative List of Orphan Products Designated or Approved through 2005. Food and Drug Administration website. Available at http://www.fda.gov/orphan/designat/list.htm. Accessed on April 10, 2005.

57. Traumatic Brain Injury Incidence and Distribution. Centers for Disease Control and Intervention website. Available at http://www.cdc.gov/node.do/id/0900f3ec8000dbdc/aspectId/A0400020. Accessed February 10, 2005.

58. American Heart Association: Heart Disease and Stroke Statistics 2005. Update. American Heart Association website. Available at http://www.americanheart.org/downloadable/heart/1105390918119HDSStats2005Update.pdf. Accessed March 15, 2005.

59. Traumatic Brain Injury in the United States, A Report to Congress: CDC Estimates of Traumatic Brain Injury-Related Disability. Centers for Disease Control and Prevention. Available at http://www.cdc.gov/doc.do/id/0900f3ec800101e8. Accessed March 19. 2005.

60. National Institute of Neurological Disorders and Stroke: Justification of Appropriation Estimates Fiscal Year 2002. National Institutes of Health website. Available at http://www.ninds.nih.gov/news_and_events/congressional_testimony/2002_appropriation_justification.htm. Accessed March 15, 2005.

61. Life Extension: Intracerebral Hemorrhage. Available at http://www.lef.org/protocols/prtcl-101.shtml. Accessed March 15, 2005.

62. LifeTrac Stroke Statistics. Available at http://www.lifetractech.com/product_info_stroke_stats.htm. Accessed on April 13, 2005.

63. Mayberg M, Batjer H, Dacey R, Diringer M, et al. Guidelines for the Management of Subarachnoid Hemorrhage. 1994. Available at http://www.americanheart.org/presenter.jhtml?identifier=1192. Accessed on April 13, 2005.

64. List of all Orphan Products Designation and Approval. Available at http://www.fda.gov/designat. Accessed September 28, 2005.

65. http://www.fda.gov/cder/guidance/3646dft.htm#P627_88719

66. Kahn K. Implementation of HIPAA's Privacy Rules. April 23, 2003. Available at http://www.modrall.com/articles/article_122.html. Accessed on February 12, 2005.

67. Premier Glossary of Terms: HIPAA Compliance. Available at http://www.premierinc.com/all/hipaa/glossary/index.jsp. Accessed on February 12, 2005.

68. HIPAA Security, IHE Guidelines Help Assure Compliance. Aunt Minnie website. November 26, 2004. Available at http://www.auntminnie.com/index.asp?Sec=sup&Sub=bai&Pag=dis&ItemId=64070. Accessed on February 16, 2005.

69. ICH M4. Common Technical Document. International Conference on Harmonization website. Available at http://www.ich.org/UrlGrpServer.jser?@_ID=276&@_TEMPLATE=254. Accessed on March 28, 2005.

70. Mathieu M, Whisenand T, Spaulding A, Kane K. The Biological License Application (BLA). Parexel International Corporation. Available at http://www.barnettinternational.com/RSC_ImgUploads/BIODEV3%20Sample%20Pages.pdf. Accessed on March 28, 2005.

71. New Drug Application (NDA) Process. Food and Drug Administration website. Available at http://www.fda.gov/cder/regulatory/applications/nda.htm. Accessed on April 2, 2005.

72. NDA Review Process. Food and Drug Administration website. Available at http://www.fda.gov/cder/handbook/nda.htm. Accessed on April 5, 2005.

73. FDA Advisory Committees. Food and Drug Administration website. Available at http://www.fda.gov/oc/advisory/default.htm. Accessed on April 8, 2005.

74. Harder K, Perry V. Preparing for an FDA Advisory Committee Meeting. March 2001. Available at http://www.devicelink.com/mddi/archive/01/03/005.html. Accessed on April 9, 2005.

75. Guidance for Industry: Accelerated Approval Products, Submission of Promotional Material. Food and Drug Administration website. Available at http://www.fda.gov/cber/gdlns/accpromdft.pdf. Accessed on April 12, 2005.

76. Managing the Risks from Medical Product Use. What are Risks and What are FDA's Roles in Managing Risk. Food and Drug Administration website. Available at http://www.fda.gov/oc/tfrm/Part1.html. Accessed on March 2, 2005.

Index